Environmental Governance in Latin America

# Environmental Governance in Latin America

Edited by

Fábio de Castro

Barbara Hogenboom

Michiel Baud

Except where otherwise noted, this work is licensed under a Creative Commons Attribution 3.0 Unported License. To view a copy of this license, visit http://creativecommons.org/licenses/by/3.0/

Selection, introduction and editorial matter © Fábio de Castro,
Barbara Hogenboom and Michiel Baud 2016
Individual chapters © Respective authors 2016
Afterword © Eduardo Silva 2016

The authors have asserted their rights to be identified as the authors of this work in accordance with the Copyright, Designs and Patents Act 1988.

Open access:

 Except where otherwise noted, this work is licensed under a Creative Commons Attribution 3.0 Unported License. To view a copy of this license, visit http://creativecommons.org/licenses/by/3.0/

First published 2016 by
PALGRAVE MACMILLAN

Palgrave Macmillan in the UK is an imprint of Macmillan Publishers Limited, registered in England, company number 785998, of Houndmills, Basingstoke, Hampshire RG21 6XS.

Palgrave Macmillan in the US is a division of St Martin's Press LLC,
175 Fifth Avenue, New York, NY 10010.

Palgrave Macmillan is the global academic imprint of the above companies and has companies and representatives throughout the world.

Palgrave® and Macmillan® are registered trademarks in the United States, the United Kingdom, Europe and other countries.

DOI 10.1057/9781137505729
E-PDF ISBN 9781137505729
E-PUB ISBN 9781137505736
Hardback ISBN 9781137505712
Paperback ISBN 9781137574084

A catalogue record for this book is available from the British Library.

A catalog record for this book is available from the Library of Congress.

# Contents

| | |
|---|---|
| List of Figures and Tables | vii |
| Preface | viii |
| List of Contributors | xi |

**Introduction:** Environment and Society in Contemporary Latin America  1
*Fábio de Castro, Barbara Hogenboom and Michiel Baud*

## Part I  Setting the Stage

1  Origins and Perspectives of Latin American Environmentalism  29
   *Joan Martinez-Alier, Michiel Baud and Héctor Sejenovich*

2  Social Metabolism and Conflicts over Extractivism  58
   *Joan Martinez-Alier and Mariana Walter*

3  Indigenous Knowledge in Mexico: Between Environmentalism and Rural Development  86
   *Mina Kleiche-Dray and Roland Waast*

## Part II  New Politics of Natural Resources

4  The Government of Nature: Post-Neoliberal Environmental Governance in Bolivia and Ecuador  113
   *Pablo Andrade A.*

5  Changing Elites, Institutions and Environmental Governance  137
   *Benedicte Bull and Mariel Aguilar-Støen*

6  Water-Energy-Mining and Sustainable Consumption: Views of South American Strategic Actors  164
   *Cristián Parker, Gloria Baigorrotegui and Fernando Estenssoro*

7  Overcoming Poverty Through Sustainable Development  186
   *Héctor Sejenovich*

## Part III  New Projects of Environmental Governance

8  Forest Governance in Latin America: Strategies for
   Implementing REDD                                              205
   *Mariel Aguilar-Støen, Fabiano Toni and Cecilie Hirsch*

9  Rights, Pressures and Conservation in Forest Regions of
   Mexico                                                         234
   *Leticia Merino*

10 Local Solutions for Environmental Justice                      257
   *David Barkin and Blanca Lemus*

11 Community Consultations: Local Responses to Large-Scale
   Mining in Latin America                                        287
   *Mariana Walter and Leire Urkidi*

**Afterword**: From Sustainable Development to Environmental
Governance                                                        326
*Eduardo Silva*

Index                                                             336

# Figures and Tables

## Figures

| | | |
|---|---|---|
| 2.1 | Latin America physical trade deficit in million tonnes, 1970–2008 | 62 |
| 2.2 | Argentina's physical and monetary external trade flows, 1970–2009 | 64 |
| 2.3 | Physical trade balance of Colombia, 1990–2011 | 65 |
| 2.4 | Domestic extraction in Argentina, 1970–2009 | 67 |
| 2.5 | Domestic extraction in Latin America by major category of material, 1970–2008 | 71 |
| 8.1 | Latin American countries in relation to their participation in REDD and the phased approach | 210 |
| 9.1 | National annual budget of CONAFOR according to different forest-related projects in Mexico (in million pesos), 2001–2008 | 241 |

## Tables

| | | |
|---|---|---|
| 2.1 | General conversion factors of gross ore versus metal content and ore concentrate | 72 |
| 4.1 | Income capture in Bolivia and Ecuador | 124 |
| 4.2 | Environmental administration in Bolivia and Ecuador | 129 |
| 6.1 | Reference cases | 167 |
| 6.2 | Overview of signifying content in the discourse models | 171 |
| 9.1 | Different uses of forest by community residents in Mexico | 242 |
| 9.2 | Indices of forest communities' performance | 244 |
| 11.1 | Mining consultations in the context of active mining conflicts, 2002–2012 | 293 |
| 11.2 | Guatemalan wave of preventative consultations against mining activities, 2005–2012 | 297 |

OPEN

# Preface

This book is the result of the collaborative research project Environmental Governance in Latin America (ENGOV) funded by the European Union (EU). For four years, a team of experts from ten Latin American and European academic institutions investigated how environmental governance is currently being shaped in Latin America. In this joint effort, we were driven by our concerns about widespread ecological degradation, poverty and injustice, as well as by our curiosity about the ways in which the emergence of new political regimes and elites, and innovative steps by communities and social organizations, affects governance practices and nature–society relations. To understand the possibilities and obstacles for sustainable and equitable natural resource use, a range of case-studies were carried out in Argentina, Chile, Uruguay, Brazil, Bolivia, Peru, Ecuador, Colombia, El Salvador, Costa Rica, Nicaragua, Guatemala and Mexico. Although some of the research topics and cases are not included in this volume, their findings have contributed to the discussions and theoretical reflections in the overall analysis.

The ENGOV project has been simultaneously challenging and inspiring. The theme of environmental governance is a huge academic enterprise because it addresses complex social relations, practices and views influencing how societies perceive nature and use natural resources. Combining methods and theories from different fields of the social sciences is a prerequisite which in practice is fairly demanding. Furthermore, by encompassing political, economic, cultural and environmental changes, formal as well as informal arrangements, and cross-scale connections, the study of environmental governance can easily become a 'mission impossible'. Arguably this is even more the case for contemporary Latin America, with its variety of local and national conditions facing rapid-paced changes. Finally, collaborating in an international research consortium of ten institutional partners and more than 25 researchers from different disciplines, schools of thought and generations has also proved to be both daring and rewarding. The fact that we spoke in different academic languages and idiom accents was not only a hurdle to tackle during our group discussions, but also forced us to learn from each other's approaches and convictions, and the foundations on which these are based. As a typical governance process, next to misunderstandings, dissonances and unbridgeable differences, the exchange of different insights and perspectives proved to bring about refreshing debates and new understandings, nuances and agreements.

Without the ambition to provide a full overview of the environmental governance in Latin America, we have tried to identify key fields for research, with an emphasis on new trends or structural problems that deserve more academic attention. The new insights from each piece of research contributed to the development of analytical frameworks to analyse the multiple interconnected processes shaping environmental governance in the region. This volume is the result of this intricate, collaborative exercise.

For the realization of this book, several people and institutions have been indispensable. It would not have been possible without the extensive support of the EU. Financed under the Seventh Framework Programme, ENGOV enabled the consortium to develop important new research on environmental governance in Latin America and the Caribbean, resulting in a long list of publications. We are particularly thankful for the professional guidance of Philippe Keraudren and Cristina Marcuzzo of the Social Sciences and Humanities division of the Research and Innovation Directorate General.

We would also like to thank the institutions participating in ENGOV for their financial and administrative support, including their directors and the employees who directly assisted the project: Consejo Latinoamericano de Ciencias Sociales (CLACSO), Institut de Ciència i Tecnología Ambientals, Universitat Autònoma de Barcelona (ICTA-UAB), Institute de Recherche pour le Développement (IRD), Centre for Development and the Environment, University of Oslo (SUM-UiO), Centro de Desenvolvimento Sustentável, Universidade de Brasília (CDS-UnB), Universidad Autónoma Metropolitana, Unidad Xochimilco (UAM-Xoc), Instituto de Estudios Avanzados, Universidad de Santiago de Chile (IDEA-USACH), Instituto de Investigaciones Gino Germani (IIGG) and Universidad Andina Simón Bolívar, Sede Quito (UASB-SQ). We are grateful to our colleagues from CLACSO, and in particular to Fernanda Saforcada and Guadalupe Rudy, for their continuous support during the project. We also thank the University of Amsterdam, which hosts our own Centre for Latin American Research and Documentation (CEDLA) and was very supportive of ENGOV, in particular Jan Jacob Sikkema and Bea Krenn. At CEDLA, the solid project support by Leontien Cremers requires a special mention. Her accurate and cheerful involvement, including the preparation of the Index of this volume, has made a difference both for CEDLA's ENGOV coordination team and for all the consortium members. We would also like to thank María Barrachina for kindly granting permission to use her photograph on the front cover. We are also most grateful to the members of ENGOV's international advisory board, who have offered insightful comments on the draft chapters: Anthony Bebbington (Clark University and University of Manchester), Alberto Cimadamore (University of Bergen), Edward F. Fischer (Vanderbilt University), Barbara Göbel (Ibero-Amerikanisches Institut), Leticia Merino Pérez (Universidad Nacional Autónoma de México), Pedro Roberto Jacobi (Universidade de São Paulo) and Eduardo Silva (Tulane University). In addition, we are grateful to all the scholars and students who have contributed to the discussions at different ENGOV meetings.

Last but not least, we are very grateful to the key project researchers, not only for the chapter they have contributed but also for their critical input to other draft chapters and their commitment to the ENGOV project. With them, we hope this book will inspire both researchers engaged in the environmental governance debate in Latin America and young scholars and non-academic readers interested in understanding the complex society–nature relations in the contemporary world.

Fábio de Castro, Barbara Hogenboom and Michiel Baud

 Except where otherwise noted, this work is licensed under a Creative Commons Attribution 3.0 Unported License. To view a copy of this license, visit http://creativecommons.org/licenses/by/3.0/

# Contributors

**Mariel Aguilar-Støen** is a political ecologist and senior researcher at SUM-UiO, Norway.

**Pablo Andrade A.** is a political scientist and professor at UASB-SQ, Ecuador.

**Gloria Baigorrotegui** is an industrial engineer and junior researcher at IDEA-USACH, Chile.

**David Barkin** is an economist and professor at UAM-Xoc, Mexico.

**Michiel Baud** is a historian and director of CEDLA, and professor of Latin American Studies at UvA, The Netherlands.

**Benedicte Bull** is a political scientist and professor at SUM-UiO, Norway.

**Fábio de Castro** is a political ecologist and assistant professor of Brazilian Studies and Human Ecology at CEDLA, UvA, The Netherlands.

**Fernando Estenssoro Saavedra** is a historian and senior researcher at IDEA-USACH, Chile.

**Cecilie Hirsch** is a human geographer and PhD candidate at SUM-UiO, Norway.

**Barbara Hogenboom** is a political scientist and Associate Professor of Political Science at CEDLA, UvA, The Netherlands.

**Mina Kleiche-Dray** is a historian and senior researcher at IRD, France.

**Blanca Lemus** is a physician specialized in labour and environment, and a visiting researcher at UAM-Xoc, Mexico.

**Joan Martinez-Alier** is an economic historian and professor of Economic History and Institutions in the Department of Economics and Economic History at UAB, Spain.

**Leticia Merino** is an anthropologist and professor at the Instituto de Investigaciones Sociales at UNAM, México.

**Cristián Parker** is a sociologist and director of IDEA-USACH, Chile.

**Héctor Sejenovich** is a political economist and senior researcher at IIGG and professor of Social Sciences and Environment at UBA, Argentina.

**Eduardo Silva** is a political scientist and Lydian Chair professor in the School of Liberal Arts at Tulane University, New Orleans, USA.

**Fabiano Toni** is a political scientist and associate professor at CDS-UnB, Brazil.

**Leire Urkidi** is an environmental scientist and researcher at the EKOPOL at EPV/EHU, Spain.

**Roland Waast** is a sociologist and engineer at the École Polytechnique de Paris, France.

**Mariana Walter** is a political ecologist and postdoctoral researcher at ICTA-UAB, Spain.

OPEN

# Introduction: Environment and Society in Contemporary Latin America

*Fábio de Castro, Barbara Hogenboom and Michiel Baud*

## Introduction

Societal change in Latin America is intimately related to nature and natural resources. In this resource-rich region, nature–society relations provide both opportunities and challenges in achieving more fair, equitable and sustainable development. Nearly half of the world's tropical forests are found in the region, next to several other natural biomes, which together carry a wealth of biodiversity. It holds one-third of the world's freshwater reserves and one-quarter of the potential arable land. And despite five centuries of extractive activities to serve global markets, the region still holds large volumes of important mineral reserves, including oil, gas, iron, copper and gold (Bovarnick, Alpizar and Schnell, 2010). On the other hand, this "biodiversity superpower" has seen a fast rate of biodiversity loss, increasing ecosystem degradation and one-third of the world's carbon emissions, mostly a result of the expansion of extractive activities and land-use change (UNEP, 2012). Together, these economic and ecological developments affect a large number of different social groups in all Latin American countries, primarily in rural areas but also in cities. Next to mobilizations and conflicts that attract national and international attention, there are numerous local socioenvironmental tensions that lead to longstanding economic problems and social injustice. Although these tensions have been part of the region's history, the accelerated pace of change, the spatial scale of impact, and the widening of social and conservation demands all point to the urgency of Latin America's current environmental challenges (Baud, Castro and Hogenboom, 2011).

Since Latin America's insertion into the world system, the extraction of natural resources has been central to its economic, social and political development. This has led to continuous tensions and antagonisms about access to natural resources, the distribution and use of revenues, and the distribution, compensation and prevention of environmental and social costs (Alimonda, 2011). In Latin America, issues of poverty, inequality and environmental protection are thus closely intertwined. Despite academic studies showing the risks of being a global provider of foodstuffs, energy, metals and environmental services without appropriate institutional arrangements, not much progress has been made in successfully tackling problems of underdevelopment (Bunker, 1988), impoverishment/marginalization (Martinez-Alier, 2002), inequality (Therborn, 2011), accumulation by dispossession (Harvey, 2003), and disempowerment and dependency in rural communities (Painter and Durham, 1995).

After a long history of elite capture and foreign exploitation of Latin American mines, agrarian lands and, later, oil and gas resources, social and political forces started to push forward reforms such as the nationalization of oil and metals, and the distribution of land in the twentieth century. Nevertheless, access to resources, revenues and power remained unequally distributed at local, national and international levels. The neoliberal regimes of the late twentieth century went against previous redistributive policies (Liverman and Vilas, 2006). This period was marked by greater attention to both environmental protection and decentralized decision-making (Larson, 2003). However, restricted funding and liberalized markets limited the potential to break with historically established patterns.

This new environmental, social and institutional context also changed environmental governance in Latin America. Both in rural and urban areas, poor citizens became more vulnerable due to environmental degradation and the increased intensity and frequency of climate disasters, including droughts, flooding, hurricanes and glacier retreat (Rios and Veiga, 2010). In many countries, especially in South America, a new phase of widespread civic discontent and mobilization of groups against exclusion, poverty, inequality and technocratic policies started in the 1990s (Harris, 2003). While many groups only called for socioeconomic redistribution, indigenous movements, landless farmers and environmental organizations also demanded different policies towards land and nature (Carruthers, 2008; Urkidi and Walter, 2011; Latta and Whitmann, 2012).

Since the turn of the twenty-first century, Latin America has experienced radical developments that have changed the dynamics of environmental governance. As will be discussed in greater detail later in this chapter, democratic elections resulted in a number of leftist governments that promised inclusionary development and more participatory decision-making. Their reforms included a more prominent role of the state in the extraction of non-renewable resources and the redistribution of revenues. At least symbolically, attention to the environment also increased. The new regimes and their policies have thus attempted to combine measures geared towards the reduction of poverty and social exclusion with policies that enhance national control over natural resources and improve environmental protection. Simultaneously, the global commodity boom brought extra revenues and foreign investments, thereby intensifying resource extraction and leading to problems of environmental degradation and more intense environmental conflicts (Fernández Jilberto and Hogenboom, 2010; Hogenboom, 2012).

Institutional adaptations played an important role in these transformations, as illustrated by the debate about the global sustainable development model. The narrative of social justice and the plural development model, established in the 1990s with strong participation by civil society organizations, was gradually replaced by narratives of institutional fixes and technological innovations (Mol, 2003). This led to a new model, framed as the Green Economy, which shifted the focus from social and political questions about deepened environmental citizenship and justice to a more technological and economic approach focused on the commodification of nature.[1] As a result, the model of participation through citizenship has gradually been reframed by participation through compensation, as installed by the post-neoliberal state in the context of an urbanized region.

This volume seeks to analyse the features, dynamics and direction of contemporary environmental governance in Latin America. Building on various local and national cases, it presents formal and informal practices of management concerning renewable and non-renewable natural resources. It also shows how rights to nature are perceived, contested and reshaped in the context of rapid social, institutional and environmental changes on multiple scales. It combines elements of power relations, diversity, complexity and dynamics in socioenvironmental systems in order to tackle this process through a cross-scale, multiactor and dialectical perspective (Robbins, 2012). One particular strength of this political ecological approach is the explicit emphasis on the social and institutional dynamics that shape social interactions and natural resource use patterns (Zimmerer and Bassett, 2003). Moreover, it takes into account the multiple conceptualizations of and claims over nature as part of a contested sphere, which we denominate "environmental governance".

The three parts of this book address the changing context, social interactions and institutional adaptations in contemporary nature–society relations in Latin America. Part I introduces the socioenvironmental context through a focus on the historical legacy of Latin American environmentalist thinking, the increasing pressure on the region's environment due to the global demand for its natural resources, and the rich ecological knowledge within local communities. These chapters set the stage to analyse the recent transformations of nature–society relations in the region. Part II addresses the politics of nature, raising issues related to the role of powerful actors – the state, elite and corporations – and their interactions in shaping discourses and practices regarding natural resource use. These processes are explored through the analysis of new policy models deployed by post-neoliberal governments, the role of new and old elites and their interactions, the narratives around the water–energy–mining nexus by contesting actors, and strategies for poverty alleviation. In Part III, new and emerging forms of environmental governance that tackle issues of participation, autonomy and environmental security are examined. The analysis of the implementation of REDD+ (reducing emissions from deforestation and forest degradation), the controversial international compensatory scheme to prevent climate change, addresses how participatory mechanisms have become invited spaces of selected legitimized groups while the bottom-up initiatives of community-based autonomous economies and local consultations to mining projects that address the struggles for effective inclusion, wellbeing and justice emerged from resistance movements.

In general, this volume aims to understand environmental governance in Latin America by looking into the ways in which historical legacies and current socioenvironmental contexts are driving new social interactions and institutional adaptations among multiple actors. The chapters cover a range of Latin American countries, mostly based on empirical data from multiple contexts, actors and production systems, and focus on transnational, national or subnational processes. Together they provide an overview of current regionwide trends, and a variety of themes and approaches to environmental governance, which feeds lively and sometimes heated debates in academia as well as in civil society and policy-making circles.

## Environmental governance as a field of inquiry

Environmental governance offers an analytical perspective that combines socioenvironmental research with development-oriented governance research (Lemos and Agrawal, 2006). Socioenvironmental research addresses the interplay between environmental and social change. In this context, as in this introductory chapter, the social dimension is broadly defined, also encompassing cultural, economic, political and institutional relations. Governance research addresses the way in which society organizes itself in order to solve its dilemmas and create new opportunities. Until the 1980s, social scientists working in Latin American countries focused on concepts of governability as the region faced unstable political conditions and structural challenges such as inequality, violence, corruption and limited citizenship. However, the growing emphasis on formal institutions and market-driven mechanisms of neoliberal governance quickly attracted the attention of social scientists to a perspective of governance as a social process that influences the level of governability (Kooiman, 2003). This perspective criticized the normative perspective of "good governance" introduced by the World Bank in the seminal report *Governance and Development* (1992). According to this document, the solution to overcome underdevelopment should be self-governance. The World Bank proposed a roadmap to achieve so-called good governance based on three pillars: a "small state" through deregulation; "market incentives" though privatization and liberalization; and "participation" through decentralization and non-governmental organizations (NGOs). Subsequent World Bank reports further elaborated this international agenda, stressing in a rather technocratic approach, the need for effective state institutions to achieve development in a global context of liberalized markets (Demmers, Fernández Jilberto and Hogenboom, 2004). Alternatively, social science scholars use (environmental) governance to emphasize social relations and, in particular, the tension between conservation and development goals in order to understand the interplay among social, institutional and environmental change.

The environmental governance research builds on a range of theoretical schools, including new institutionalism (Ostrom, 1990; Young, 1999; Biermann and Pattberg, 2008), sociopolitical studies (Kooiman et al., 2005; Lemos and Agrawal, 2006) and sociocultural approaches (Cleaver, 2002; Alimonda and Gandásegui, 2006; Castro, 2008; Gudynas, 2011). Despite their different theoretical and methodological stands (see Castro, 2013), they all address social behaviour towards natural resources as a complex arrangement of formal and informal interactions among state and non-state actors across different scales, driven by ecological and social factors. In this book we follow a similar approach and define environmental governance as the process of formulating and contesting images and designs, and implementing procedures and practices that shape the access, control and use of natural resources among different actors.

In recent decades, environmental governance in Latin America has undergone major transformations. We observe multiple layers of governance, mediated by formal and informal social interactions, which have gradually evolved over time. Nevertheless, a particular arrangement has typically dominated discourses and practices at the national level. As of the 1940s, state-centred governance mode increasingly dominated most of the region. Particularly during the period of military dictatorship, decision-making processes were based on bureaucratic authoritarian regimes and top-down procedures controlled by a technocratic elite and grounded in a strong nationalist discourse of state sovereignty.

In the 1990s, most Latin American countries underwent a societal change through democratization, political decentralization and neoliberal restructuring. Civil government and electoral democracy were (re-)established and the former exclusionary governance gave way to electoral forms of political representation. At the same time, the role of the state was limited by far-reaching structural adjustment policies imposed by international institutions, in particular the International Monetary Fund, the World Bank and the Inter-American Development Bank (Liverman and Villas, 2006). Self-governance mode, as conceptualized by the World Bank, calls for a small role of national states, and reliance mainly on market-based mechanisms such as privatization, self-designed corporate conduct guides (e.g. corporate social responsibility (CSR)) and voluntary mechanisms (certification and compensation schemes). While promising environmentally and socially sound initiatives, the market-based approach to self-governance primarily sought to improve the image of transnationally operating companies vis-à-vis their shareholders and to consequently ease their insertion into host countries (Lyon, 2009).

At the same time, self-governance mode, as conceptualized by political scientists (e.g. Ostrom, 1990), includes mostly local governance systems shaped through collective action to regulate access to and use of natural resources. This governance mode, long overlooked by policy-makers, became visible through a large number of community-based management studies (see McCay and Acheson, 1990; Berkes and Folke,

1998) and was brought to the attention of society at large by environmental justice movements that built on socioenvironmental discourses and political connections with transnational activism networks (Keck and Sikkink, 1998). While self-governance through collective action became important in more remote areas during this period (Schmink and Jouve-Martín, 2011), in areas of large-scale economic production a type of self-governance based on market-based mechanisms thrived, leading to a wave of natural resource privatization in the region. As these two governance systems collided, local social relations were disrupted (Bebbington, 2012), and local elites and transnational corporations were strengthened (Larson, 2003; Perreault, 2005). This led to an intensification of local conflicts that often had national and global repercussions (Walter and Martinez-Alier, 2012). Combined with other political and social demands, environmental conflicts contributed to major political transformations and may be considered to have been instrumental in the election of left-leaning parties in many Latin American countries.

As part of this struggle for resources, participatory governance mode emerged in the 2000s as an alternative to the previously proposed monolithic governance modes. This was part of the project to deepen democracy and citizenship by the new Latin American governments. Grounded in discourses of social justice, equity and poverty alleviation, participation of civil society organizations has become a central element of environmental governance in the region. Instead of state-, community- or market-based governance, participatory governance is based on partnerships among relevant actors to set goals and to design and implement initiatives. Participatory governance ranges from co-management models, in which state and local communities develop a sustainable plan for traditional territories (Castro, 2012), to more complex arrangements that include multistakeholders and multiscale institutions, such as that of climate governance. Here, governments, transnational social movements and transnational corporations are engaged in the shaping of an international institutional arrangement that combines semilegal agreements to tackle climate change and related environmental issues, such as emission targets, Agenda 21 and the Convention on Biological Diversity (Biermann and Pattberg, 2008).

Participatory environmental governance therefore takes place in a contested political space where different actors struggle to strengthen their positions. More than a new governance mode, it represents a new layer in hybrid governance models composed by state-centred, market-based and local-based mechanisms. To what extent participation can actually be fostered, inequalities diminished and the environment protected in this complex arrangement depends on the way different images of nature–society relations are negotiated, how problems are prioritized, and how compatible the proposed solutions are with the social, institutional and environmental context. In this respect, Latin America has recently experienced some interesting new trends.

## Recent trends in Latin American environmental governance

Environmental governance in Latin America is a contradictory process. The dominating discourse of participatory governance in several Latin American countries is accompanied by increasing socioenvironmental conflicts.[2] In the centre of this contradiction are the changes to the socioenvironmental context observed in the last decade. The impressive economic and social progress of the 2000s and the new approaches to poverty alleviation, redistribution and sovereignty were supported by large segments of the population. However, social programmes were usually based on increased public revenues from extractive activities, both through booming global commodity markets and through higher national taxes and royalties (Hogenboom, 2012). As many countries deepened their dependence on the extractive use of natural resources, this prompted a "reprimarization" of the economy. As soon as these tendencies became evident, the problems and contradictions of (neo)extractivism and the possibilities for post-extractivist development strategies became the subject of vivid debates in countries such as Ecuador (*Ecuador Debate*, 2011), Bolivia (Radhuber, 2014), Argentina (Giaracca and Teubal, 2013) and Peru (Alayza and Gudynas, 2011). Critics of extractivism point to the new partnerships between the national state and transnational corporations, which simultaneously reinforced state-centred and market-based principles of governance. Despite the increasing implementation of impact assessments and prior consultations, the involvement of local stakeholders in decision-making processes remains very limited (Schilling-Vacaflor, 2012). Grassroots organizations, human rights activists and environmentalists accordingly denounce the imposition of top-down arrangements. Next to the limited influence of civil society, and especially of marginalized groups, they call attention to the increasing criminalization of social mobilization against large-scale projects of mining, oil and gas extraction, hydroelectricity or infrastructure.

These processes reinforced the longstanding tension between the commodification of nature and the "safeguard of nature" (Silva, 2012).

On the one hand, governments and corporations are receiving support from the urban population to further the expansion of extractive activities in order to fulfil urgent societal needs. On the other hand, rural communities, indigenous organizations and environmentalists stress the relevance of nature for ecological sustainability, social reproduction and cultural notions of belonging rooted in local cosmologies. The implications for the safeguard of nature and local communities in the region have been complex and contested. Facilitated by national policies, large companies are attracted to resource-endowed areas to supply the increasing global demand for commodities. The expansion of extractive activities has deepened the pressure on the natural environment and its local residents. This has become particularly clear in the Amazon, where the rapid expansion of a range of large- and small-scale activities (Dijck, 2014) threatens the livelihoods of indigenous and other communities, sparking numerous conflicts and violent clashes (Alimonda, Hoetmer and Saavedra Celestino, 2009; Gavaldà i Palacin, 2013; Vásquez, 2014). However, Maristella Svampa (2011) also notes that due to a convergence between indigenous communitarian views and environmental discourses, an interesting ecoterritorial turn in socioenvironmental struggles has come about.

The frequency and intensity of socioenvironmental conflicts indicate that, in the context of democracy and post-neoliberal development models, major dilemmas between conservation and development remain. For the solution of these dilemmas, a range of proposals and actions have been brought forward that are meant to bring actors together to find new forms of more consensual environmental governance. The existing proposals can be categorized as one of two contrasting models.

On the one hand we can distinguish a tendency that we call *neodesarrollismo* (new developmentalism). This refers to mainly business-like proposals that rely on institutional engineering, technological modernization and market-based mechanisms to bring about efficient and sustainable use of natural resources. This model tends to dominate policy circles in most Latin American governments. It is closely related to the globally dominant environmental governance model known as the Green Economy. Grounded in neoinstitutionalism, the model relies on institutional fixes to fine-tune market-based incentives in order to drive collaborative behaviour and sustainable practices (UNEP, 2011). The Green Economy model assumes that shortcomings such as asymmetric relationships, injustices and unsustainable behaviour can turn into more equitable and sustainable outcomes through proper institutional design (Biermann, 2007). By relying on institutional engineering, solutions are based on apolitical means such as innovation of technology (de Mol, 2003) and "green" consumptive behaviour (Dobson, 2003). The pragmatism of this approach finds fertile ground among elite groups because it addresses the dilemmas around equity, sustainable development and conservation from within the capitalist market-based structure. Its advocates rely on market-based incentives and compensation schemes, such as REDD and payment for ecosystem services (PES), as mechanisms to replace state regulation, minimize conflict-related costs and improve corporate image. The model also fits well into the institutional ethos of a technocratic state apparatus, which tends to rely on blueprint institutional designs. Finally, it satisfies part of the environmentalist agenda, including several international environmental NGOs such as the World Wide Fund for Nature (WWF), Conservation International and the Nature Conservancy. These transnational organizations have gradually moved towards an agenda of compensation schemes and market-based incentives in order to promote sustainable behaviour among corporations, states and local communities (Hall, 2012).

On the opposite side, we find a number of proposals that envision a radically different model of production and environmental governance, brought together under the label of *Buen Vivir* ("good living"). This tendency includes a range of alternative conceptions of nature and human–nature relations that depart from indigenous ideas about the relationship between human production and the environment and rights of nature (Gudynas, 2011). The proposals recommend a bottom-up and unorthodox environmental governance perspective, which calls for the transformation, or even the end, of the hegemonic capitalist model that is considered to be the very source of environmental degradation and injustice. Their advocates argue that *neodesarrollismo* and its connection with the Green Economy only mean a repackaging of old development models to maintain unequal power relations on multiple scales. Instead of the technocratic belief in "institutional deficiencies" that only need to be fixed, they consider these deficiencies to be the very foundation of asymmetric relationships and environmental degradation (Alimonda, 2011). They argue that institutional fixes will hardly be effective in solving socioenvironmental problems unless the unequal power relations between different social groups and the basic foundations of the market-based economy are properly addressed (Gudynas, 2009). Grounded in discourses of wellbeing, civil rights and a plural state, advocates leaning towards this narrative argue that capitalism is limited to tackling issues of justice, equity and sustainability, and they call for alternative models of heterodox economy, such as degrowth (Russi et al., 2008) and the solidarity economy (Barkin and Lemus, 2011), or local practices such as agroforestry (Altieri and Toledo, 2011) and community-based management systems (Bray, Merino and Barry, 2005).

The *Buen Vivir* model has provoked two kinds of criticism. On the one hand, some observers consider the anti-market basis of these ideas to be unfeasible and unrealistic. In their view it is impossible in today's world not to participate in the market economy. Other observers focus on the governments that want to implement these ideas, such as those of Bolivia and Ecuador. They criticize the lack of clarity in the concept of *Buen Vivir* and highlight the contradictions that its supposed implementation engenders (Bretón Solo de Zaldívar, 2013). They argue that, in practice, these ideas serve as an excuse for continuing developmentalist and extractive models.

It is clear that both *neodesarrollismo* and *Buen Vivir* have their flaws and contradictions. In practice, we can see that most governments in Latin America today combine elements of both models. Indeed, we can speak of a mixed governance model, in which governments and other actors eclectically use different models to implement their practices or to formulate their demands. In this way multilayered and flexible institutional arrangements are continuously constructed and reconstructed through a process of hybridization and bricolage (Cleaver, 2002).

To understand projects of environmental governance in Latin America today, we need to start from the fact that they emanate from different actors who have particular historical experiences and use a variety of local, national and global discourses. These projects at the same time present a number of often contradictory goals and proposals. In the last instance they aim to find solutions or create new opportunities for this predicament of a balance between productive activities, societal equality and environmental policies. In the remainder of this introduction, we will try to shed light on the consequences of these complex proposals for environmental governance.

## Environmental governance as a social process

Environmental governance is thus embedded within a historical, environmental and social context that is continuously shaped by political struggles, environmental change and contested values of nature over time (Miller, 2007). Environmental attributes, such as availability and distribution of renewable and non-renewable natural resources, influence access to production territories by different stakeholders (see Haarstad, 2012). Social attributes – such as consumption patterns, poverty and inequality levels, democracy and citizenship, cultural diversity, and economic growth – are some of the driving factors underlying the actions of Latin American societies to shape multiple patterns of the exploitation and protection of nature (Latta and Wittman, 2012). In particular, institutional arrangements that define the "rules of the game" – which include both formal and informal practices and mechanisms mediating social-environment relations on multiple scales – are based on different sets of principles, values and images of nature, conservation and development.

To understand how environmental governance takes place in the region, we have to look at the intricate and heterogeneous environmental, social and institutional arrangements in Latin America (see Helmke and Levitsky, 2006). Changes in the social, institutional and environmental context continuously reshape the set of opportunities and constraints for different actors, triggering new social interactions and institutional adaptations.

In these highly complex and dynamic processes, multiple actors make use of elements of different, often contrasting, discourses to legitimate their proposals or projects. To disentangle and unpack the practical and discursive contradictions of today's environmental governance in Latin America, we identify three analytical lines that are reflected throughout this book. First, perceptions, values and discourses are important because they show the variety of images of nature, environmental problems and possible solutions among different social groups. Second, social interactions further give shape to people's actions and relations towards decision-making processes. And third, institutional change and adaptations are the result of concrete efforts to deal with these different and often conflicting images and a multitude of social interactions.

### Perceptions, values and discourses

Perceptions and values are fiercely contested by different actors according to their representations of nature. The contestation over values, principles and knowledge sources guiding the way nature is conceptualized is one of the key elements of environmental governance. The way nature conservation is framed directly influences how environmental dilemmas are problematized, how solutions are designed and how priorities and trade-offs between conflicting goals are set. The more actors are engaged in environmental governance, the more complex and heterogeneous the images become. The central question is how these complex dynamics lead to specific forms of environmental governance, and maybe even more importantly, how these forms can be directed towards social inclusion and environmental sustainability.

As argued by Martinez-Alier, Baud and Sejenovich (Chapter 1), Latin America has a long epistemological and political tradition in relation to the balance between human production, natural resources and the environment. This academic perspective goes in the same direction as indigenous cosmologies, in which nature is an integrated part of their lives. By using a range of illustrative examples, Kleiche-Dray and Waast (Chapter 3) describe in detail how cultural practices are intimately related to production and food systems. Similarly, Barker and Lemus (Chapter 10) explain how cultural perspectives of nature form the core concept of indigenous peasant communality.

While indigenous and peasant communities tend to perceive nature as important for symbolic meanings and for sustaining their livelihoods, extraction-oriented images connect nature to the interests of exploiting its resources and generating revenues. The latter images have been especially advocated by national governments and large companies. Interestingly, although Andean governments today also use the symbolic indigenous images of *Pachamama* and *Buen Vivir* in their discourses, their meaning has been reframed (see Teijlingen and Hogenboom, 2014). The governments have adapted such images to a political agenda in which nature mainly serves to support national development. This leads to the coexistence of seemingly competing images and discourses, such as *Buen Vivir* with the idea of the so-called *país minero* (mining country), as explained in detail by Andrade (Chapter 4).

Parker, Baigorrotegui and Estenssoro (Chapter 6) demonstrate how the discourses of private companies resemble those of the national Latin American governments. Through multiple – and often contrasting – discourses, large private companies strive to defend their interests, to confront contested political contexts and to legitimate their projects. However, while national governments define the control of natural resources as an element of national sovereignty, corporate actors interpret the dilemmas of environmental governance as transcending national boundaries, such as in the case of the fictitious United Republic of Soybeans, the agricultural area covering parts of Argentina, Brazil, Paraguay and Bolivia that is controlled by the world's largest food companies (see Grain, 2013).

Environmentalists' images of nature also transcend national interests and boundaries, and often pit them against national governments and interest groups. However, their views contrast with indigenous communities or companies by defining nature as a biophysical entity, characterized by its ecological function of biodiversity repository and carbon sink with direct implications in regulating the global climate. By using metaphors such as "Earth's lung" or "carbon sink", or superlatives such as megabiodiversity spots, biomes such as the Amazon are usually emphasized over other ecosystems, as shown by the REDD+ case described by Aguilar-Støen, Toni and Hirsch (Chapter 8).

In sum, whether a lifestyle, a commodity or a biological stock, nature's multiple images and values create dissonance among stakeholders' perceptions of nature-related problems and possible solutions. At the core of this dilemma is the struggle over meanings of nature, conservation, development and participation. The consequences of these different perceptions and the contradictions within existing discourses become apparent in concrete social interactions.

## Social interactions

Social interactions are the propeller of environmental governance. Through their ambitions to deepen democracy and foment popular participation, often in response to social demands and mobilization, Latin American governments have expanded the range of actors and interests involved in environmental governance. Even though these ambitions may have often been confined to discourse and rhetoric, they have opened political spaces for more varied and dynamic social interactions. As a result, decisions regarding environmental dilemmas in Latin America today involve a range of actors that may hold multiple political and identity positions. These positions may be strategically shifted according to new opportunities and constraints that emerge from changes in the socioenvironmental context. Because they concern concrete decisions that present technical, economic and political choices and ambiguities, social interactions are dynamic and constantly swing between the opposites of cooperative or accommodating to conflictive and resisting relations. In this intricate social interaction, the struggle to participate and control the decision-making process is a central element of environmental governance.

It is interesting to note that the relevance of participation for effective solutions to economic, social and conservation challenges is no longer questioned by the elite groups. As Chapter 6 shows, even the most conservative and market-oriented stakeholders acknowledge the importance of the inclusion of local or marginalized groups. In fact, participation has become a central element in official documents drafted by government agencies, corporations, donors and multigovernmental agreements. However, the participation of local communities has been framed in terms of them being recipients of compensatory benefits decided by other legitimated actors.

In the case of mining consultations, Walter and Urkidi (Chapter 11) argue that companies try to demobilize local participation with technological solutions and false promises. Through top-down procedures, they only give local populations the opportunity to be informed in order to legitimize their activity. In the case of REDD+, Chapter 8 argues that projects are dominated by "invited" actors who decide which knowledge tools, goals and models are legitimized. What remains for the local populations is some compensation in the form of money or material facilities. Despite the different territorial and political contexts, both of these chapters demonstrate the dangers of framing participation as a distribution of compensatory measures.

The reframing of participation through compensation has emerged from coalitions between the state and other elite groups. Chapter 4 and Bull and Aguilar-Støen's Chapter 5 focus on state-business coalitions for the expansion of extractive industries. The former focuses on the political and economic agenda of the state based on natural resources, while the latter describes how this process has driven new forms of political interactions between the state and the new and old elite. Chapter 8 focuses on the NGOs, experts and state coalition for the expansion of protected areas.

The unfulfilled promises of participatory policies combined with the increased exploitation of natural resources in many Latin American regions have fuelled socioenvironmental conflicts almost at the same pace as the implementation of participatory initiatives. According to Martinez-Alier and Walter (Chapter 2), these conflicts concentrate on the distribution of the ecological debt and basically emerge from the unequal exchange of material between different parts of the world. In addition, as Sejenovich (Chapter 7) shows, dominant production processes have high social and environmental costs. To end poverty and realize sustainable development, social rights as well as ecological limits need to be fully integrated into governance processes. In recent years, some progress has been made in this direction. To regain their protagonism in environmental governance, various local communities have developed and designed bottom-up decision-making processes to defend their local interests and to keep their autonomy in shaping their livelihood strategies (see chapters 9, 10 and 11).

These bottom-up solutions are built on environmental justice networks and peasant and indigenous movements, an instrumental strategy in the struggle for access and control over natural resources in Latin America (Carruthers, 2008). They struggle to empower themselves through a discourse of human–nature interdependency and territorial autonomy. In this process, local actors try to scale down the decision-making process. Chapter 10 argues that locally developed economic models are the only way to liberate subalterns from their marginalized position in the capitalist structure. Chapter 2 shows how local communities organize themselves around glocal (global–local) networks in order to reclaim their political position within the capitalist structure. At the implementation level, a myriad of initiatives have been observed on the ground. Local communities draw on their local knowledge and institutions in order to develop new strategies to tackle new challenges. In some cases they have actively designed their own decision-making systems to counter the manipulative consultations carried out by private companies, as described in detail in Chapter 11. In other cases, communities have engaged in commercial activities by building on their social capital to develop their technical and entrepreneurial capacity (see Merino's Chapter 9).

In sum, the increasing tension between environmental justice and post-neoliberal policies is characterized by a dynamic reshaping of strategies among contesting actors. This central element of environmental governance drives new institutional adaptations based on discourse, relationships and practices on the ground.

### Institutional change and adaptation

Institutional adaptations involve strategies developed by different actors to increase their ability to be included or to define the "rules of the game" in environmental governance. These adaptations comprise formal and informal mechanisms, and range from discourse reshaping and new communication strategies to innovative initiatives, technologies and knowledge integration. Latin America has been the stage for two key forms of institutional adaptation among different contesting actors: the reshaping of environmental discourse and the rescaling of environmental governance.

Generally, dominant actors have reframed their discourses in order to fit their interests and objectives into a "green growth" agenda. Corporations favour models based on technological innovation while leftist governments argue for the expansion of extractive activities in order to reach social objectives. The ideologies and discourses of the new so-called post-neoliberal governments in Latin America have greatly influenced the adaptations of environmental governance. By framing natural resources as a national wealth to solve inequality problems, they have strengthened the state's political position vis-à-vis the transnational corporate sector. This has allowed them to acquire a more central position in the governance of natural resources and to impose stronger conditions for the exploitation of natural resources. The increased income from taxes and royalties on natural resource use have allowed for a redistribution of benefits among different stakeholder groups, resulting in decreasing poverty and income inequality in the region, even though the problem of structural poverty still needs to be resolved (see Chapter 7).

Among several actors, gradual shifts may be observed in environmental attitudes, mechanisms and practices. The state has been instrumental in reformulating procedures for the socioenvironmental assessment of extractive industries and infrastructure expansion, decision-making processes and control over environmental conflicts. To prevent further legislative restrictions, and in response to social pressures, corporations have become proactive in the development of a discourse in which they hold a key role in solving societal problems. This discourse has materialized through the CSR framework, which promises to reconcile their productive activities with social and environmental demands. Many researchers and environmentalists, on the other hand, have adapted to the new context by claiming their "expert" role as knowledge-holder of the technical information that is necessary to design better policies.

These different discursive strategies mediate the institutional changes promoted by contesting actors. At the national level, Chapter 4's analysis of the state in Andean countries reveals the strong role of the recentralization of environmental governance as a key strategy of post-neoliberal states in order to subsidize the accomplishment of their social policies. Chapter 5 offers several examples in which elite groups try to ensure their access to land and natural resources through different means (see also Otero, 2010; Borras et al., 2012; Harstaad, 2012). In some other cases, however, different governmental levels may compete for control of the decision-making process. The REDD+ implementation process provides an illustrative example of tensions between different governmental levels in the attempt to recentralize or decentralize the funding scheme to compensate forest-protection initiatives. In the current "race" for the implementation of REDD+ in Brazil, state governments have built state-level coalitions in order to bypass national governments and reach out to different international funding schemes (Chapter 8).

Politically less powerful actors also strive to rescale decision-making processes in order to overcome undesirable policies and developments, structural constraints or environmental degradation. Chapter 11 describes the efforts of local communities to build up both glocal connections and coalitions with local governments in order to have control over consultations and decide about the implementation of mining projects in Latin America. According to Chapter 10, the scaling down of environmental governance to the local level is fundamental in safeguarding the self-determination of local communities. Chapter 9 argues that social capital and institutional strength in communities are key factors for the protection of forest commons and for local capacities to face traditional and emergent pressures on forest ecosystems.

The extent to which local communities and social movements succeed in bringing about institutional change partly depends on their interactions with other actors. In this respect it is also important to point out that social actors (the state, corporations, communities, etc.) are not homogeneous entities. They may consist of various groups with different power, interests and positions, which may shift over time. Local governments, for example, occasionally confront central governments by developing alliances with local communities or other state agencies. Also, experts from corporations, governments and environmental organizations may take very different stances on energy efficiency, production technologies and social responsibilities, despite the fact that they work in the same sector or country (see the analysis of views and discourses of strategic actors in Chapter 6). In some cases, environmentalists support local communities against development policies that promote the expansion of infrastructure and extractive industries in fragile ecosystems (Chapter 11). In other cases, they may favour compensatory schemes in conservation policies, regardless of the criticism raised by environmental justice movements (Chapter 8).

In sum, while the central state has repositioned itself in processes of environmental governance of Latin America, institutional adaptation to the new contexts, discourses and demands has come from a range of (contesting) actors, and the interactions among them, across multiple scales. Overall, elite groups have tried to adjust some of their discourses and practices in order to partly comply with new demands and regulations, without having to give up their prominent position. Simultaneously, various marginalized groups have attempted to strike back by (re-)establishing and (re)appropriating local decision-making processes in order to regain their autonomy. To what extent these institutional adaptations may lead to structural transformations in environmental governance remains to be seen.

## Environmental governance in the making

Environmental governance is a social arena of multiple demands, goals and images of nature, in which priorities and trade-offs are negotiated according to the interests of those who are able to influence decision-making. In Latin America, several social and institutional arrangements through which environmental governance takes place are currently changing. Trends such as the repositioning of the national state (Chapter 4), the emergence of new elite groups (Chapter 5) and the development of new mining technologies (Chapter 6) are largely supportive of the increasing resource extraction for global markets, which is a cause for numerous environmental conflicts in the region (Chapter 2). At the same time, however, new communication means (Chapter 11), knowledge exchanges (chapters 3 and 9), increased attention for social rights (Chapter 7) and strengthened bottom-up organizations (chapters 9, 10 and 11) create opportunities for marginalized groups to counter top-down political and economic processes that greatly affect the lives of people who have limited voice.

Whether new trends in Latin America's environmental governance will prove to have transformative implications depends on how relevant actors are involved in the process. In this respect, the contributions to this book reveal profound tensions between the compensatory approaches favoured by governments and corporations (chapters 4, 5, 6, 7 and 8), and the participatory proposals and practices of socioenvironmental analysis, political decision-making and economic production that are championed by local communities and activists (chapters 2, 3, 9, 10 and 11). Although compensation can be a means for dealing with social and environmental debts and injustices, an overly strong emphasis on local "damage control", financial reparation and social projects not only legitimizes practices that threaten the integrity of fragile ecosystems but also jeopardizes a protagonist role of local communities in environmental governance. While a second generation of environmental justice movements is taking a lead in struggles over resource-related meanings and rights (Chapter 2), compensatory policies gain space in Latin America in the context of resource-based economic growth and poverty reduction (chapters 4 and 7).

The tension between participatory and compensatory approaches is in practice often not so evident or clear-cut. Take, for instance, the political visibility of injustices and the institutionalization of rights granted to marginalized groups, especially indigenous peoples, since the 1990s. While meaningful progress has undoubtedly been made, this is partly overshadowed by neoliberal and post-neoliberal institutional adaptations that give greater power to corporations and the state, and more room to expansionary large-scale production and infrastructure projects that tend to threaten the livelihoods of some of the same marginalized groups. By the same token, participation, formerly defined as full involvement of local groups in decision-making over socioenvironmental change, has been reframed to include marginalized groups mainly as co-beneficiaries through compensation schemes. Paradoxically, as state agencies more actively promote participatory initiatives, local populations may in fact be less actively involved in decision-making. And especially when coalitions between the state and corporations foster the expansion of natural resource exploitation (chapters 2, 4 and 5), the genuine participation and empowerment of local communities has been limited, and in some cases protests have even been criminalized in the name of progress and national security (Chapter 11; see also Taylor, 2011; Saguier, 2012; Zibechi, 2012).

In addition to economic and social compensation, the fast transformation of rural areas reveals a trend towards territorial compensation, in which some protected areas are supposed to make up for the vast areas where large-scale productive or extractive activities are basically given a free hand (Castro, 2014; see also Zimmerer, 2011). The expansion of protected areas (e.g. parks, reserves and ethnic communities) by national governments is primarily aimed at protecting forests, coinciding with national and international climate change and biodiversity policies (Chapter 8; see also Castro, 2013). In many cases, the expansion of these activities and infrastructure takes place in environmentally and socially sensitive areas, and forces peasants and traditional communities to fight for their autonomy, food and land security. Meanwhile, from this ongoing territorial reconfiguration, new inequalities, injustices and vulnerabilities emerge. While productive territories become gradually more concentrated in the hands of elite groups, secluded protected areas where land-use activities are limited by market constraints and restrictive rules are allocated to the rural poor.

Finally, this book's collection of studies shows that in order to tackle the current and emerging socioenvironmental problems in Latin America, three main challenges must be urgently addressed: first, the political challenge of promoting democracy and citizenship in a public space that is safeguarded for effective participation in the agenda-setting and negotiation of conflicting interests; second, the social challenge of ensuring the improvement of wellbeing through food and land security, social reproduction and self-determination of marginalized groups; and third, the environmental challenge of protecting ecological integrity, carbon emission mitigation and adaptation to climate change.

## Notes

1. See, for example, The Economics of Ecosystems and Biodiversity (TEBB) – www.teebweb.org/.
2. See http://www.engov.eu/bd_justicia_ambiental_es.php.

## References

Alayza, A. and Gudynas, E. (eds) (2011) *Transiciones. Post extractivismo y alternativas al extractivismo en el Perú* (Peru: RedGE and CEPES).
Alimonda, H. (ed.) (2011) *La Naturaleza Colonizada: Ecología Política y Minería en América Latina* (Buenos Aires: Ediciones Ciccus, CLACSO).
Alimonda, H. and Gandásegui, M.A. (2006) *Los Tormentos de la Materia: Aportes para una Ecología Política Latinoamericana* (Ciudad de Buenos Aires: CLACSO).
Alimonda, H., Hoetmer, R. and Saavedra Celestino, D. (eds) (2009) *La Amazonía Rebelde. Perú 2009* (Buenos Aires: CLACSO).
Altieri, M.A. and Toledo, V.M. (2011) "The Agroecological Revolution in Latin America: Rescuing Nature, Ensuring Food Sovereignty and Empowering Peasants", *Journal of Peasant Studies* 38(3): 587–612.
Barkin, D. and Lemus, B. (2011) "La Economia Ecológica y Solidaria: Una Propuesta frente a nuestra Crisis", *Sustentabilidades* 5: 1–13.
Baud, M., Castro F. and Hogenboom, B. (2011) "Environmental Governance in Latin America: Towards an Integrative Research Agenda", *European Review of Latin American and Caribbean Studies* 90: 79–88.
Bebbington, A. (ed.) (2012) *Social Conflict, Economic Development and Extractive Industry: Evidence from South America* (London: Routledge).
Berkes, F. and Folke, C. (eds) (1998) *Linking Social and Ecological Systems: Management Practices and Social Mechanisms for Building Resilience* (Cambridge: Cambridge University Press).
Biermann, F. (2007) "'Earth System Governance' as a Crosscutting Theme of Global Change Research", *Global Environmental Change* 17(3–4): 326–337.
Biermann, F. and Pattberg, P. (2008) "Global Environmental Governance: Taking Stock, Moving Forward", *Annual Review of Environment and Resources* 33: 277–294.
Borras, Jr. S., Franco, J.C., Gómez, S., Kay, C. and Spoor, M. (2012) "Land Grabbing in Latin America and the Caribbean", *The Journal of Peasant Studies* 39(3–4): 845–872.
Bovarnick, A., Alpizar, F. and Schnell, C. (eds) (2010) *The Importance of Biodiversity and Ecosystems in Economic Growth and Equity in Latin America and the*

*Caribbean: An Economic Valuation of Ecosystems* (United Nations Development Programme).
Bray, D.B., Merino L. and Barry, D. (eds) (2005) *The Community Forests of Mexico. Managing for Sustainable Landscapes* (Austin: University of Texas Press).
Bretón, V. (2013) "Etnicidad, Desarrollo y 'Buen Vivir': Reflexiones Críticas en Perspectiva Histórica", *European Review of Latin American and Caribbean Studies* 95: 71–95.
Bunker, S.G. (1988) *Underdeveloping the Amazon. Extraction, Unequal Exchange and the Failure of the Modern State* (Chicago: University of Chicago Press).
Carruthers, D.V. (ed.) (2008) *Environmental Justice in Latin America: Problems, Promise, and Practice* (Boston: MIT Press).
Castro, F. (2012) "Multi-Scale Environmental Citizenship: Traditional Populations and Protected Areas in Brazil", in A. Latta and H. Wittman (eds), *Environment and Citizenship in Latin America: Natures, Subjects, and Struggles*, pp. 39–58 (New York: Berghahn Books).
Castro, F. (2013) "Crossing Boundaries in Environmental Governance". *Analytical Framework Report* (ENGOV), http://www.engov.eu/documentos/AFR_WP10_D10_1.pdf.
Castro, F. (2014) "Ethnic Communities: Social Inclusion or Political Trap?", in F. Castro, P. van Dijck and B. Hogenboom (eds), *The Extraction and Conservation of Natural Resources in Latin America: Recent Trends and Challenges*. Cuadernos CEDLA 27 (Amsterdam: CEDLA).
Castro, J.E. (2008) "Water Struggles, Citizenship and Governance in Latin America", *Development* 51: 72–76.
Cleaver, F. (2002) "Reinventing Institutions: Bricolage and the Social Embeddedness of Natural Resource Management", *The European Journal of Development Research* 14(2): 11–30.
Demmers, J., Fernández Jilberto, A.E. and Hogenboom, B. (eds) (2004) *Good Governance in the Era of Global Neoliberalism: Conflict and Depolitization in Latin America, Eastern Europe, Asia and Africa* (New York: Routledge).
Dijck, P. van (ed.) (2014) *What Is the Future for Amazonia? Socio-Economic and Environmental Transformation and the Role of Road Infrastructure*, Cuadernos CEDLA 28 (Amsterdam: CEDLA).
Dobson, A. (2003) *Citizenship and the Environment* (Oxford: Oxford University Press).
*Ecuador Debate* (2011) "Problemas y Perspectivas del Extractivismo" (Tema Central) 82: 45–135.
Fernández Jilberto, A.E. and Hogenboom, B. (eds) (2010) *Latin America Facing China: South-South Relations Beyond the Washington Consensus* (Oxford: Berghahn Books).
Gavaldà i Palacin, M. (2013) *Gas Amazónica. Los Pueblos Indígenas frente al Avance de las Fronteras Extractivas en Perú* (Barcelona: Icaria).
Giarracca, N. and Teubal, M. (eds) (2013) *Actividades Extractivas en Expansión ¿Reprimarización de la Economía Argentina?* (Buenos Aires: Antropofagia).
GRAIN (2013) "The United Republic of Soybeans: Take Two", *Against the Grain*. Published online on 2 July 2013. http://www.grain.org/article/entries/4749-the-united-republic-of-soybeans-take-two, date accessed 12 December 2014.
Gudynas, E. (2011) "Buen Vivir: Today's Tomorrow", *Development* 54(4): 441–447.

Gudynas, E. (2009) "Ciudadanía Ambiental y Meta-Ciudadanías Ecológicas: Revisión y Alternativas en América Latina", *Desenvolvimento e Meio Ambiente* 19: 53–72.
Hall, A. (2012) *Forests and Climate Change. The Social Dimensions of REDD in Latin America* (Cheltenham and Northampton: Edward Elgar).
Harris, R.L. (2003) "Popular Resistance to Globalization and Neoliberalism in Latin America", *Journal of Developing Societies* 19(2–3), 365–426.
Harvey, D. (2003) *The New Imperialism* (Oxford: Oxford University Press).
Haarstad, H. (ed.) (2012) *New Political Spaces in Latin American Natural Resource Governance* (New York: Palgrave Macmillan).
Helmke, G. and Levitsky, S. (eds) (2006) *Informal Institutions and Democracy: Lessons from Latin America* (Baltimore: Johns Hopkins University Press).
Hogenboom, B. (2012) "Depoliticized and Repoliticized Minerals in Latin America", *Journal of Developing Societies* 28(2):133–158.
Keck, M.E. and Sikkink, K. (1998) *Activists Beyond Borders: Advocacy Networks in International Politics* (Ithaca: Cornell University Press).
Kooiman, J. (2003) *Governing as Governance* (Los Angeles: Sage).
Kooiman, J., Bavinck, M., Jentoft, S. and Pullin, R. (eds) (2005) *Fish for Life: Interactive Governance for Fisheries* (Amsterdam: Amsterdam University Press).
Larson, A.M. (2003) "Decentralization and Forest Management in Latin America: Towards a Working Model", *Public Administration and Development* 23: 211–226.
Latta, A. and Wittman, H. (2012) *Environment and Citizenship in Latin America: Natures, Subjects and Struggles* (Oxford: Berghahn Books).
Lemos, M.C. and Agrawal, A. (2006) "Environmental Governance", *Annual Review of Environment and Resources* 31: 297–225.
Liverman, D.M. and Vilas, S. (2006) "Neoliberalism and the Environment in Latin America", *Annual Review Environment and Resources* 31: 327–363.
Lyon, T. (2009) "Environmental Governance: An Economic Perspective", in M.A. Delmas and O.R. Young (eds), *Governance for the Environment: New Perspectives*, pp. 43–68 (Cambridge: Cambridge Academic Press).
Martinez-Alier, J. (2002) *The Environmentalism of the Poor: A Study of Ecological Conflicts and Valuation* (Cheltenham: Edward Elgar Publishing).
McCay, B.J. and Acheson, J.M. (eds) (1990) *The Question of the Commons: The Culture and Ecology of Communal Resources* (Tucson: University of Arizona Press).
Miller, S.W. (2007) *An Environmental History of Latin America* (Cambridge: Cambridge University Press).
Mol, A. (2003) *Globalization and Environmental Reform: The Ecological Modernization of the Global Economy* (Boston: MIT Press).
Ostrom, E. (1990) *Governing the Commons: The Evolution of Institutions for Collective Action* (Cambridge: Cambridge University Press).
Otero, G. (ed.) (2010) *Food for the Few: Neoliberal Globalism and Biotechnology in Latin America* (Austin: University of Texas Press).
Painter, M. and Durham, W.H. (eds) (1995) *The Social Causes of Environmental Destruction in Latin America* (Ann Arbor: University of Michigan Press).
Perreault, T. (2005) "State Restructuring and the Scale Politics of Rural Water Governance in Bolivia", *Environment and Planning* A 37: 263–284.
Radhuber, I.M. (2014) *Recursos naturales y finanzas públicas. La base material del Estado plurinacional de Bolivia* (La Paz: Plural Editores).

Rios, S.P. and Veiga, P.M. (2010) "Tackling Climate Change in Latin America and the Caribbean: Issues for an Agenda", *Integration and Trade* 3: 55–70.

Robbins, P. (2012) *Political Ecology: A Critical Introduction*, 2nd edn (Chichester: Blackwell Publishing).

Russi, D., Gonzales-Martinez, A.C., Silva-Macher, J.C., Giljum, S., Martinez-Alier, J. and Vallejo, M.C. (2008) "Material Flows in Latin America: A Comparative Analysis of Chile, Ecuador, Mexico, and Peru, 1980–2000", *Journal of Industrial Ecology* 12(5–6): 704–720.

Saguier, M. (2012) "Socio-Environmental Regionalism in South America: Tensions in New Development Models", in P. Riggirozzi and D. Tussie (eds), *The Rise of Post-Hegemonic Regionalism*, pp. 125–146 (Dordrecht: Springer).

Schilling-Vacaflor, A. (2012) "Democratizing Resource Governance Through Prior Consultations? Lessons from Bolivia's Hydrocarbon Sector", GIGA Working Paper, No. 184 (Hamburg: German Institute of Global and Area Studies).

Schmink, M. and Jouve-Martín, J.R. (2011) "Contemporary Debates on Ecology, Society, and Culture in Latin America", *Latin American Research Review* 46, special issue.

Silva, E. (2012) "Environment and Sustainable Development", in P. Kingstone and D.J. Yashar (eds), *Routledge Handbook of Latin American Politics*, pp. 181–199 (New York: Routledge).

Svampa, M. (2011) "Extractivismo Neodesarrollista y Movimientos Sociales. ¿Un Giro Ecoterritorial hacia Nuevas Alternativas?", in Grupo Permanente de Trabajo sobre Alternativas al Desarrollo, *Más Allá del Desarrollo* (Quito: Fundación Rosa Luxemburg and Abya Yala).

Taylor, L. (2011) "Environmentalism and Social Protest: The Contemporary Anti-Mining Mobilization in the Province of San Marcos and the Condebamba Valley, Peru", *Journal of Agrarian Change* 11(3): 420–439.

Teijlingen, K. van and Hogenboom, B. (2014) "Development Discourses at the Mining Frontier: Buen Vivir and the Contested mine of El Mirador in Ecuador", ENGOV Working Paper Series, No. 15.

Therborn, G. (2011) "Inequalities and Latin America: From the Enlightenment to the 21st Century", *DesiguALdades.net*, Working Paper Series, No. 1.

UNEP (2012) *Latin America and the Caribbean. Global Environment Outlook*, http://www.unep.org/geo/pdfs/geo5/GEO5_report_full_en.pdf, date accessed 2 December 2014.

UNEP (2011) *Towards a Green Economy: Pathways to Sustainable Development and Poverty Eradication*. www.unep.org/greeneconomy, date accessed 23 October 2014.

Urkidi, L. and Walter, M. (2011) "Dimensions of Environmental Justice in Anti-Gold Mining Movements in Latin America", *Geoforum* 42(6): 683–695.

Vásquez, P.I. (2014) *Oil Sparks in the Amazon: Local Conflicts, Indigenous Populations and Natural Resources* (Athens, GA: University of Georgia Press).

Walter, M. and Martinez-Alier, J. (2012) "Social Metabolism, Ecologically Unequal Exchange and Resource Extraction Conflicts in Latin America", *Analytical Framework Report* (ENGOV).

World Bank (1992) *Governance and Development* (Washington, DC: World Bank).

Young, O. (1999) *Governance in World Affairs* (Ithaca: Cornell University Press).

Zibechi, R. (2012) "Latin America: A New Cycle of Social Struggles", *NACLA Report on the Americas* 45(2): 37–49.

Zimmerer, K.S. (2011) "'Conservation Booms' with Agricultural Growth? Sustainability and Shifting Environmental Governance in Latin America, 1985–2008 (Mexico, Costa Rica, Brazil, Peru, Bolivia)", *Latin American Research Review* 46: 82–114.

Zimmerer, K.S. and Bassett, T.J. (eds) (2003) *Political Ecology: An Integrative Approach to Geography and Environment-Development Studies* (New York: Guilford Press).

 Except where otherwise noted, this work is licensed under a Creative Commons Attribution 3.0 Unported License. To view a copy of this license, visit http://creativecommons.org/licenses/by/3.0/

# Part I
# Setting the Stage

OPEN

# 1
# Origins and Perspectives of Latin American Environmentalism

Joan Martinez-Alier, Michiel Baud and Héctor Sejenovich

## Introduction

The debate on the socioenvironmental challenges faced by Latin America has a long history. This history is crucial to understanding Latin American perspectives on environmental governance and, above all, to understanding the specific characteristics which determine these perspectives. Traditional debates on environmental governance tend to see the Western debates on nature and environment as determining views and perspectives on a global scale. The suggestion is that Latin American environmental debates were directed by the changing views in the industrialized world. This chapter, however, suggests that Latin America has developed its own strands and perspectives on environmental issues which were emerging from its peculiar historical position. A focus on the specific, and to a large extent autonomous, knowledge development on nature and environment allow us to understand the determining roots of Latin American ideas on environmental governance.

Latin American environmental ideas are closely connected to an environmental history since the Spanish Conquest, which was characterized by a dramatic drop in population and a series of export booms driven by one commodity after another. An early case in point may be the exportation of guano from Peru that amounted to about 11 million tons over 40 years, from 1840 to 1880, and was based on the exploitation of indentured Chinese workers (Gootenberg, 1993). In the last decades of the nineteenth century and in the beginning of the twentieth century, the entire Latin American region experienced a dramatic boom in agriculture for exportation. New crops such as coffee, cacao and banana, along with more traditional goods such as sugar, changed the economic and ecological context of much of Latin America as well as the lives of large sectors of its population. The agrarian frontier expanded, and large territories, often in the interior of the new republics, were deforested and occupied by new forms of agriculture. The expansion of coffee cultivation in Antioquia, Colombia, and of cacao in the interior of Ilhéus in the north-east of Brazil have been iconic examples, just like rubber and henequen in southern and south-eastern Mexico, the banana belt in Central America, Colombia and Ecuador, and the occupation of the Pampas in Argentina and southern Brazil (for a number of examples, see Topic, Marichal and Frank, 2006). Cuban sugar export increased from 1 million tons per year around 1900 to 3 million tons by 1920, causing dramatic deforestation on the island (Funes Monzote, 2004a, 2004b). This sacrifice was unaccounted for in the modernizing ideology of the time, epitomized by Arango Parreño's slogan of 1770, "*sin azúcar no hay país*" ("without sugar, no country") (Moreno Fraginals, 1978).

This expansion of the agrarian frontier was accompanied by ideologies of progress, the incorporation of new business elites, and a strong dependence on the international market. With the Chilean triumph in the Pacific War (1879–1883) and the incorporation of Antofagasta and Tarapacá, Chile became the world's principal producer of the mineral saltpetre. The exportation of this sodium nitrate increased until 1914 and remained constant until the crisis of 1929, oscillating between 1.5 and 3 million tons per year (Miller and Greenhill, 2006). This provoked an economic boom like the country had not experienced before.

In the beginning of the twentieth century, the oil industry in Venezuela and Mexico began to grow, causing ecological and social disasters at a scale unknown at the time (Santiago, 2006). This process continues today: the calculation (in tons) of primary materials that are exported (West and Schandl, 2013) reveals a multiplication of four, from 1970 to 2010.[1] As an example, Venezuela exports roughly 120 million tons of oil per year.

Recently, with the expansion of the Chinese economy, the extraction of natural resources (not only minerals and oil but also agrarian products, such as soy) has grown at an extraordinary rate. The Government of Uruguay is considering exporting 18 million tons of iron ore per year under the Aratirí project. Meanwhile, Chile exports 5 million tons of copper per year, which requires the removal of land, enormous production of slag and a large input of energy. Colombia exports almost 100 million tons of coal per year; Brazil annually exports 400 million tons of soy and iron ore. There are signs that the recent economic bonanza from primary exports is coming to a halt in 2015, reinforcing the critiques from the "post-extractivist" school. However, this might be only a temporary situation. New supplies of energy and materials from Latin America will find markets, and domestic and foreign demand.

## The beginning

The population of the American continent suffered an enormous drop during the Spanish colonization. The population was drastically reduced by the exploitation to which it was subjected, but the "Great Dying", as it was called by Eric Wolf (1982: 133ff), was primarily due to the spread of infectious diseases. From an estimated 140 million people in the year 1500, only 40 million were registered 60 years later (Tudela, 1990; also Sánchez-Albornoz, 1984). The American population, which had a size comparable to that of Europe at the time, dropped some 80%. This historical process is unparalleled in other continents with the exception of Australia and a few other places in the world (e.g. the Canary Islands, Hawaii) that have experienced a similar phenomenon. The decrease in the native population – and its slow substitution by an immigrant population in the neo-European (as they were called by Crosby, 2004) and also later in the humid tropics – should be understood as a biological as well as a military process. The conquistadores arrived in new territories in search of riches. They had little mercy for the native population and, unwittingly but also relentlessly, they contaminated it with new fatal illnesses.

However, the depopulation in the first century after the colonization can not only be attributed to the arrival of Hernán Cortés and Francisco Pizarro and their troops in the former Mexican and Andean empires (or even before they arrived, as death travelled fast). The archaeology of the Amazon today confirms the existence of population densities much greater than those during several centuries following the conquest. There had already been collapses of empires, and perhaps also of populations before the Spanish Conquest, such as in the Mayan territory, but what happened in the American demography after 1492 had no precedent on a continental scale and throughout the history of mankind.

Today's low population density in Latin America (with local exceptions such as El Salvador and Haiti) negates one of the principle arguments in ecological thinking, namely, that population density is the key problem of environmental degradation. Nowhere in Latin America is there an issue of overpopulation as in Europe (with densities of up to 300 people per square kilometre in Germany, Italy and England) or in India and Bangladesh. In Latin America, population increase later became an explicit policy of modernist governments. In this sense, the famous remark by Argentinian Juan Bautista Alberdi in 1852, "to govern is to populate", is symbolic of the mindset of the Latin American elites of that time. Much later, during the time of the military dictatorship (1964–1986), the Brazilian state – in its geopolitical delirium – called for an increase in birth rate in order to populate the Amazon against foreign threats.

Ecology and demographics thus changed rapidly in the context of early colonization. Under the rule of one single dynasty – the Habsburgs – for the first 200 years, the Spanish American territories saw enormous ecological and demographic changes. Invasive species arrived (Melville, 1999), whereas the expansion of modern mining methods (modern in technology and scale) in regions such as Potosí, Zacatecas and also Minas Gerais led to a great decrease in population and enormous pollution by mercury (Machado Araoz, 2014). In a later stage, the frontiers of silver and gold extraction and – almost always at the same time – of deforestation moved to those of sugarcane in the Caribbean and the north-east of Brazil, and later the regions that produced and exported coffee, rubber, wood such as mahogany and quebracho, meat, banana, soy, copper, oil and coal, iron ore and bauxite (Brannstrom, 2004).

## Conservationist environmentalism

Despite the anthropogenic changes that happened before and after 1492, Latin America managed to conserve immense biological diversity in many of its diverse ecosystems. The Amazon had scarcely been touched before the rubber whirlwind at the end of the nineteenth century. This enormous biological richness attracted the attention of European explorers such as Alexander von Humboldt (1769–1859), the renowned Prussian scientist. Without his explorations of this part of the world that came to be known as the "Neotropics", biogeography, the study of the geographical distribution of plants and other life forms, would not have been developed in the same way. His intention, which he never accomplished, was to return to Latin America once it had become independent and to direct an academy with scientific correspondents from Mexico to Patagonia.

On 29 July 1822, when he was in Paris, Humboldt wrote a letter to Simon Bolívar introducing him to the young mining experts, Jean Baptiste Boussingault and Mariano de Rivero. Some years later, in his *Memoria sobre el Guano de los Pájaros* (1827), Mariano de Rivero remembered how Humboldt had given samples of guano to Fourcroy and Vauquelin who analysed the chemical elements of this fertilizer. Still later, Mariano de Rivero regretted that Peru had not durably invested the revenues from guano exports in a policy that we now call "weak sustainability" (Alcalde Mongrut, 1966). This renewable product was exported at such a rate that it led to its depletion. It should have been invested in businesses that could have generated permanent income. This proposal is similar to that which was later proposed by Uslar Pietri in Venezuela in 1936, baptized as the "sowing of the oil" (*sembrar el petróleo*) (Martínez-Alier and Roca, 2013: 116–117).

Humboldt described the geology, volcanoes, biogeography and the richness of species of the American territories that he visited between 1799 and 1805. Later – and largely due to Darwin – Latin America came to hold a privileged role in the science of biological evolution. Darwin's explanation of the origin of species owes much to his trip to America during the Beagle mission (from 1831 to 1836) to collect materials. He came up with ideas that eventually, after his crucial stay in the Galápagos, led him to express his astonishment at the number of endemic species, given that the islands had only come to exist in a geologically recent period. By observing finches and variations in the size and form of their beaks (which ecotourists continue to discuss today), he concluded that only one race of such birds had arrived and established itself on the archipelago, and that new species had arisen through adaptation to specific food sources.

South America was therefore crucial to the history and evolution of biology as well as the history of agrarian chemistry and the development of the idea of "social metabolism". By 1840, Liebig, Boussingault and other scientists, based on the analysis of Peruvian guano and other fertilizers, determined that plants need three principal nutrients – phosphor, potassium and nitrogen – and that agriculture should evolve from a system of plundering to one of restitution (McCosh, 1984: 81–82). The fertilizing properties of guano were known by the historic inhabitants of Peru but had not been described or analysed in chemical terms. Guano had global importance – it was exported as a fertilizer but also served and strongly influenced the minds of the agrarian chemists (Gootenberg, 1993; Cushman, 2013).

In the course of the nineteenth century, conservationist environmentalism increased. Most intellectuals and politicians lived in parts of Latin American cities which were somewhat removed from the environmental destruction caused by mining and by the agro-export model. Gradually, however, urban populations also started to be confronted by issues of pollution and environmental destruction in their own habitat. This was most directly the case with dirty water, sanitation and infectious diseases, which alarmed urban elites. The growth of cities also led to environmental destruction and deforestation to which they could not close their eyes. Warren Dean presented some impressive estimates about urban-led deforestation in Brazil. He calculated that a city such as Rio de Janeiro consumed at least 270,000 tons of firewood every year in the 1880s (almost 20% provided by mangroves). For the construction of a small brick house, 37 tons of firewood may have been needed. This would mean that the buildings of the city of Rio de Janeiro by 1890 cost the deforestation of 200 square kilometres (Dean, 1995: 196–197). He may have overstated his case and exaggerated the importance of wood as the principal source of energy for Brazil's urban growth (Brannstrom, 2005), but there is no doubt that the relentless progress promoted by Latin American elites came at the cost of rapid deforestation.

These developments led to a plethora of environmental research. The distinct biomes of the Americas have all had their iconic researchers. The dry tropical forest of the Chaco was studied by the great ecologist Jorge Morello (1932–2013). He sponsored excellent collective research at the University of Buenos Aires, on the Pampas and the Chaco, and also on the coastal areas and the conurbation of Buenos Aires (e.g. Morello and Matteucci, 2000). He occupied the post of director of National Parks for a short time under the government of Raúl Alfonsín. In the ecological and political history of Argentina, the logging of red quebracho for railroad ties and the export of tannin for tanneries (by the British company La Forestal) in Santa Fe and in the Chaco during the first 40 years of the twentieth century played a notable role. In Argentina there has been active conservationism since the end of the nineteenth century, responsible for the creation of various national parks in different ecosystems. The dedication of Maximina Monasterio to the study of the Andean *páramo* has been similar to that of Jorge Morello in the Chaco. Born of a Galician refugee family in Argentina, educated and graduated with a doctorate in ecology in France, with long sojourns in Bolivia and exiled to Venezuela in 1966, she has been a crucial figure in research on and education about the Andean highlands from Venezuela to Ecuador. Monasterio studied, in her own words, "from the *frailejones* to the potatoes" (i.e. both the "wild" and the agricultural biodiversity of the highlands) (Monasterio, 2003). Today the ecosystemic services provided by the *páramos* are common knowledge – as sources of water for the people in the lowlands and their livestock. Thus in Colombia the biodiversity research institute (Instituto de Investigación de Recursos

Biológicos) "Alexander von Humboldt" is currently in charge of delimiting and protecting the *páramo* ecosystems, and in this way of preventing coal mining in such areas.

In Mexico, Arturo Gómez Pompa, a biologist at the National Autonomous University of Mexico (Universidad Nacional Autónoma de México (UNAM)) and of the same generation as Morello and Monasterio, studied the ecology of tropical forests and ethnobotany (see http://www.agomezpompa.org). He was one of the most prominent voices in denouncing deforestation in south-east Mexico. He is also known for having discovered the chocolate tree in the Mayan jungle. The idea of the cultivated jungle (or the "cultured jungle", as Philippe Descola (1986) called the Amazonian Achuar forest) became very important in Latin American conservationism.

Conservationism in Latin American is a consequence of foreign influence but it also has its own local tradition. It uses universal and more or less strict instruments, such as the Constitution of the National Parks, the inclusion of wetlands and marshes in the list of the international Ramsar Convention, and the Biosphere Reserves sponsored by UNESCO. The natural reserves have sometimes been protected by the support of international conservationism. However, many countries rightly stress the importance of their own national scientists and public policy-makers in the designing of conservationist policies. In Peru, the forest engineer Marc Dourojeanni played an important role in establishing protected areas – around 1970 during the administration of Velasco Alvarado – to save both the vicuña in the Andean highlands and the Amazonian forests (Dourojeanni, 1988, 1990). In Mexico the conservation efforts of figures such as Enrique Beltrán and Miguel Angel de Quevedo (Simonian, 1995) are still well remembered 100 years later. In Ecuador, Nicolás Cuví has highlighted the figure of Acosta Solís, botanist and conservationist, with one foot in his country and the other in the USA (Cuví, 2005). The latter's research on the remnants of the quinine tree (the tree that is on the shield of the Republic of Peru) became suddenly relevant by the Second World War when the US troops were fighting in the Pacific tropics and were threatened by malaria.

More than a century ago, part of the Amazon suffered from the onslaught of the rubber boom, which had a significant negative impact on indigenous populations. Another principal threat is perhaps the global climate change that could convert the rainforest into savannah. Meanwhile, the Atlantic Forest in Brazil, the forests of southern Mexico and Central America, like the forests of southern Chile and Argentina, were largely destroyed in the twentieth century by grazing, agricultural crops and monocultures of trees such as pine and eucalyptus. José Augusto Pádua has explained how the statesman José Bonifacio predicted the destruction of the coastal forests as early as the moment of Brazilian independence. Conservationists such as Alberto Torres (born in 1865 on a plantation in Rio de Janeiro that was already in decline because of soil erosion) also publicly deplored the forest destruction in the march of extractivist civilization towards the interior (Pádua, 2002, 2010; see also Drummond, 1997).

It is noteworthy to mention that, in the conservation movement of 80 years ago, there was already a major controversy. Ciriacy-Wantrup suggested that "conservationism itself may not mean non-use". This Berkeley economist anticipated an economic approach to sustainability. His major book was published in 1952 and its translation (by Edmundo Flores, an agricultural economist), published in Mexico in 1957, had an important impact on the region (Ciriacy-Wantrup, 1957).

In summary, there is a Latin American conservationist tradition with deep historic roots. It found scientific support in the sciences of biogeography and conservation biology, and also, later, in the economics of natural resources and the study of watersheds. Different from the popular environmentalism and the agroecology and post-development movements that we shall analyse below, this conservationist trend has had powerful support in the North, among organizations such as the International Union for Conservation of Nature (IUCN), the WWF and other international institutions, such as the US Resources for the Future, and the Smithsonian.

## Agroecology and post-developmentalism

The agroecological pride of the Andean and Mesoamerican regions (with authors such as Chilean Miguel Altieri and Mexican Victor Toledo) (Altieri and Toledo, 2011) has roots that are even older than conservationism, but it did not manifest itself significantly until the 1970s and 1980s. A good example of this new visibility was the Andean Project for Peasant Technologies (PRATEC) in Peru, which was established by dissident agronomists from the school of La Molina. In this school they had learned the technological simplification as the result of the focus on the main export crops, sugar and cotton, that included the elimination of native varieties of coloured cotton. They reacted against this teaching (Proyecto SEINPA, 1990) and were critical of the notion of uniform "development". They were responsible for the first edition in

Spanish in 1996 of *The Development Dictionary* edited by Wolfgang Sachs, a post-developmentalist classic (Sachs, 1981). They began to research and apply the agrarian epistemologies of the indigenous inhabitants of the Sierra, expressed in the conservation and use of many varieties and species of seeds.

Latin American environmentalism is different from that of the USA as it has drawn significantly from ancestral agricultural practices and respect for indigenous knowledge. There is a line from the agroecological studies and practices of the influential agronomist from Chapingo, Efrain Hernández Xolocotzi (1913–1991), whose career (in the USA and in Mexico) culminated in a substantial and competent school of Mexican ethnoecologists, to the peasant movement in Mexico which manifests itself in the twenty-first century under the motto "without maize, no country" (*sin maíz no hay país*) (Esteva and Marielle, 2003). Victor Toledo (*La Jornada*, 5 August 2014) asserts that the indigenous agrarian Mesoamerican civilization survives and persists: "These indigenous populations are the principle opponents to the industrial civilization model." Indigenous agriculture and agroforestry are major sources of Latin America environmentalism.

In order to understand traditional Latin American agricultural systems, it is necessary to enter into a "dialogue of knowledges", if not a rejection of Western thought. The communities whose situation and practices have been studied by anthropologists and agronomists bring to the table their own perspectives and knowledge to guide the research, an idea that Robert Chambers of Sussex University (Chambers, 1983) developed from Paulo Freire and Orlando Fals Borda, important Latin American intellectuals. This dialogue of knowledges is also shared by environmentalists in other contexts, such as in Funtowicz and Ravetz's doctrine of "post-normal science", which supports and even requires an "extended peer review" in situations of technological uncertainty and of urgent decisions (Funtowicz and Ravetz, 2000).

Even more radically, political ecologist Héctor Alimonda explains that environmental degradation is caused by "persistent colonialism". He writes: "Over five centuries, entire ecosystems were destroyed by the implementation of monoculture export crops" (2011: 22). "Colonialism" is also useful for interpreting the environmental crisis in terms of the loss of indigenous knowledge and cultures, true "epistemicides" (Sousa Santos' word) that cannot be compensated by either Western science or by a dialogue of knowledges.

Patterns of economic and environmental sustainability in pre-Hispanic societies, which we know from archaeology or which have survived with many changes, express the social values of these societies. They are more useful for the period in which we live because they question the illusion of universal, uniformizing development. Arturo Escobar (1995, 2010) and Gustavo Esteva (who met with Ivan Illich in 1983) have been outstanding thinkers in the field of post-developmentalism, previous or parallel to the discussion of degrowth, *décroissance* or "prosperity without growth" in Europe.[2] They have deep roots in the Latin American mindset (or *Abya-Yala*, as it is sometimes called) but they also find inspiration in Ivan Illich, Cornelius Castoriadis and André Gorz, political ecologists of the 1970s, and in authors from India, such as Ashish Nandy and Shiv Visvanathan.

In Ecuador, the political debate after 2007 has introduced the concept of Sumak Kawsay, *Buen Vivir*, possibly after many hundreds or thousands of years of verbal usage. Since the year 2000, the concept has been revisited in articles and theses by Quechua intellectuals such as Carlos Eloy Viteri. Viteri comes from the Amazonian village of Sarayaku, which prevented a local oil-extraction project, and his ideas have been heavily influenced by this situation. Sumak Kawsay was converted into a national objective included in the Ecuadorian constitution of 2008, introduced under the presidency of Alberto Acosta in the constituent assembly (Hidalgo-Capitán et al., 2014).

Beyond disputes over the merits of these constitutional developments, the fact is that putting Sumak Kawsay central is very different from saying that the main objective being pursued is economic growth or even sustainable development. Sumak Kawsay is something similar to a solidary and ecological economy, which had already existed and needed to be recovered. It is a concept related to "post-developmentalism".

## Governments and international organizations: "Our own agenda"

Since the last decades of the nineteenth century, there have been voices of scientists as well as writers criticizing the indiscriminate use of natural resources, but they were never heard amid the obsession with the modernity of the time (Baud, 2013). In the second half of the twentieth century, the critique became more coherent and politically articulate. Although it occurred in the context of a global debate, it showed a markedly Latin American perspective and influenced the creation of what is now called an "environmental institutionalism" with new ministries, laws and regulations. Since Rachel Carson published *The Silent Spring* in 1962, and especially since the Meadows Report to the Club of

Rome in 1972, international environmentalism has taken off. At first this debate was scarcely considered by Latin American governments or by the Economic Commission for Latin America and the Caribbean (Comisión Económica para América Latina y el Caribe (ECLAC/CEPAL)). For them the problem of underdevelopment and poverty was the bigger issue, and their main objective was to augment the productive capacity of the region and to consolidate its economic expansion. Nevertheless, in those decades, all national governments created legal and administrative structures for natural resource management. It is important to note the creation of the United Nations Environment Programme (UNEP) at a worldwide level and furthermore the active participation of the Regional Office for Latin America and the Caribbean, which from 1975 onwards promoted courses and debates in all Latin American countries, effectively training university professors, NGOs, and personnel from natural resources and environment administrations.

With the support of UNEP, the Spanish Iniciativa de Copenhague para Centroamérica y México (CIFCA) was created and a multitude of courses and seminars were organized in Latin America and Europe. In 1980 the Latin American governments and universities decided to create their own Environmental Education Network. The Argentinian economist Héctor Sejenovich and the Colombian philosopher Augusto Angel Maya elaborated a plan for training and research. All countries had an office from the Environmental Education Network (Red de Formación Ambiental), in large part with governmental organizations but also with NGOs. In Europe a debate was initiated by Sicco Mansholt, president of the European Commission, who converted to the "growth below zero" doctrine upon reading the Meadows Report. This European debate, which involved the participation of André Gorz, Edgar Morin, Herbert Marcuse and other early ecological thinkers, was published in Santiago de Chile in 1972 and in Buenos Aires in 1975 with the spectacular title *Ecology and Revolution* (Marcuse, 1975). However, the book does not seem to have been influential, perhaps because of Latin America's military-led neoliberal backlash at the time.

In fact, the first articulated response to the environmental problems in Latin America came in the 1970s from the Bariloche Foundation in Argentina which in 1976 published the report *Catastrophe or New Society? Latin American World Model* (Herrera et al., 1976). In this report, various specialists such as Gilberto Gallopin developed a new environmental model for Latin America, in which the idea of the scarcity of natural resources was basically rejected. Gudynas (1999: 110) observes that these ideas were considered a direct attack on the idea of development and progress for Latin America. As a logical consequence, the reaction to the Meadows Report was negative, as is evident in the writings of Amilcar Herrera and Helio Jaguaribe (1973; see also Estenssoro Saavedra, 2014, cap. 7). The general conviction was that Latin American natural resources were abundant and that it was necessary to exploit them in order to develop the region. The Bariloche group emphasized two issues: the low population density of Latin America and its enormous and unknown ecological potentials. Latin American diplomats started to reject notions of "limits to growth" and believed that Latin America could resolve its problems of poverty and development, and at the same time achieve a more sustainable model, drawing also on the world's solidarity. This line of thought was very clear in Brazil, where the national ideology focused on the Amazon (Garfield, 2013). Before the Stockholm Conference of 1972, João Augusto de Araujo Castro, Brazilian diplomat of the United Nations, had asked for "a worldwide compromise on development" from and towards the poor countries. He talked of "a contamination of opulence and a contamination of poverty" (Estenssoro Saavedra, 2014: 129).

Since the mid-1970s, under the influence of Ignacy Sachs (who was a university professor in Paris and travelled to Mexico and Brazil), the notion of "ecodevelopment" spread (e.g. Sachs, 1981, 2008), long before sustainable development would triumph in the rhetoric of the Brundtland Report of 1987. Various Latin American authors, from within official organisms or as consultants or university professors, and people involved in activism – including Enrique Leff, Vicente Sánchez, Victor Toledo and Augusto Angel Maya – were inspired by the idea of ecodevelopment. As part of the actions of UNEP, and along with the participation of the University of Tehran (under the direction of Mohammad Taghi Fharyar), a network of ecodevelopment projects was established. In 1976 the first Symposium on Ecodevelopment was hosted at UNAM, organized by Enrique Leff.

In October 1974, UNEP organized a famous conference in Cocoyoc, Mexico. It was here that the so-called Charter of Obligations and Rights of the States was proclaimed. Above all else, Article 30 about environmental governance was important: "The protection, the preservation and the betterment of the environment for current and future generations is the responsibility of all States. They should try to establish their own environmental and development policies in accordance with this responsibility. The environmental policies of all States should promote and not adversely affect the current and future potential of development of developing countries."

In the 1970s and 1980s, ministries of the environment were created in various countries. The influence of UNESCO's Man and Biosphere (MAB) programme was evident, generating new interdisciplinary activity. An example is the reference to urban ecology and human settlements by Martha Schteingart at the Colegio de México (Schteingart y Graizbord, 1998). In economic management, Héctor Sejenovich proposed that to minimize degradation and waste it is necessary to take all costs into account, including those of the reproduction of nature (research, regeneration, control and management), and also all the potential benefits, for an integrated management of resources or, rather, an integrated management of the natural patrimony. The Latin American Council of Social Sciences (El Consejo Latinoamericano de Ciencias Sociales (CLACSO)) formed a working group on environment and development in 1978, led by Sejenovich (Estenssoro Saavedra, 2014, cap. 8). In Colombia, in the National Institute of Renewable Natural Resources and Environment (Instituto Nacional de los Recursos Naturales Renovables y del Ambiente (INDERENA)), Julio Carrizosa and Margarita Merino de Botero (who would later represent South America in the Brundtland Commission) began to take action. No less important was Anibal Patiño, whose early work addressed environmental problems in the Cauca Valley in Colombia (Patiño, 1991).

Environmental issues arrived at CEPAL in the form of a book edited by Osvaldo Sunkel and Nicolo Gligo, *Estilos de desarrollo y medio ambiente en la América Latina* (1981), published after developing activities for more than one year along with the UNEP Regional Office. They emphasized the notion of the ecosystem, the understanding that all of us are part of the same ecosystem and that there is a direct relationship between that which happens in society and in nature (Sunkel and Gligo, 1981). In his contribution to the book, Raúl Prebisch (who, as an economist, had been oblivious to environmental issues during his long and brilliant career) observed from the periphery that "the environmental crisis was generated by the centre's irrational capitalist development model". He also mentioned the danger of excessive carbon dioxide emissions from rich countries. However, the book found little response within CEPAL, despite the efforts of Axel Dourojeanni and Nicolo Gligo himself. CEPAL has not been a leader of environmental thought in Latin America. Nowadays the economic crisis of "extractivism" (the rapidly deteriorating terms of trade in 2014–2015, partly because of excessive global investment in the extractive industries) has caught CEPAL by surprise, just as both the neoliberal and the national-popular governments.

Back in the 1980s, the UNEP Regional Office discussed several other issues around the binary development and environment. One of the questions addressed the roles that the small producers and large business owners play in the deterioration of nature. Some sustained that, as peasants were obliged to occupy lands of lesser quality at the agricultural frontier, they generated soil degradation. However, other indicators exist that support the view that the processes of degradation and dilapidation were caused by large landowners.

Later, in response to the Brundtland Report of 1987, another study called *Our Own Agenda* was elaborated by UNEP and Inter-American Development Bank (IDB), and coordinated by the hydraulic engineer Arnaldo Gabaldón (the Venezuelan minister of the environment) (Gabaldón, 1994).[3] Gilberto Gallopin, Vicente Sánchez and other expert authors participated, proposing to the governments, to the NGOs and to society at large that the agenda be incorporated into the Rio meeting of 1992. Part of this work was published in more accessible language by Sejenovich and Panario (1996). All of this contributed, on the one hand, to the United Nations' Agenda 21 and, on the other hand – within civil society – to the various alternative Treaties of NGOs in Rio 1992. At the official conference, the Convention on Climate Change and the Biodiversity Convention were signed by all countries (with the sole exception of the USA). At that time, a prominent Latin American representative was Jose Lutzenberger, who had published the ecological manifest, *End of the Future?* (*Fim do Futuro?*) in 1976. As Brazilian minister of the environment, Lutzenberger asked in 1992 that the World Bank not lent any more money to Brazil (Hochstetler and Keck, 2007: 74ff). He was forced to resign.

In parallel meetings to Rio 1992, popular environmentalism emerged in a very public and urgent fashion. In fact, 1,500 organizations from all over the world met to debate the treaties that the governments were discussing, and effectively drafted alternative treaties that were much more exigent, including one about "ecological debt" (Alternative Treaty, n. 13). Despite all of this, the anti-environmentalist prejudice in Latin American official circles continued for decades, until today. Instead of using Chico Mendes (assassinated in December 1988) as a symbol of popular Latin American environmentalism, an international official conflict evolved over the interpretation of the struggle of rubber tappers against deforestation. Fearing initiatives that would internationalize the Amazon, so as to not passively let Brazil destroy it, the president of Brazil conspicuously left a public meeting.

In conclusion, from Stockholm in 1972 until Rio+20 in 2012, Latin American governments have emphasized that the solution to the environmental problem does not consist of halting economic growth, but rather that the main and ultimate solution resides in changing the unequal distribution of power and wealth in the world, and by stimulating distinct styles of development in accordance with each ecological and social reality at national and continental levels (Estenssoro Saavedra, 2014: 155). At the governmental level there was, and is still, a lack of a sense of urgency about the continuing destruction of biodiversity and about climate change (the concentration of carbon dioxide in the atmosphere rose from 360 ppm to 400 ppm between 1992 and 2012). Empathy for popular ecology has also been missing. Neither peasant agroecology nor post-developmentalism nor popular environmentalism – as discussed below – has been part of Latin America's official "own agenda".

## Popular environmentalism

Governmental and international debates over new environmental policies occurred at the same time that a debate emerged in civil society which quickly grew stronger. Influenced by the new ideas of Liberation Theology and different social movements in the region, a widely shared critique of the economic growth models in Latin America would give voice to a popular environmentalism, or the environmentalism of the poor. It drew from the ideas of two important Latin American thinkers. Paulo Freire emphasized social and environmental justice, local knowledge, the morality of political decisions, and respect for the planet and its diverse habitats. These ideas led some to adopt a fundamental rejection of capitalism; others regarded it as an agenda that was more cultural and moral, and which could present an alternative to materialist developmentalism. The other thinker with great influence in the debate was the Uruguayan writer Eduardo Galeano. In his 1971 book *Open Veins of Latin America* (*Las Venas Abiertas de América Latina*), he presented a ferocious critique of the extractivist logic throughout all of Latin America's history. The book became an iconic text in the debates over the consequences of extractive capitalism and the social and ecological destruction in the region. In recent years another Uruguayan, Eduardo Gudynas (2009), attracted many followers for his elaboration of "post-extractivism". Meanwhile, Maristella Svampa leads a flourishing group of Argentinean authors doing excellent political ecology research with an "anti-extractivist" agenda (Svampa, 2011, 2013, 2015), as do Gian Carlo Delgado in Mexico (Delgado Ramos, 2000) and Mario A. Pérez Rincón in Colombia (Pérez-Rincón, 2006, 2014).

In the 1970s and 1980s, nationalist-popular political parties (in the style of Peronismo in Argentina and the American Popular Revolutionary Alliance (Alianza Popular Revolucionaria Americana (APRA) in Peru, before their incongruent neoliberal moments with presidents Menem and Alan García) had protested against the insertion of Latin America in the world economy as provider of raw materials and with episodes of terrible indebtedness. And they were joined by other political currents. For example, the influential Argentinian economist Aldo Ferrer of the Radical Party presented a well-argued plea for "living within our means" in 1983 (Ferrer, 1983). This has been replaced in recent times by a "commodity consensus" (or a new "Beijing consensus") at an official level.

Beyond the government and international debates directed towards new public environmental policies and beyond university research, a popular environmentalism developed with greater force encompassing movements that are sometimes purely reactive and that, in general, do not aspire to achieve political influence per se. Instead they emerged as a reaction to specific environmental problems, which are often local but have worldwide importance. In this sense, one can see Latin American agroenvironmentalism as an international movement that is not only defensive but one that also makes propositions that show the "productive ecological rationality" about which Enrique Leff speaks (Leff, 2004).

Much of the resistance manifested in popular environmentalism did not create permanent alternatives but was rather linked at one point or another to specific places of mineral extraction or investment projects. The protests in Mexico in the 1980s against the nuclear plant in Laguna Verde present a now distant example. There have been many instances of resistance to dams, which lasted for decades and eventually led to nothing. The local movement in Ecuador against copper mining in Intag is a current example. They resisted and succeeded against Mitsubishi in 1995 and against Ascendant Copper (of Canada) in 2006, and developed productive alternatives such as the trade of organic coffee and ecotourism. After these victories, in 2014 it suffered the ravages of President Correa's policies ("we shall leave extractivism behind through more extractivism") in alliance with the state-owned company Codelco of Chile.[4]

Popular environmentalism, otherwise known as the environmentalism of the poor and indigenous, is above all the expression of a "moral economy" that confronts commodification and manifests itself in the commodity-extraction frontiers (Martínez-Alier, 1992, 2005). The peasant and indigenous populations protest against the extractive industries of minerals and biomass, using distinct languages of valuation. They succeed in halting conflictive projects in perhaps 20% of the cases, according to the inventories of the EJOLT (Environmental Justice Organizations, Liabilities and Trade) Project (www.ejatlas.org). Sometimes they demand monetary compensation for the damage inflicted or for that which they are going to suffer; other times they argue in terms of inalienable territorial rights, they appeal to Convention 169 of the International Labour Organization (ILO), or they argue that landmarks that are going to be destroyed (hills, rivers, lakes) are sacred. They oppose the loss of common goods and natural resources that they need to live and survive. Not only in the countryside but also in the city there are groups of relatively poor citizens who, without being "card-carrying" environmentalists, protest when they lose green areas of public use, demand space for pedestrians or cyclists, and practise urban horticulture.

Today, this Latin American popular environmentalism congregates in (virtual) networks of information and agitation such as those of the Observatory of Mining Conflicts in Latin America (Observatório de Conflitos Mineiros da América Latina (OCMAL)) and the Latin American Observatory of Environmental Conflicts (Observatorio Latinoamericano de Conflictos Ambientales (OLCA)), both based in Chile. There are parallels and connections (through international networks such as Oilwatch, the World Rainforest Movement (WRM), the Vía Campesina and Latin American Coordination of Rural Organizations (Coordinadora Latinoamericana de Organizaciones del Campo (CLOC)) with resistance movements in India and Africa, and there are also similarities with the movement for environmental justice in the USA. Networks such as the MAB (Movement of People Affected by Dams/Movimento dos Atingidos por Barragens) in Brazil and MAPDER (Movement of those Affected by Dams and in Defence of Rivers/Movimiento Mexicano de Afectados por las Presas y en Defensa de los Rios) in Mexico (which oppose dams) are also connected with international movements. This popular environmentalism has made itself visible in a great number of local conflicts that have arisen in recent decades. In Latin America, in almost half of the cases collected in the Environmental Justice Atlas (www.ejatlas.org), the indigenous or African-American populations participate as actors in such ecological-distributive conflicts. There are also new networks of statistical political ecology (Pérez Rincón, 2014).

Popular environmentalism does not only have indigenous roots; religion was also important. The book by Brazilian theologian Leonardo Boff, *Ecology: Cry of the Earth, Cry of the Poor* (1996), stands out along with the leadership of former priest Marco Arana in Peru in the movement and political party Tierra y Libertad (Land and Liberty), founded after several years of resistance in Cajamarca against the Yanacocha Mine. Previously there was a movement called Movement of Priests for the Third World, which played an important role in the slums (*villas miserias*) in Argentina and in general with the poor. It was harshly repressed and obliged to dissolve itself, but it reappeared 20 years later in the agrarian leagues of north-eastern Argentina, forming environmental movements in the fight against the soy production that invades the Chaco forest. Alongside this process emerged a non-governmental network called Doctors of the Fumigated Towns (Médicos de los Pueblos Fumigados por Glifosato), which supports the substantial movement called Let's Stop Fumigating (Paremos de Fumigar), with emblematic activists such as Sofia Gatica in Córdoba (Goldman Prize) of the Mothers of Ituzangó (Madres de Ituzangó) movement.[5] In Brazil, the active presence of the Pastoral da Terra is noted in land conflicts in the north of the country (Porto et al., 2013).

The term "ecological debt" was first used in 1991 by Latin American organizations that were opposed to the loss of the ozone layer and to climate change (Robleto and Marcelo, 1992), and it was applied a little later to the results of ecologically unequal trade and instances of "biopiracy". There are other slogans or expressions, such as "water is worth more than gold" (*el agua vale más que el oro*), "water justice" (*justicia hídrica*), "living rivers" (*ríos vivos*), "climate justice" (*justicia climática*), "tree plantations are not forests" (*las plantaciones no son bosques*) (Carrere and Lohman, 1996), "food sovereignty" (*soberanía alimentaria*, from Vía Campesina) and, more recently, "energy sovereignty", which were born in or have been spread across the continent. Environmental justice associations also ask for an international criminal court for environmental damages and an international convention about "ecocide". This is truly very distant from the rhetoric of the "green economy" deployed by the United Nations in the Rio+20 conference of June 2012, not to mention the super-oxymoron of "green growth".

One of the important elements of the environmental justice movement is the word "biopiracy", introduced in 1993 by Pat Mooney (of the Rural Advancement Foundation International (RAFI), which is today Action Group on Erosion, Technology and Concentration (ETC)), and spread on a worldwide scale by Vandana Shiva, frequent visitor to

Latin American countries. In Latin America, Carlos Vicente, author of numerous books on the subject, coordinates the Action for Biodiversity Network. What started as allegations by environmental justice activist organizations against biopiracy has now been converted into legal actions of some governments or court cases in megadiverse countries. In Peru, as in Brazil, the state authorities now speak of "biopiracy". Even the Brazilian minister of the environment, Izabella Teixeira, said in March 2012 – after having fined some companies – that opportunities to advance in the economic valorization of biodiversity should be avoided so as not to "disguise biopiracy actions".[6]

In the regulation of investment projects, advances have been made in imposing a process of public audience for environmental impact assessments (EIAs), which are crucial moments in many socioenvironmental conflicts (Wagner, 2014). The EIAs sometimes provide a setting of participation or of struggle, and allow advancement towards participatory environmental governance. In Tambogrande, Peru, the refusal of the population to participate in a rigged EIA public audience was a step towards a referendum or popular consultation in 2002.[7]

Environmental conflicts do not only consist of local populations on one side and corporations on the other. Local and international NGOs participate, along with state representatives, in a multitude of conflicts not only over the administrative management of the EIAs or granting of mining or oil concessions, but also through other legal channels (with spectacular cases, such as the recent suspension of the Barrick Gold Pascua Lama project in Chile, after investments of thousands of millions of dollars), including court cases. Legislative authorities also sometimes intervene in favour of environmentalism, such as in the prohibition of open-pit mining by various provincial legislatures in Argentina (Wagner, 2014). Mediation bodies can also intervene, such as the ombudsman (*Defensoría del Pueblo*) in Peru and Bolivia. However, in other instances, quite often the police, military and private security forces protected by the state intervene against popular environmentalists. Although there is a consensus between neoliberal and national-popular governments in attributing environmentalism to foreign influences and interpreting it as a phenomenon of "full bellies", it is impossible to ignore the numerous outbreaks of bottom-up environmental mobilizations all over Latin America and the hundreds of victims killed in environmental conflicts in Mexico, Honduras, Guatemala, Colombia, Peru, Brazil and other countries documented by Global Witness, by the OCMAL inventories, the Oswaldo Cruz Foundation (Fundação Oswaldo Cruz (FIOCRUZ)) map of Brazil (Porto et al., 2013), and the EJ Atlas (www.ejatlas.org).

## A Latin American ecosocialism?

In the 1980s, new ideas about socioecological politics in Latin America emerged. Authors such as Victor Toledo, Enrique Leff, José Augusto Pádua and Ivan Restrepo formulated more radical ideas about the political context of environmental governance. Augusto Angel Maya's explicit message (1996, 2002) was to avoid interpreting environmental problems as exclusively ecological or technological. He understood the environment as an object of study in all the scientific disciplines, from the natural sciences and technologies to sciences that study human behaviour.

Beginning in the 1980s, activist groups such as the Political Ecology Institute (Instituto de Ecología Política) in Chile, Censat in Colombia, Ecological Action (Acción Ecologica) in Ecuador (composed of young female biologists), REDES (Amigos de la Tierra Uruguay/Friends of the Earth) in Uruguay, FASE (Federação de Órgãos para Assistência Social e Educacional/ Federation of Organizations for Social and Educational Assistance) in Brazil with Julianna Malerba, and others have emerged. There is a strong Latin American environmental thinking that enumerates, and denounces the multitude of environmental conflicts that the growth of the social metabolism brings with it. Some 20 years later, these views have not only been expressed in writings and manifestos of social actors and alternative thinkers of post-developmentalism, of agroecology and of popular environmentalism, but also in some national constitutions, in the discourses of government officials and even by some ministers.

After the defeat in 2005 of the US plans to promote the Free Trade Area of the Americas (FTAA), new leftwing, progressive governments emerged with the electoral victories of Evo Morales in Bolivia (2005) and Rafael Correa in Ecuador (2006). In the following years it even seemed that an international "official" environmental leadership could arise from South America. The Ecuadorian Constitution of 2008, for example, has been a very important symbol of environmental thinking in Latin America, with the presence of Alberto Acosta – ex-president of the Constituent Assembly – in a multitude of forums. Another example was the radical speech of Ecuador representative Fánder Falconí, at the failed climate change conference in Copenhagen in 2009, when he made reference to the ecological debt or climate debt of the North with the South. He compared the poor countries with "passive smokers" and he defended the Yasuni Ishpingo-Tambococha-Tiputini (ITT) initiative to "leave the oil below ground" in front of more than 150 presidents of state and leaders of government.[8]

The contradictions of the new leftwing governments, which had to choose between environmental protection and economic growth, became clear when only a few weeks later Falconí resigned as minister of foreign affairs because of President Correa's refusal to take the Yasuni ITT initiative forward. In Cochabamba, Bolivia, in April of 2010, a large meeting was held after the failure of the United Nations meeting in Copenhagen, attempting to position Evo Morales as an environmental leader of the South, but neither he nor his vice president, García Linera (who believes that environmentalism is a luxury for the rich), was in favour of concrete measures regarding environmental protection. They went rather for the exploitation of the Amazon as in the plan for the TIPNIS (Isiboro Secure National Park and Indigenous Territory/Territorio Indígena y Parque Nacional Isiboro Secure) highway. The Bolivian ambassador to the UN, Pablo Solón, was alone in the insistence on the responsibility of the developed countries for climate change in December of 2010 in Cancún in one more ineffectual climate conference.[9]

The inability of Latin American governments to take on environmentalism as a main issue, and even more the repression and "criminalization" of popular environmentalism, is opening up space for a political environmentalism that is opposed to neoliberal as much as it is to the national-popular governments. Both share the "commodities consensus" (Svampa, 2013). This is leading to a mature Latin American environmentalist political thinking, albeit incipient, proposing new principles of international environmental governance, and also criticizing extractivism and environmentally unequal trade in the defence of the rights of nature, the human right to water, and the integral and sustainable management of resources for the benefit of local livelihoods.

In support of ecosocialism, Enrique Leff in *Ecology and Capital* (1986) and James O'Connor (in the first issue of the journal *Capitalism, Nature, Socialism* (1988)) explained that the growing social and environmental costs caused by economic growth are also the catalysts for an explosion of environmental protest (Leff, 1986, 2012). Currently we see a major global process of dispossessing indigenous and peasant lands by private or state enterprises: expropriating mangroves by the shrimp industry, and land-grabbing for tree plantations and agrofuels, for megamining and dams, and for the extraction of gas and oil. These are neocolonial processes of appropriating natural resources and territories where new actors, such as Chinese companies, appear. There is also much resistance in urban areas, including recycling cooperatives of "scavengers" of urban waste, who play a very important and under-recognized role.

The Latin American Network of Recyclers and Urban Reclaimers has come into existence which has attained notable success in places such as Bogotá under the leadership of Nohra Padilla, who won the 2014 Goldman Prize for grassroots environmentalism.

## Conclusion

A common element of Latin American environmentalist thought (absent in Europe and also in India, for example) is the awareness of the demographic disaster brought about by the European Conquest. This led to a perhaps justified disdain for Malthusian approaches in the region. The environmentalism of Paul Ehrlich with his focus on the "population bomb" was never successful in Latin America, where the population density is generally low (in comparison with Europe, East Asia and South Asia). Since the beginning of the 1970s, there has been a profound discussion among Latin American governments and on the part of the UNEP Regional Office to establish a shared environmental position. The 1972 Meadows Report, *The Limits of Growth*, garnered a general rejection in official circles in Latin America. It was emphasized that the problem was not the finite supply of resources but rather their distribution. However, 40 years after this debate, we have indeed found that today there are "planetary boundaries" of resources and sinks. Current world trends are negative in regard to the loss of biodiversity and climate change. Above and beyond this initial negative reaction in the 1970s and 1980s from official circles, and the search for a "Latin America agenda" of its own, we have identified a set of environmental ideas and practices that have emerged in Latin America and which in part coincide and in part diverge from other continents:

- awareness of the demographic disaster after the conquest and a widespread rejection of the Malthusian approach to the problem of overpopulation;
- an agroenvironmental pride, especially present in Mesoamerica and the Andes (and absent in the USA);
- a shared admiration by European and Latin American science (since 1800, with Alexander von Humboldt) for the great biological richness of the continent in its diverse ecosystems, together with conservation programmes implemented since the nineteenth century;
- a keen awareness of global political and economic inequality, and the consequent plundering of natural resources in the region; this

awareness runs from the time of colonial exploitation through to today;
- the rejection by Latin American governments – since Stockholm in 1972 – of the idea of limits to growth, defining an agenda that proposed distinct "styles of development" but eventually accepting a confusing notion of "sustainable development";
- from the 1980s onwards, a growing number of socioenvironmental conflicts that gave way to "popular environmentalism" with networks of activists that denounce the extraction of natural resources and the destruction of the commons;
- the validity of ancient indigenous worldviews, the celebration of *Pachamama* that is recognized in the constitutions of Bolivia and Ecuador, the respect for nature in Afro-American communities, and the contributions of liberation theology; also, on a cultural level, the presence of ecology in twentieth-century literature.

There is clearly a Latin American conservationist environmentalism that is common with other continents: a shared admiration of European science (which is also American science) since Humboldt because of the enormous biodiversity of Latin America's many diverse ecosystems, which were only partially explored. The extraordinary biological richness of not only the Amazonian rainforest but also of other ecosystems (such as the Atlantic forest in Brazil, mangroves and coral reefs, the Andean highlands, the tropical dry forests, the Pantanal, and other wetlands and marshes) are seen as a promise of the economic potential that is not yet confirmed and, on the other hand, periodically leads to protests against "biopiracy".

Conflicts around the extraction and export of natural resources are increasing in Latin America. The resistance against the exploitation of nature has led to the growth of popular environmentalism, to environmental justice movements, to protests against climate injustice and water injustice, and to the defence of the commons. Latin American politicians and public administrators have basically ignored this movement of the environmentalism of the poor, but they have not suppressed it.

Recently, however, there have been signs of an emerging post-extractivist and post-developmentalist environmentalism that attack impartially both the neoliberal and the national-popular governments. Some would call it ecosocialism. This political environmentalism is very distinct from that of European green parties that focus on "ecoefficiency". Post-extractivism is intellectually powerful but still politically weak, although it seems much reinforced by the declining terms of trade of 2014–2015. This movement attempts to include new concrete proposals for continental and international governance, such as oil and open-pit mining moratoria, campaigns against dams and against the "green deserts" of pine and eucalyptus trees, and the defence of peasant seeds. Rather than the objective of economic growth or development, it proposes an objective of *Buen Vivir* and also to give rights to nature (as in the 2008 Constitution of Ecuador). The Latin American concept of "ecological debt" has been very fruitful and has provoked important debates, as has the emphasis on the human right to water, supported by Bolivia on the experience of the Cochabamba "water wars" of 2000. Latin America is at a crossroads where various critical political and economic theories are seeking a point of convergence with environmentalism, which will give it the opportunity to present a real alternative to extractivism. One of the crucial challenges will be to transfer these debates to the new circles of politicians and policy-makers. This has been a permanent challenge in Latin American environmental history, but today it has a renewed intensity.

## Notes

1. Chapter 2 gives statistics on the social metabolism.
2. For Esteva's analysis of the meanings of "development", see https://desarrolloxxi.files.wordpress.com/2010/05/desarrollogustavoesteva1.pdf
3. See Garcia-Guadilla (2013) for an interesting account of "neoextractivism" and its conflicts in today's Venezuela.
4. www.http://codelcoecuador.com/news/ and Rafael Correa, Discurso para la XIV Cumbre Iberoamericana, Veracruz, Mexico, 8 December 2014: "Debemos hacer uso del extractivismo para salir de él".
5. See, for instance, https://noticiasdeabajo.wordpress.com/2012/07/30/informe-del-primer-encuentro-nacional-de-medicos-de-pueblos-fumigados/
6. See http://www.bbc.co.uk/mundo/noticias/2012/03/120323_biopirateria_brasil_lp.shtml
7. See Chapter 11 about local referenda or popular consultations against mining investments.
8. See https://mail.uevora.pt/pipermail/ambio/2009-December/015749.html, taken from the webpage of the Ministry of Foreign Relations of Ecuador.
9. Chapter 4 compares post-neoliberal environmental governance in Ecuador and Bolivia.

## References

Alcalde Mongrut, A. (1966) *Mariano de Rivero* (Lima: Editorial Universitaria).
Alimonda, H. (ed.) (2011) *La Naturaleza Colonizada. Ecología Política y Minería en América Latina* (Buenos Aires: Ciccus/CLACSO).

Altieri, M. and Toledo, V.M. (2011) "The Agroecological Revolution in Latin America: Rescuing Nature, Ensuring Food Sovereignty and Empowering Peasants", *Journal of Peasant Studies* 38(3): 587–612.
Ángel Maya, A. (1996) *El Reto de la Vida* (Bogotá: Ecofondo).
Ángel-Maya, A. (2002) *El Retorno de Icaro* (Bogotá: PNUMA).
Baud, M. (2013) "Ideologies of Progress and Expansion: Transforming Indigenous Culture and Conquering Nature in Latin America, ca. 1870", in *The Emergence of New Modes of Governance of Natural Resources Use and Distribution in Latin America and Ecuador*, ENGOV Working Paper No. 4, 7–25.
Boff, L. (1996) *Ecología: Grito de la Tierra, Grito de los Pobres* (Buenos Aires: Lumen).
Brannstrom, C. (ed.) (2004) *Territories, Commodities and Knowledges: Latin American Environmental Histories in the Nineteenth and Twentieth Centuries* (London: Institute of the Americas).
Brannstrom, C. (2005) "Was Brazilian Industrialisation Fuelled by Wood? Evaluating the Wood Hypothesis, 1900–1960", *Environment and History* 11: 395–430.
Carrere, R. and Lohman, L. (1996) *Pulping the South. Industrial Tree Plantations and the World Paper Economy* (London: Zed Press).
Chambers, R. (1983) *Rural Development: Putting the Last First* (London: Longman).
Ciriacy-Wantrup, S.V. (1957) *Conservación de los Recursos: Economía y Política* (trad. Edmundo Flores) (Mexico: Fundación de Cultura Económica).
Crosby, A.W. (2004) *Ecological Imperialism. The Biological Expansion of Europe 900–1900*, revised edition (Cambridge: Cambridge University Press).
Cushman, G.T. (2013) *Guano and the Opening of the Pacific World. A Global Ecological History* (Cambridge: Cambridge University Press).
Cuví, N. (2005) "Misael Acosta Solís y el Conservacionismo en el Ecuador", *Scripta Nova*, IX (191). http://www.ub.edu/geocrit/sn/sn-191.html
Dean, W. (1995) *With Broadax and Firebrand. The Destruction of the Brazilian Atlantic Forest* (Berkeley: University of California Press).
Deléage, J.P. (1994) *Historia de la Ecología* (Barcelona: Icaria).
Delgado Ramos, G.C. (ed.) (2010) *Ecología Política de la Minería en América Latina* (México: UNAM).
Denevan, W.M. (1980) "Tipología de Configuraciones Agrícolas Prehispánicas", *América Indígena* XL(4): 619–652.
Descola, P. (1986) *La Nature Domestique: Symbolisme et Praxis dans l'Écologie des Achuar* (Paris: Ed. de la Maison des Sciences de l'Homme).
Dore, E. (1994) "Una Interpretación Socio-Ecológica de la Historia Minera Latinoamericana", *Ecología Política* 7: 49–68.
Dourojeanni, M.J. (1988) *Si el Árbol de la Quina Hablará* (Lima: Fundación Peruana para la Conservación de la Naturaleza, ProNaturaleza).
Dourojeanni, M.J. (1990) *Amazonia ¿Qué hacer?* (Iquitos: Centro de Estudios Teológicos de la Amazonía, CETA).
Drummond, J.A. (1997) *Devastação e Preservação Ambiental no Rio de Janeiro* (Niterói: Editora da Universidade Federal Fluminense).
Escobar, A. (1995) *Encountering Development: The Making and Unmaking of the Third World* (Princeton: Princeton University Press).
Escobar, A. (2010) "Latin America at a Crossroads. Alternative Modernizations, Post-Liberalism, or Post-Development?", *Cultural Studies* 24(1): 1–65.

Estenssoro Saavedra, F. (2014) *Historia del Debate Ambiental en la Política Mundial 1945–1992. La Perspectiva Latinoamericana* (Santiago de Chile: Instituto de Estudios Avanzados, Universidad de Santiago de Chile).
Esteva, G. and Marielle, C. (eds) (2003) *Sin Maíz no hay País: Páginas de una Exposición* (México: Consejo Nacional para la Cultura y las Artes, Dirección General de Culturas Populares e Indígenas).
Falconí, F. (2013) *Al Sur de las Decisiones. Enfrentando la Crisis del Siglo XXI* (Quito: El Conejo).
Farah, H.I. and Vasapollo, L. (eds) (2011) *Vivir bien, ¿Paradigma no Capitalista?* (La Paz: Cides-UMSA).
Ferrer, A. (1983) *"Vivir con lo Nuestro" – Para Romper la Trampa Financiera y Construir la Democracia* (Buenos Aires: El Cid Editores).
Funes Monzote, R. (2004a) *De Bosque a Sabana: Azúcar, Deforestación y Medio Ambiente en Cuba, 1492–1926* (Mexico: Siglo XXI).
Funes Monzote, R. (2004b) "Deforestation and Sugar in Cuba's Centre-East: The Case of Camagüey, 1898–1926", in C. Brannstrom (ed.), *Territories, Commodities and Knowledges. Latin American Environmental Histories in the Nineteenth and Twentieth Centuries* (London: Institute of the Americas), 148–170.
Funtowicz, S.O. and Ravetz, J.R. (2000) *La Ciencia Posnormal. Ciencia con la Gente* (Barcelona: Icaria).
Gabaldón, A.J. (1994) "Del Informe Brundtland a Nuestra Propia Agenda", in M.P. García-Guadilla and J. Blauert (eds), *Retos al Desarrollo y la Democracia: Movimientos Ambientales en América latina y Europa* (Mexico: Nueva Sociedad con Fundación Friedrich Ebert de México), 27–36.
Gallopin, G. (ed.) (1995) *El Futuro Ecológico del Continente. Una Visión Prospectiva de la América Latina*, 2 vol. (Tokyo/México: Editorial de la UNU/Fondo de Cultura Economica).
García-Guadilla, M.P. (2013) "Neo-Extractivismo, Neo-Rentismo y Movimientos Sociales en la Venezuela del Siglo XXI: Conflictos, Protestas y Resistencia", XXXI International Congress of the Latin American Studies Association (LASA), Washington D.C.
Garfield, S. (2013) *In Search of the Amazon: Brazil, the United States, and the Nature of a Region* (Durham and London: Duke University Press).
Gligo, N. and Morello, J. (1980) "Notas sobre la Historia Ecológica de América Latina", *Estudios Internacionales* 13(49): 112–148.
Gligo, V.N. and Sunkel, O. (eds) (1980) *Estilos de Desarrollo y Medio Ambiente en la América Latina*, 2 vol. (Mexico: Fondo de Cultura Económica).
Gootenberg, P. (1993) *Imagining Development: Economic Ideas in Peru's "Fictitious Prosperity" of Guano, 1840–1880* (Berkeley: University of California Press).
Grillo, E. et al., (1988) *Ciencia y Saber Campesino Andino* (Lima: PRATEC). http://pratecnet.org/wpress/wp-content/uploads/2014/pdfs/Ciencia%20y%20saber%20campesino%20andino.pdf
Gudynas, E. (1999) "Concepciones de la Naturaleza y Desarrollo en América Latina", *Persona y Sociedad* 13(1): 101–125.
Gudynas, E. (2009) *El Mandato Ecológico. Derechos de la Naturaleza y Políticas Ambientales en la Nueva Constitución* (Quito: Abya Yala).
Herrera, A.O., Scolnik, H.D. and Chichilnisky, G. (1976) *Catastrophe or New Society? A Latin American World Model* (Ottawa: International Development Research Centre).

Hidalgo-Capitán, A.L., Guillén García, A. and Guazha, N.D. (eds) (2014) *Antología del Pensamiento Indigenista Ecuatoriano sobre Sumak Kawsay* (Universidad de Huelva y Universidad de Cuenca).
Hochstetler, K. and Keck, M.E. (2007) *Greening Brazil. Environmental Activism in State and Society* (Durham and London: Duke University Press).
Jaguaribe, H. (1973) "El Equilibrio Ecológico Mundial", *Pensamiento Político* 12(46): 235–254.
Leff, E. (1986) *Ecología y Capital* (Mexico: Siglo XXI).
Leff, E. (2001) *Los Problemas del Conocimiento y la Perspectiva Ambiental del Desarrollo*, 2nd ed. (Mexico: Siglo XXI).
Leff, E. (2004) *Racionalidad Ambiental. La Reapropiación Social de la Naturaleza* (Mexico: Siglo XXI).
Leff, E. (2006) *Aventuras de la Epistemología Ambiental. De la Articulación de las Ciencias al Diálogo de Saberes* (Mexico: Siglo XXI).
Leff, E. (2012) "Latin American Environmental Thinking: A Heritage of Knowledge for Sustainability", *Environmental Ethics* 34(4): 431–450.
Machado Aráoz, H. (2014) *Potosí, el Origen. Genealogía de la Minería Contemporánea* (Buenos Aires: Mardulce).
Marcuse, H. (1975) *Ecología y Revolución* (Buenos Aires: Editorial Nueva Visión).
Martinez-Alier, J. (1991) "Ecology and the Poor: A Neglected Issue of Latin American History", *Journal of Latin American Studies* 23(3): 621–639.
Martinez-Alier, J. (1992) *De la Economía Ecológica al Ecologismo Popular* (Barcelona: Icaria).
Martinez-Alier, J. (2002) *The Environmentalism of the Poor: Environmental Conflicts and Languages of Valuation* (Cheltenham: Edward Elgar).
Martinez-Alier, J. and Roca Jusmet, J. (2013) *Economía Ecológica y Política Ambiental*, 3rd ed. (Mexico: Fondo de Cultura Económica).
McCosh, F.W.J. (1984) *Boussingault: Chemist and Agriculturist* (Dordrecht: Reidel).
Melville, E. (1999) *Plaga de Ovejas. Consecuencias Ambientales de la Conquista de México* (México: Fondo de Cultura Económica).
Miller, R. and Greenhill, R. (2006) "The Fertilizer Commodity Chains: Guano and Nitrate, 1840–1930", in S. Topic, C. Marichal and Z. Frank (eds), *From Silver to Cocaine: Latin American Commodity Chains and the Building of the World Economy, 1500–2000* (Durham: Duke University Press), 228–270.
Monasterio, M. (2003) "Del Frailejón a la Papa...entre la Conservación y la Agricultura. Una Apuesta Permanente por el Reencuentro entre Ecología y Sociedad en el Escenario de los Páramos Andinos. Entrevista por Neyllana Salas", *Fermentum* 36: 153–173.
Morello, J. and Matteucci, S. (2000) "Biodiversidad y Fragmentación de los Bosques en la Argentina", in S. Matteucci, O. Solbrig, J. Morello and G. Halffter (eds), *Biodiversidad y Uso de la Tierra. Conceptos y Ejemplos de Latinoamérica* (Buenos Aires: EUDEBA).
Moreno Fraginals, M. (1978) *El Ingenio. Complejo Económico-Social Cubano del Azúcar*, 3 vol. (La Habana: Editorial de Ciencias Sociales).
Pádua, J.A. (2002) *Um Sopro de Destruição: Pensamento Político e Crítica Ambiental no Brasil Escravista, 1786–1888* (Rio de Janeiro: Jorge Zahar).
Pádua, J.A. (2010) "European Colonialism and Tropical Forest Destruction in Brazil', in J.R. McNeill, J.A. Pádua and M. Rangarajan (eds), *Environmental History: As If Nature Existed* (Delhi: Oxford University Press), 130–148.

Palerm, A. (1978) *Obras Hidráulicas Prehispánicas en el Sistema Lacustre del Valle de México* (México: INAH).
Patiño, R.A. (1991) *Ecología y Compromiso Social: Itinerario de una Lucha* (Bogota: Activistas Ecológicos).
Pérez-Rincón, M.A. (2006) "Colombian International Trade from a Physical Perspective: Towards an Ecological 'Prebisch Thesis' ", *Ecological Economics* 59(4): 519–529.
Pérez-Rincón, M.A. (2014) "Conflictos Ambientales en Colombia: Inventario, Caracterización y Análisis", in L.J. Garay (ed.), *Minería en Colombia: Control Público, Memoria y Justicia Socio-Ecológica, Movimientos Sociales y Posconflicto*. (Bogotá: Contraloría General de la República), 253–319.
Porto, M.F., Pacheco, T. and Leroy, J.P. (eds) (2013) *Injustiça Ambiental e Saúde no Brasil: O Mapa de Conflitos* (Rio de Janeiro: Fiocruz).
Robleto, M.L. and Marcelo, W. (1992) *La Deuda Ecológica, una Perspectiva Sociopolítica* (Santiago de Chile: Instituto de Ecología Política).
Sachs, I. (1981) *Ecodesenvolvimento: Crescer sem Destruir* (São Paulo: Vértice).
Sachs, I. (2008) *La Troisième Rive: À la Recherche de l'Écodéveloppement* (Paris: Nouvelles François Bourin).
Sachs, W. (ed.) (1996) *Diccionario del Desarrollo: Una Guía del Conocimiento como Poder* (Lima: PRATEC).
Sánchez-Albornoz, N. (1984) "The Population of Colonial Spanish America", in L. Bethell (ed.), *The Cambridge History of Latin America, Vol. II: Colonial Latin America* (Cambridge: Cambridge University Press), 1–36.
Santiago, M.I. (2006) *The Ecology of Oil. Environment, Labor, and the Mexican Revolution, 1900–1938* (Cambridge: Cambridge University Press).
Schteingart, M. and Graizbord, B. (eds) (1998) *Vivienda y Vida Urbana en la Ciudad de México* (Mexico: El Colegio de México).
Sejenovich, H. and Panario, D. (1996) *Hacia Otro Desarrollo: Una Perspectiva Ambiental* (Montevideo: Redes, Nordan).
Sevilla Guzmán, E. and Martinez-Alier, J. (2006) "New Rural Social Movements and Agroecology", in P. Cloke, T. Marsden and P. Mooney (eds), *Handbook of Rural Studies* (London: Sage), 468–475.
Simonian, L. (1995) *Defending the Land of the Jaguar: A History of Conservation in Mexico* (Austin: University of Texas Press).
Sunkel, O. and Gligo, N. (eds) (1981) *Estilos de Desarrollo y Medio Ambiente en la America Latina* (Mexico: Fondo de Cultura Económica).
Svampa, M. (2011) "Modelos de Desarrollo, Cuestión Ambiental y Giro Eco-Territorial", in H. Alimonda (ed.), *La Naturaleza Colonizada. Ecología Política y Minería en América Latina* (Buenos Aires: CLACSO), 181–215.
Svampa, M. (2013) "Consenso de los Commodities y Lenguajes de Valoración en América Latina", *Nueva Sociedad* 244: 30–46.
Svampa, M. (2015) "Commodities Consensus: Neoextractivism and Enclosure of the Commons in Latin America", *South Atlantic Quarterly* 114(1): 65–82.
Topic, S., Marichal, C. and Frank, Z. (eds) (2006) *From Silver to Cocaine. Latin American Commodity Chains and the Building of the World Economy, 1500–2000* (Durham: Duke University Press).
Tudela, F. (ed.) (1990) *Desarrollo y Medio Ambiente en América Latina* (Madrid: MOPU).

Vavilov, N.I. (1992) *Origin and Geography of Cultivated Plants* (Cambridge: Cambridge University Press).
Von Humboldt, A. (1980) *Cartas Americanas* (Caracas: Biblioteca Ayacucho).
Wagner, L. (2014) *Conflictos Socio-Ambientales: La Megaminería en Mendoza 1884–2011* (Quilmes: Editorial Universidad Nacional de Quilmes).
West, J. and Schandl, H. (2013) "Material Use and Material Efficiency in Latin America and the Caribbean", *Ecological Economics* 94: 19–27.
Wolf, E.R. (1982) *Europe and the People Without History* (Berkeley and Los Angeles: University of California Press).

 Except where otherwise noted, this work is licensed under a Creative Commons Attribution 3.0 Unported License. To view a copy of this license, visit http://creativecommons.org/licenses/by/3.0/

OPEN

# 2
# Social Metabolism and Conflicts over Extractivism

*Joan Martinez-Alier and Mariana Walter*

## Introduction

The natural resource conflict dimension of environmental governance is usually centred on the social and political aspects of production systems and has hardly addressed the biophysical features of the natural resources themselves. Here we aim to address renewable and non-renewable resource-extraction conflicts in Latin America in the context of a changing global social metabolism and increasing demands for environmental justice (M'Gonigle, 1999; Sneddon, Howarth and Norgaard, 2006; Gerber, Veuthey and Martínez-Alier, 2009; Martinez-Alier et al., 2010). "Social metabolism" refers to the manner in which human societies organize their growing exchanges of energy and materials with the environment (Fischer-Kowalski, 1997; Martinez-Alier, 2009). In this chapter we use a sociometabolic approach to examine the material flows (extraction, exports, imports) of Latin American economies and furthermore look into the socioenvironmental pressures and conflicts that they cause. Sociometabolic trends can be appraised using different and complementary indicators. For instance, the Human Appropriation of Net Primary Production (HANPP) measures to what extent human activities appropriate the biomass available each year for ecosystems (Haberl et al., 2007). Other examples are indicators that study virtual water flows, the energy return on investment (EROI) or a product life cycle. Each indicator provides information on different aspects of our economic performance.

In this chapter we will address the economy-wide material flow analysis (MFA) in more detail. The MFA is "a consistent compilation of the overall material inputs into national economies, the material accumulation within the economic system and the material outputs to other economies or to the environment" (EUROSTAT, 2001: 17). MFA aims to complement the system of national accounting with a compatible system of biophysical national accounts, using tonnes per year as the key unit of measurement. Such methodology provides a picture of the physical dimension of the economy, where the total turnover of energy and materials of the socioeconomic system can be analysed historically or cross-sectioned through the accounting of input flows (tonnes of biomass, fossil fuels, construction minerals, etc.) or output flows (tonnes of materials exported, waste or pollutant generated). Focusing on the input side by taking into account all materials that enter into the national economy allows for an acknowledgment of the physical dimension of foreign trade and can determine the amount of all outputs transferred to the environment (Gonzalez-Martinez and Schandl, 2008). While MFA presents some limitations regarding, for instance, the qualitative differences between materials (i.e. toxicity, environmental or social context of extraction), it offers a picture of the overall evolution of the pressures exerted by an economy to extract renewable and non-renewable resources.

A social metabolic approach acknowledges that inputs into the economy ultimately become outputs from the economy in the form of waste (except for the part that accumulates as a stock, as in buildings). The main output in volume from rich economies (apart from wastewater) is carbon dioxide from the burning of fossil fuels, the excessive production of which is a main source of climate change. Solid wastes produced by the economy are disposed of locally (in landfills or incinerators), or sometimes exported to distant regions or countries. All goods circulate through "commodity chains" (Raikes, Friis Jensen and Ponte, 2000) – that is, from cradle to grave or from point of extraction to waste disposal. Ecological distribution conflicts occur at different stages as peasant or tribal groups, national or multinational companies, national governments, local or international NGOs, and consumer groups are all stakeholders.

Economic change generally occurs for the benefit of some groups and at the expense of other existing or future groups (Hornborg, 2009). Externalities can be positive (like the free environmental services provided by a forest) or negative. Negative externalities are not seen here as market failures but rather as (provisional) cost-shifting successes (Kapp, 1950). Optimistic views regarding ecological modernization, the "dematerialization" of the economy (Stern, 2004), are confronted with the reality of increased inputs of energy and materials into the world economy, thereby increasing the production of waste and ecological distribution conflicts.

Ecological distribution conflicts are struggles over the burdens of pollution or over the sacrifices made to extract resources, and they arise from inequalities of income and power (Martinez-Alier and O'Connor, 1996; Douguet, O'Connor and Noel, 2008). The concept of ecological distributive conflicts is born of the intersection between the fields of ecological economics and political ecology, which links the emergence of environmental conflicts in the global South with the growth of the metabolism of societies in the global North (which includes parts of China). Political ecology focuses on the exercise of power in environmental conflicts. In other words, the question is: Who has the power to impose decisions on resource extraction, land use, pollution levels, biodiversity loss, and more importantly, who has the power to determine the procedures to impose such decisions (Martinez-Alier, 2001, 2002; Robbins, 2004)?

Ecological distribution conflicts emerge from the structural asymmetries in the burdens of pollution and in the access to natural resources that are grounded in unequal distributions of power and income, and in social inequalities of ethnicity, caste, social class and gender (Martínez-Alier, 1997; Martinez-Alier et al., 2011). As processes of valuation surpass economic rationality in attempts to assign market prices and chrematistic costs to the environment, social actors mobilize for material and symbolic interests (of survival, identity, autonomy and quality of life), beyond strictly economic demands of property, means of production, employment, income distribution and development (Leff, 2003). Sometimes the local actors claim redistribution, leading to conflicts that are often part of, or lead to, larger struggles of gender, class, caste and ethnicity (Agarwal, 1994; Robbins, 2004). Hence the concept of "environmental justice" is important. It was born in the USA (Bullard, 1990) and it has gained growing acceptance in extractive industries, water use and waste-disposal conflicts all over the word (Urkidi and Walter, 2011). Not all conflicts are born from immediate metabolic needs. Demand for certain commodities such as gold arises in part from the search to have an investment outlet that furthermore allows for speculation. Other metals, such as copper, can also be stored and used as guarantees for speculative loans. The fact remains that both energy-carriers (coal, gas, oil) and metallic minerals are inputs for the industrial economy and that their use, in total, grows more or less in proportion to the growth of the economy.

In this chapter, we analyse the material flows of Latin American countries and their implications in terms of socioenvironmental conflict. First, we present an overview of recent material-flow studies conducted in this region. Second, we examine in further detail the socioenvironmental pressures exerted by the extraction of renewable and non-renewable materials. We propose a classification of extractive conflicts based on the commodity at stake. With this double approach we address the process of growing primarization of Latin American economies, its trends and some of its drivers, while simultaneously exploring the local pressures and conflicts that this process is fostering. At the macroeconomic level, we point to the paradox that the large physical exports are unable, or scarcely able, to finance the imports so that many countries are falling into commercial deficits.

## Latin American sociometabolic trends

Different indicators can be used to analyse Latin American sociometabolic features and trends. Here we consider recent MFA studies conducted on Latin American economies and discuss their implications in terms of socioenvironmental pressures and injustices. MFAs have been conducted in most Organisation for Economic Co-operation and Development (OECD) countries, but only recently has research been conducted in the Latin American region and some of its countries in particular, such as Argentina (Perez-Manrique et al., 2013), Colombia and Ecuador (Russi et al., 2008; Vallejo, Pérez Rincón and Martinez-Alier, 2011; West and Schandl, 2013; Samaniego, Vallejo and Martinez-Alier, 2014). MFAs conducted on the overall region indicate that there was a four-fold increase in material flows between 1970 and 2008 for domestic consumption and also for exports. The Latin American economy has certainly not become "dematerialized" – one could compare such trends with other geographical regions, such as Europe, where the rate of increase in material extraction has been much lower, or with India, which has a lower rate of material extraction per capita than Latin America and which is not a net exporter in physical terms (Singh et al., 2012). Such physical indicators are useful for characterizing the economic structure of countries and regions.

Latin American economies, and particularly South American economies, have a persistent and increasing physical trade deficit (West and Schandl, 2013). The physical trade balance (PTB) is the difference between the number of tonnes of materials that are imported by an economy and the number of tonnes that are exported. The monetary trade balance (MTB) is the difference between how much is paid for the imports and how much is earned by exports in monetary terms. Exports in tonnes are larger than imports in tonnes, resulting in a

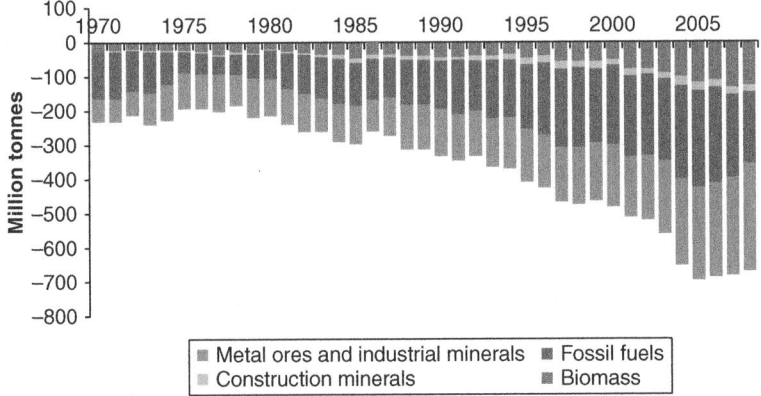

*Figure 2.1* Latin America physical trade deficit in million tonnes, 1970–2008
Source: UNEP and CSIRO, 2013.

"deficit" in the same sense that would be applied to a tree plantation that grows less than the harvest rate. Figure 2.1 presents a yearly PTB of the Latin America region (including Mexico) per type of material from 1970 and 2008. Note in Figure 2.1 the increased physical trade deficit for metal ores and industrial minerals, which reflects the growing pressure to extract and export these materials. While one tonne of uranium is, of course, environmentally very different from one tonne of sand and gravel, or one tonne of cellulose from one tonne of shrimp, our aim here is to show trends within broad material categories, where the shift in the composition by commodities is not that important. Later we take a closer look at the different commodities within the categories of biomass and metal ores.

There are internal and external pressures to increase the extraction of materials, for domestic use and for export. Such increasing pressures to extract materials displace the commodity frontiers (Moore, 2000) to new territories often inhabited by peasant and indigenous groups, who complain accordingly as we signal in further detail in the next section (Conde and Walter, 2014). In regard to external trade, trends point to a structural persistence of an "ecologically unequal exchange". This concept challenges the argument that exports from developing nations foster economic growth and development, and points to the physical and socioenvironmental trade-offs at play (Hornborg, 1998; Muradian and Martinez-Alier, 2001; Bunker, 2007). Studies in this field highlight how poor countries are exporting goods at prices that do not take into account local externalities or depletion of natural resources, in exchange for the purchase of expensive goods and services from richer regions. One can measure ecologically unequal trade in terms of the inequality of various dimensions, such as hours of labour, hectares of land, tonnes of materials, water footprints, and joules or calories. When all or most indicators point in a similar direction, then we can state that there has been an unequal exchange (Hornborg, 2006). Ecologically unequal exchange arises from the structural fact that the metropolitan regions or countries require increasing amounts of energy and materials at cheap prices for their metabolism.

The terms of trade are persistently negative for South America as a whole and for most countries individually (one tonne of imports is always more expensive than one tonne of exports, from two to five times) in the very long term. However, the terms of trade improved somewhat in the first decade of the twenty-first century, fuelling a wave of optimism regarding economic growth but later deteriorating again (Samaniego, Vallejo and Martinez-Alier, 2014). Currently, the large physical exports can scarcely pay for the imports in most South American countries. A large physical trade deficit does not imply a positive MTB, and, on the contrary, recent LA trends point to simultaneous physical and monetary deficits. Either in 2013 or 2014, or in both years, there were commercial deficits in Brazil, Colombia, Ecuador, Peru and other countries. While Argentina's commercial surplus has been much reduced, there is now a need to finance commercial deficits (Samaniego, Vallejo and Martinez-Alier, 2014). For Argentina, our analysis of the external trade over a long period (1970–2009) shows (Figure 2.2) small monetary surpluses since the end of the 1990s (in 2001–2002 the surplus increased because the economic crisis violently reduced imports). Such small monetary surpluses almost disappeared in 2013–2014. From a physical point of view, Argentina has exported increasing amounts (in tonnes) since the early 1990s (between three and four times its imports in tonnes), thus demonstrating structurally negative terms of trade.

We do not enter into a detailed study here of the physical structure of external trade in the sense of looking at its biomass, mineral and fossil-fuel components (Perez-Manrique et al., 2013; West and Schandl, 2013). We point out, however, that Argentina exports – like Brazil – large amounts of biomass. In comparison, another large South American country, Colombia, does not export large amounts of biomass products but it does export large amounts of coal. The PTB of Colombia shows long-term trends that are not very different from those of Argentina,

64  *Social Metabolism and Conflicts*

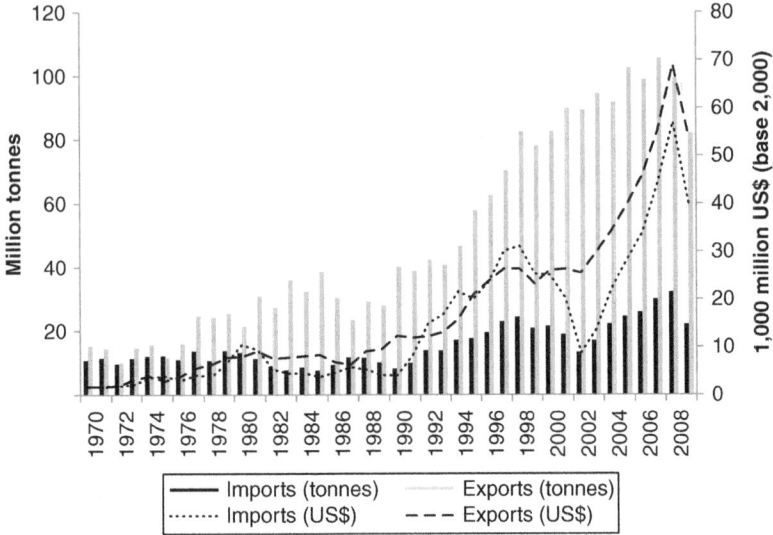

*Figure 2.2* Argentina's physical and monetary external trade flows, 1970–2009
Source: Walter et al. (2013).

namely, physical exports exceed physical imports by a factor of no less than three (Figure 2.3). It must be noted that Colombia's large physical exports (which entail large unpaid socioenvironmental liabilities) are now unable to pay for the imports. As Figure 2.3 shows, in 2011, Colombia exported about 120 million tonnes and imported about 30 million tonnes, leaving a physical trade deficit of more than 90 million tonnes. This is for a country of more than 45 million inhabitants. Argentina, with a population of about 40 million, has reached exports of about 100 million tonnes and imports of about 30 million tonnes (Perez-Manrique et al., 2013). Similar trends, with slight differences, are identified in Brazil, Ecuador and Peru. Growing exports in tonnes (of different commodities) are not succeeding in improving the MTBs due to the negative terms of trade (Vallejo, Pérez Rincón and Martinez-Alier, 2011; Pérez-Rincón, 2014; Samaniego, Vallejo and Martinez-Alier, 2014).

To conclude this section, the critiques against extractivism have a double economic foundation. Domestic extraction and exports increase as they are driven by internal and external demand. Raw materials-based economies incur disproportionate environmental costs, which are not factored into the price of commodities (Rice, 2007; Jorgenson, 2009; Roberts and Parks, 2009). Moreover, exhaustion of resources is

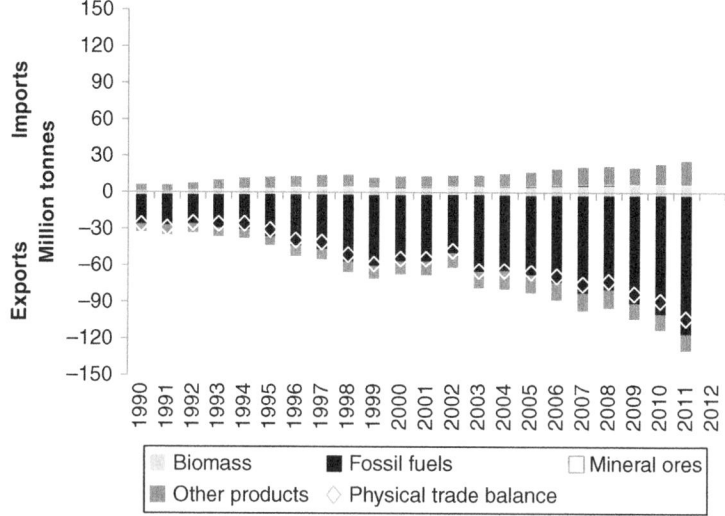

*Figure 2.3* Physical trade balance of Colombia, 1990–2011
Source: Samaniego et al. (2014) based on COMTRADE, DANE.

renamed as "production" and it sustains periodic periods of bonanza. Outside demand does increase because of the metabolic needs of the world industrial economy. The recent growth of Asian economies, and China in particular, is exacerbating the primarization of Latin American economies by boosting the pressure to extract environmentally sensitive resources (Muradian, Walter and Martinez-Alier, 2012). Recently, an absurd situation has been reached: not only are the environmental costs of the booming extractive activities not accounted for, and the exhausted resources not replenished, but, moreover, the great excess of physical exports over imports is not able to pay for the imports. The commercial deficits will have to be compensated for by foreign investments or other forms of debt, which in due course will produce repayments to foreign countries. These are becoming key drivers that strengthen extraction trends, thereby expanding the commodity frontiers and reaching areas of high biodiversity and cultural value – the land of indigenous and peasant communities.

## Extractive conflicts in Latin America

As pointed out in the previous section, there is an ongoing boom in the extraction of commodities in Latin America, and a large share of these materials is exported. This boom has been related to an increase in the number of extractive conflicts, which we frame as "ecological distribution conflicts". In order to elucidate the connections between sociometabolic trends and extractive conflicts, we propose a typology based on the commodity at stake. For each commodity type we will briefly explain some key features and illustrate with examples. Each commodity has its particularities and, as a result, different typologies could be proposed. We don't claim that the one used here is the best or the only possible one, but we use it as a guiding tool to distinguish key trends and features. We propose a classification that distinguishes between biomass (crops, plantations, fisheries) and minerals (metal ores, fuels, industrial, construction materials).

Within this typology, other subclassifications could be considered. For instance, from a social metabolism point of view, another distinction can be made between precious materials and bulk commodities when considering metallic minerals or biomass products (Wallerstein, 1974). Precious materials, such as diamonds, gold or shrimp, have a high economic value per unit of weight but are physically not necessary as inputs for the metabolism of the importing countries, compared with "bulk commodities", such as oil, gas, copper, iron, wood or soyabeans. This distinction does not mean that gold does not play an important social and economic role in the world of jewellery-making, in the world of love and marriage (as in India) or in the world of financial investments (Ali, 2006), but the difference stands in the point of view of the metabolism of the importing economies. Moreover, this difference is also related to different drivers for extraction and the related socioenvironmental pressure exerted.

### Biomass

Extractive conflicts related to biomass involve a range of activities, including soy, oil palm and timber production, plantations, fisheries, and mangrove destruction and other deforestation. We could also include related conflicts such as those over the use of glyphosate (for the production of genetically modified organisms, such as soy) and over the implementation of projects for Reducing Emissions from Deforestation and Forest Degradation (REDD).

Let us consider here the case of Argentina (Perez-Manrique et al., 2013). As shown in Figure 2.4, biomass is the predominant material flow of this economy. On average, biomass represents 70% of all materials extracted in the country from 1970 to 2009, of which 71% comprise fodder for livestock (forage, silage, grazing and by-products),

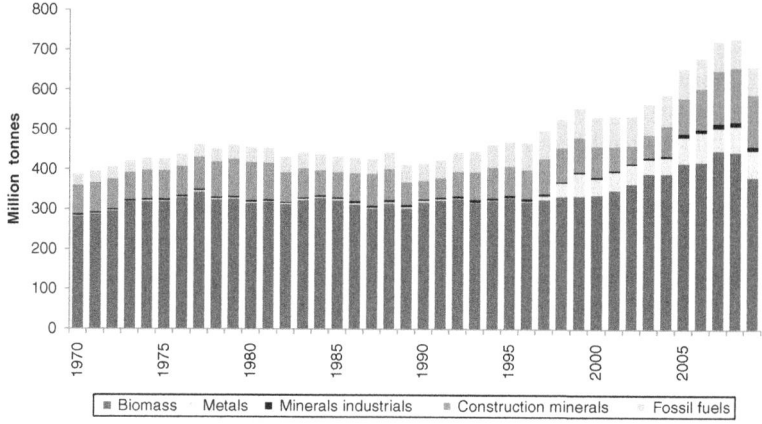

*Figure 2.4* Domestic extraction in Argentina, 1970–2009
Source: Walter et al. (2013).

2% fishing and forestry biomass, and 27% crops. From 1997 to 2009, biomass extraction from primary crops increased from 50 megatonnes (Mt (1 million tonnes)) to 137 Mt, mainly for export. Soyabeans constitute the predominant flow within the primary crops. According to Pengue (2001), soyabeans (mostly genetically modified) have displaced other domestically produced crops such as cereals, roots, tubers, vegetables and melons. Indeed, during the period studied, these crops have decreased their participation in the primary crop extraction from 44% to 25% for cereals, from 6% to 2% for roots and tubers, and from 5% to 2% for vegetables and melons. From 1970 to 2009, Argentina's soyabean production jumped from 26,000 tonnes to 30.9 Mt. This growth was driven by high international prices for this commodity from the 1990s onwards, and by technological factors such as the mechanization of agriculture, and the introduction of transgenic soyabeans and chemical weeding with glyphosate (Teubal, 2006). Since the introduction of genetically modified soyabeans in Argentina in 1996, this crop represents an average of 26% of all primary crops.

The rise in crop production led to the expansion of the agricultural frontier, thereby clearing land and forest as well as displacing indigenous and rural communities. Since the 1990s, Argentina has been experiencing one of the largest processes of deforestation in the history of the country (UMSEF, 2007). This entails new issues, such as the weakening of food security, as crops are mainly exported and the production of locally consumed crops is decreasing. The growing use of agrochemicals produces water, air and soil pollution, and causes health impacts on the surrounding populations (Binimelis, Pengue and Monterroso, 2009). The harvested area of soyabeans multiplied from 38,000 hectares (Ha) in 1970 to 18 million Ha in 2009, accounting for more than half of the total agricultural land (MAGyP, 2011). The predominant biomass flow in the economy of Argentina is still grazing, foraging, silage and by-products. Nevertheless, the expansion of soyabean crops diminished the amount of land available for cattle-grazing. Millions of hectares that were in agricultural-cattle rotation have been allocated to permanent agriculture, while livestock increasingly depends on feed crops (i.e. cereal, soymeal) (Santarcángelo and Fal, 2009; PEA, 2010).

These trends have contributed to an increased number of conflicts over land in Argentina, as peasants and indigenous groups are confronted with the expansion of the soy-extraction frontier into their lands (Aranda, 2010). The expansion of the agricultural frontier has led to the clearing of lands and forest, as well as the displacement of many indigenous and rural populations (Teubal, 2006). This has resulted in various conflicts over access to land. This is the case for the inhabitants of La Primavera (Formosa, Argentina), who have been displaced by the expansion of soy production ever since 2008. Indigenous communities have been dispossessed of their lands, and the Qom people are struggling to recover 5,000 Ha (Asociación Civil Nodo Tau, 2010; García-López and Arizpe, 2010).

The increased use of chemicals in genetically modified (GM) crops has also triggered an increasing number of conflicts related to the health impacts. This is the case for the "mothers of Ituzaingó" of Cordoba, who lead a movement that is mainly composed of women who since 2001 have been demanding that the provincial government stop the air fumigation of soy fields. The spraying of large amounts of glyphosate near urban areas was causing cases of cancer (mostly in children) and birth defects induced by contamination. In 2009 the movement succeeded in forbidding the spraying of this product in urban areas (GRR, 2009). Incidentally, some invasive species such as Aleppo sorghum (or Johnsson grass) acquired resistance to glyphosate spraying, and as a result agriculture steps not only into a pesticide treadmill but also into a "transgenic treadmill" (Binimelis, Pengue and Monterroso, 2009).

Tree plantations have similarly been the subject of socioenvironmental conflicts. As analysed by Gerber (2011), industrial tree plantations for wood, palm oil and rubber production are among the fastest-growing monocultures and are currently being promoted as carbon sinks and energy producers. Such plantations are causing a large number of conflicts between companies and local populations, mostly in the tropics and subtropics. Relying on the most comprehensive literature review to date, corresponding to 58 worldwide conflict cases (drawing on the WRM database), Gerber (2011) finds that the prominent cause of resistance is related to corporate control over land that results in displacements and the end of local uses of ecosystems as they are replaced by monocultures.

Biomass conflicts related to fisheries and shrimp aquaculture are also relevant in Latin America. Let us briefly consider here the environmental injustices related to the promotion of the shrimp aquaculture industry in Central America, in the Gulf of Fonseca region of Nicaragua and Honduras on the Pacific Coast. This is one of the most densely populated areas in Central America and also one of the poorest. This regional economy depends, to a large extent, on artisanal fishing, specifically shellfish harvesting. Industrial aquaculture activities began in Honduras at the start of the 1970s and in Nicaragua in the second half of the 1980s with small-scale projects. Nowadays this activity has sharply increased. According to the Food and Agriculture Organization (FAO) of the United Nations, in 2008 production had reached 26,584 tonnes, and 14,690 tonnes in Honduras and Nicaragua, respectively. This implies an increase in total production of more than 200% in both countries over ten years (1998–2008). Most of the production is for export, mainly to the USA and to European markets. Where there were once estuaries and natural lagoons, nowadays there are large ponds for producing shrimp. In Nicaragua the surface area under production expanded from 771 Ha in 1989 to 10,396 Ha in 2009, and in Honduras from 750 Ha in 1985 to 14,954 Ha in 2000 (Mestre Montserrat and Ortega Cerdà, 2012).

What was supposed to become a source of wealth for the regional economy has disempowered local fishing communities, which have seen their access to natural resources enclosed and limited. This has triggered serious social conflicts in the region. The industrial sites are located in areas populated by poor communities that rely on the communal use of coastal resources. The main response of the shrimp industry to the theft of their product has been the armed surveillance of their lands, both private and public. This has been a common practice in Nicaragua since 2008, when an agreement was established between the Association of Aquaculturalists of Nicaragua and the armed forces. These measures have further limited the access of local communities to coastal resources, fostering conflict and further impoverishing the population, thereby increasing social marginalization and unrest. As Mestre Montserrat and Ortega Cerdà (2012) indicate, successive conflicts between security forces protecting aquaculture farms and local fishermen have caused various injuries and at least one death in Nicaragua, and twelve deaths in Honduras. Fishermen have reported cases in which navigation to their fishing grounds through the estuarine channels has been restricted, along with cases of detention and harassment – in the form of constant demands for documentation to be shown – at sea. In Honduras, people engaged in campaigns to resist the expansion of the shrimp industry into protected areas have also been detained.

In Latin America, as elsewhere, the views of social groups involved in such conflicts over biomass are expressed in different "languages", using, for example, discourses about land and territorial dispossession, territorial rights, biopiracy, consultation rights, health impacts (due to chemical use), food sovereignty, human rights (given criminalization and militarization of extractive activities) and democracy. Unsustainable biomass extraction is also linked with conflicts over the rights of nature and of future generations, as biodiversity and nature's genetic pool are affected (by reducing the diversity of crops or advancing towards high-diversity areas). Potential future conflicts could also arise as intensive agricultural practices affect the long-term quality of soils (Pengue, 2001, 2004; Binimelis, Pengue and Monterroso, 2009).

## Minerals

Mineral mining includes a range of commodities that can be grouped as metals (e.g. copper, gold, silver, iron, bauxite, uranium, nickel), mineral fuels (e.g. oil, gas, coal, shale oil), industrial minerals (e.g. phosphates, asbestos, salt) and construction minerals (e.g. sand, gravel, stones). The general stages of the mining process are shared: exploration to locate and characterize the mineral deposits, exploitation to mine the ores, mineral processing to refine the mineral, and transport to the consuming economies. However, the features and impacts of each commodity vary. Here we present some key features of the different minerals, and analyse in more detail metal and fuel minerals whose extraction is currently triggering significant debates in Latin America.

### Metal ores

The extraction boom of raw materials in Latin American has been particularly significant for metal ores (see Figure 2.5). While in 1970 the weight of industrial and metal ores accounted for 10% of the total material flows of Latin America, in 2009 it reached 25%. In fact, in 2009, industrial and metals ores were, after biomass, the second greatest material extracted and, in part, exported from the region, accounting

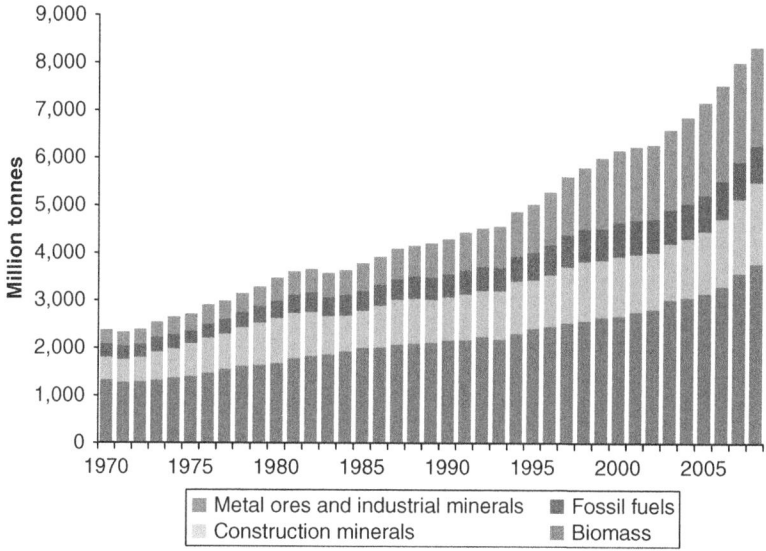

*Figure 2.5* Domestic extraction in Latin America by major category of material, 1970–2008
*Source*: UNEP and CSIRO (2013).

for 2,100 million tonnes of ores (West and Schandl, 2013). In 2012, Latin America provided 45% of the global copper output, as well as 50% of silver, 26% of molybdenum, 21% of zinc and 20% of gold (Henriquez, 2012), attracting a third of global metal-mining investments (US$210 billion) (Ericsson and Larsson, 2013). We will address with some detail metal ore extraction features and trends that are currently related to a boom of conflicts in Latin America.

One of the particularities of the metal-mining production chain is that its initial stages are characterized by low value but high environmental cost: resource extraction and then processing/refining have the highest impact. Later stages, such as assembling, are estimated to have less environmental impact but generate the majority of the economic value. This relationship represents a general trend of the impact/value curve that also applies more generally to other products that use metal ores (Giurco et al., 2010). Moreover, the socioenvironmental impacts of resource extraction increase when ore grades decline, as more waste is generated. As pressure to extract ores increases and the extraction frontier expands, reaching lower quality deposits, the environmental pressures in the stages of extraction and processing become greater (Giurco et al.,

*Table 2.1* General conversion factors of gross ore versus metal content and ore concentrate

| Metal | Gross ore/metal content | Gross ore/concentrate |
| --- | --- | --- |
| Iron | 43.32 | 81.93 |
| Copper | 1.04 | 3.33 |
| Nickel | 1.83 | 23.45 |
| Lead | 11.86 | 16.52 |
| Zinc | 8.34 | 14.50 |
| Tin | 0.24 | 0.33 |
| Gold | 0.00021 | 0.06630 |
| Aluminium | 18.98 | 67.55 |
| Silver | 0.034 | 2.552 |
| Uranium | 0.0015 | 0.3744 |

*Source*: Based on Schoer et al. (2012).

2010). Table 2.1 presents general conversion factors for the relationship between metal ores or concentrates and the gross ore that is mined. This factor is derived from the average of the annual business reports of about 160 metal mines in the world (Schoer et al., 2012).

Precious materials, such as gold, have the highest generation of overburden. As indicated in Table 2.1, to obtain 2 grams of gold, an average of 1 tonne of gross ore has to be mined. As the price per unit of precious metals is higher than for bulk metals, it becomes economically feasible to extract ore of decreasing quality or grade, entailing the processing of larger amounts of ore in open-cast mining and, as a result, generating increasing amounts of waste rock and tailings. This has also been made possible with the development of (more intensive) processing techniques that allow miners to obtain metals from decreasing ore concentrations (i.e. cyanide leaching for gold) (Bridge, 2004).

Moreover, other studies point to a worldwide decline in the quality of ore.[1] As the high-grade ores have been depleted, the mining frontier moves to lower-grade ores, with increasing environmental costs. The decline in the quality of ores has direct implications in terms of land intervention of mining activities, as larger mines (open-pit mining) have to be built and larger quantities of waste rock – especially sensitive in the case of sulphidic material that has the potential to generate acid drainages[2] – are generated (Bridge, 2004; Giurco et al., 2010; Mudd, 2010). For instance, recent studies conducted in the gold-mining sector in Australia indicate that, as ore quality decreases, the amount of water and energy used in the mining process increases significantly.

This trend overlaps with other environmental pressures, such as larger requirements of chemical inputs and larger amounts of waste (Mudd, 2007a, 2007b; Giurco et al., 2010; Prior et al., 2012).

The significance of these trends grows as we consider the expansion of the mining frontier to sensitive and critical ecosystems, such as tropical and cloud forests, or the very high mountains next to pasturelands and glaciers. These are also the homes of indigenous people. As pointed out by Bridge (2004), an increasing proportion of mineral exploration and investment expenditures during the 1990s targeted the tropical areas around the globe, reaching ecologically sensitive and/or high-value conservation areas. The International Union for Conservation of Nature (IUCN) has raised concerns related to the expansion of the mining, gas and oil frontier in World Heritage Sites, demanding protection for them (IUCN, 2011). Furthermore, recent studies led by scholars and activists are pointing to the large overlap of mining concessions with the land of peasants and indigenous people in Latin America (Bebbington, 2012b). For instance, de Echave (2009, quoted in Bebbington, 2012b) estimates that over half of Peruvian peasant communities are affected by mining projects or concessions. According to the EJOLT database (see below), in Latin America, indigenous peoples are present in over 50% of the environmental conflicts recorded to date in this registry (Pérez-Rincón, 2014). Chapter 11 on community consultations analyses in more detail some aspects of metal-mining conflicts in Latin America.

Moreover, it is important to stress that in the case of mining activities, ecoefficiency and technological approaches are limited. As the environmental impacts of mineral extraction can be reduced but not eliminated (Bridge, 2004), inputs to the mining process – such as water, energy or chemical compounds – can be reduced (per unit of production), the management of waste can be improved (e.g. better membranes to isolate waste from soil), and mining sites can be rehabilitated (e.g. revegetation). However, mineral mining necessarily modifies the environment to some degree. Moreover, operationalizing ecoefficiency in the mining sector is complicated by the fact that mining (unlike other industrial processes) is a segregative process that cannot avoid the production of large volumes of waste. This is increasingly significant considering the wider trends of declining ore qualities. Along the same vein, Giurco et al. (2010) maintain that mineral resource depletion is as much about falling resource quality (decreasing ores) and accessibility (distant and difficult to extract, with higher social and environmental costs and related conflicts) as it is about a reduction in resource quantity and availability. As follows, Prior and colleagues (2012) suggest that the "peak metal" (the time when extraction can no longer rise to meet the demand) has more to do with a carefully weighed decision that considers the social and environmental implications of continuing to extract than a question of existing metal quantities available.

In early 2014, OCMAL, a network of organizations that records large-scale metal-mining conflicts, listed 203 active conflicts affecting 308 communities. According to OCMAL (2014), the largest number of mining conflicts are found in Peru (35), Chile (35), Argentina (26), Mexico (32), Brazil (20), Colombia (12), Bolivia (9) and Ecuador (7). Central America as a whole also has many mining conflicts. The impact of large-scale metal-mining activities on water, land, health, livelihoods and rights raises concerns among communities that feel disempowered by official decision-making procedures that place a premium on ecoefficiency and pecuniary criteria. Governments and mining companies frame complaints as being politically motivated and misinformed (Walter, 2014), but such a widespread wave of complaints (and so much violence against the protestors, at least in some countries) is evidence of a vigorous grassroots social movement.

*Mineral fuels*

This category includes a diversity of commodities, such as oil, natural gas and shale-gas fracking. We could also consider energy-related conflicts related to thermoelectricity plants. Oil is the main source of energy of modern societies; it is an essential input for the exosomatic energy metabolism of contemporary rich economies (transport, industry, etc.). The growth of the world economy has relied on fossil fuels over the last century, and the oil demand and consumption have increased steadily throughout the twentieth century. However, since the 1960s, there has been a decrease in the number of new discoveries of conventional oil reservoirs. Moreover, recent discoveries reveal decreasing quality, thus implying larger economic and environmental costs (Tsoskounoglou, Ayerides and Tritopoulou, 2008). As the pressure to find and extract conventional and unconventional fossil fuels augments, the frontiers of exploration and extraction expand, reaching environmental and socially sensitive locations.

One area in Latin America where the expansion of the oil-mining frontier has strongly impacted one of the culturally and biologically most diverse regions on Earth is in the Peruvian Amazon. Orta-Martínez and Finer (2010) indicate that since the 1920s, oil exploration and extraction in this region have threatened both biodiversity and indigenous peoples, particularly those living in voluntary isolation. They argue that the phenomenon of peak oil, combined with rising demand and consumption, is pushing oil extraction into the most remote corners of the world. As modern patterns of production and consumption, and high oil prices, are forcing a new oil exploratory boom in the Peruvian Amazon, conflicts are spreading across indigenous territories, new forms of resistance appear, and indigenous political organizations are born. The expanding oil and gas frontiers are overlapping with the lands of indigenous peoples, some of whom were previously uncontacted, which fosters conflict, disease and unrest among these communities (Finer and Orta-Martínez, 2010; Orta-Martínez and Finer, 2010; Gavaldà, 2013).

An important case of struggle over the environmental injustices of oil extraction is in Lago Agrio, in the Ecuadorian Amazon. Between 1964 and 1992, Texaco's oil operations polluted the northern region of the Amazon forest in Ecuador, spanning 1 million Ha inhabited by various indigenous communities and resulting in environmental and health damage. Texaco was bought by Chevron in 2001. In 1993, local residents and indigenous communities filed a class-action lawsuit against Texaco in the District Court in New York for damages caused to their health and to the environment. For ten years the case was stalled in the US Courts, until 2003, when eventually the trial was moved to the Ecuadorian Amazon town of Lago Agrio. In 2011, in a landmark judgement, the local Sucumbios court sentenced Chevron Texaco to pay US$9.5 billion to the Frente de Defensa de la Amazonia, which would be doubled if the company did not publicly apologize. The court decision was upheld in 2012. Chevron has refused to pay and activists have tried to seize the company assets in third-party countries, such as Canada and Argentina.

*Industrial and construction minerals*

Industrial minerals include those used in industrial and agricultural processes. These minerals have different levels of toxicity and the pressures to extract them depend on their industrial uses. There are, for instance, conflicts related to the asbestos-mining in different places in Latin America. An example is the conflict of Sao Felix do Amianto in the state of Bahia (Brazil), which was open between 1939 and 1967 in the towns of Bom Jesus da Serra and Poçoes. There are many claims asking for compensation for health impacts, from workers both in the mine and in the factory.

There are also conflicts related to industrial minerals that are less toxic, such as phosphates. For instance, the Bayovar mine that is located in the north of Peru and is owned by Vale produces 5 million tonnes of phosphates per year (EJOLT, 2014).

Construction minerals are materials such as sand and gravel that are related to urbanization processes and infrastructure construction. These materials travel less than other materials because of their relatively low price per unit of weight, and for this reason they tend to be near the sites of processing and final use. As follows, conflicts over quarries are usually related to conflicts over processing plants (e.g. cement factories). An example of conflicts related to sand and gravel extraction is in Rio Tunjuelo (Bogotá, Colombia), one of the main sources of construction minerals in Bogotá. Some 50 years of extraction of sands and gravels have changed the urban landscape, shaping large holes in the ground. These holes are 30, 50 or 70 m deep and have diameters that reach several hundreds of metres. In 2002, in order to avoid the impact of a serious flood, old mining holes were used as water reservoirs to divert overflowing water from the Tunjuelo River. Flooded quarries became a source of infections and bad odours, as abandoned quarries became water oxidation ponds. Social unrest was born from the impact of abandoned quarries on water, and the environmental impacts related to the nearby processing plants. Another example is the conflict in San Juan Sacatepequez in Guatemala, where indigenous communities fostered a local consultation to stop the opening of a quarry and its processing plant on their lands. These activities were promoted by the national government without the consent of local inhabitants (EJOLT, 2014).

## Conflicts at different points in the commodity chain

The classification presented here focuses on extractive activities, but conflicts can emerge at other stages of the life cycle of a commodity. In such a way, material extraction is connected to environmental and social pressures at different localities and to social groups that exceed the specific place where extraction is occurring. We point to four key stages related to the life cycle of a (raw material) commodity where conflicts emerge: extraction, transport, processing and final disposal.

First, conflicts can arise at the site of extraction. We have previously pointed out some of the socioenvironmental pressures and conflicts directly related to extraction.

Second, the transport of raw materials to processing plants is also related to noise, dust and air pollution. This stage also includes the impacts and conflicts related to the construction of transport infrastructures, such as pipelines and ports. An example of the tensions related to these activities is the Initiative for the Integration of the Regional Infrastructure of South America (IIRSA), led by a group of Latin

American governments with the support of the Interamerican Development Bank (IDB) and the Corporación Andina de Fomento (CAF). The IIRSA initiative aims to improve the connection of Latin American economies, connecting the Pacific and Atlantic oceans to facilitate the extraction and export of Latin American raw materials. It includes the construction of hydroways, gas and oil pipelines, ports and so forth. IIRSA-related projects are giving way to numerous large conflicts in the region (Svampa, 2012; Gavaldà, 2013), as these infrastructures are reaching the lands of distant communities that are also areas of high biodiversity and landscape value.

Third, processing plants usually require energy, water and chemical substances, and can also affect the quality of soil, air, and surface and underground waters, triggering health problems and social conflict. A paradigmatic case is La Oroya in Peru. La Oroya is a mining town in the Peruvian Andes that, since 1922, has been the site of a polymetallic smelter. This has produced toxic emissions and wastes from the plant. Recently the smelter was recycling scrap metals imported through El Callao (Lima's harbour) and taken up by railway to La Oroya, which has suffered from critical levels of air pollution and is considered to be one of the most polluted places on Earth (Blacksmith Institute, 2006). Owned by the Missouri-based Doe Run Corporation, the smelter was long signalled as responsible for the dangerously high lead levels found in children's blood.

Fourth, conflicts can arise when commodities reach the end of their life cycle and are discarded. Waste generation also includes impacts on soil, air and water generated during extraction, transport and processing (e.g. mining waste ponds and landfills). Climate change could be seen as a waste-disposal conflict because we have exceeded the capacity of new terrestrial vegetation and the oceans to absorb the carbon dioxide produced, and therefore its concentration in the atmosphere has increased to 402 ppm.

## New approaches to studying environmental conflicts: A statistical political ecology

Since the 2000s, various groups have been creating online databases that register information on ongoing socioenvironmental conflicts in Latin America and beyond. These databases reflect an effort initiated by NGOs and social movements to make visible the increasing environmental injustices that communities confront. More recently, universities and research projects have also engaged in such systematization initiatives.

Some aim at mapping out environmental conflicts in one country, such as a recent inventory of over 80 conflicts in Colombia (Pérez-Rincón, 2014) and the Brazilian Mapa da Injustiça Ambiental e Saúde (Environmental Injustice and Health Map, by FIOCRUZ). In addition, there is a growing number of databases recording socioenvironmental conflicts throughout the region, including OLCA, and worldwide, such as our EJOLT project (Martinez-Alier et al., 2011). There are also databases focused on specific issues, such as tree plantations (see WRM), mining (OCMAL) and land-grabbing (Genetic Resources Action International (GRAIN)). Furthermore, there are important efforts being made to report on processes of protest and "criminalization" of activists or human rights violations in Latin America and the Caribbean (OCMAL, 2013; Toledo, Garrido and Barrera Bassols, 2013). This "criminalization of protest" refers to different processes that range from government officials and politicians who promote and apply laws that typify protest as unacceptable social behaviour and label protest as sabotage, terrorism or an obstruction of public space; to protesting organizations as illicit associations or publicly framing protestors as criminals (Saavedra, 2013); and, most dramatically, to the reality of countries such as Brazil, Mexico, Colombia and Peru, where environmental activists are being killed while defending livelihoods and nature (see the lists provided by Global Witness). The ENGOV project has created an inventory of Latin American databases and maps (available at www.engov.eu), while the global inventory by EJOLT allows us to analyse and compare different features of numerous extractive conflicts (available at www.ejatlas.org).

## Conclusion

In this chapter we have explained the main trends in the social metabolism of Latin America and have focused on one of the main indicators, the material flows. In the last 40 years the extraction of materials has increased four-fold, far more than the population. A substantial part of the extracted materials (whether biomass, fossil fuels or metal ores, although not the building materials) goes to exports. We have developed a typology of conflicts according to the commodities in question. Many grassroots environmental organizations, and also academics and state bodies, are aware that there are more ecological distribution conflicts, and they contribute to environmental governance by making them visible through inventories and maps.

In regard to external trade and economic policies, we have insisted that at present most South American economies have large physical trade deficits (in tonnes), and simultaneously they have or are about to have commercial trade deficits (in monetary terms). That is to say, the large physical exports that carry heavy ecological and social rucksacks are scarcely able to pay for the imports. In all of South America there are huge exports in volume (tonnes of oil, coal, iron ore, soyabeans, wood, copper, etc.) and yet several countries (Brazil, Colombia, Peru, Venezuela and Ecuador) have monetary commercial deficits. Remarkably, the recent "extractivist" trend happens both in countries with national-popular governments and in those with neoliberal governments. Even President Mujica of Uruguay favoured an iron-mining project with the Indian company Zamin Ferrous Metals in 2014. This project aims to export 18 million tonnes per year during the next 20 years – about 6 tonnes per inhabitant – leaving behind large environmental liabilities.

There are structurally unfavourable terms of trade for Latin American countries exporting natural resources. First, persistent physical trade deficits are recorded. We call it a "deficit" because natural resources are lost or depleted. In recent years, this trend has been accompanied by a monetary trade deficit that affects both small and large countries. Brazil had, between January and March of 2014, a trade deficit of US$6,072 million. This is the highest deficit for a quarter in 21 years, while Argentina has seen its monetary trade surplus sharply decrease between 2012 and the first quarter of 2014. Monetary trade deficits must be balanced by other income in the current account or in the capital account balance. The inflow of foreign direct investment can offset the trade deficit but it will generate income that will later leave the country. Increased indebtedness will lead to a need to export more and more, causing further environmental damage and social conflict.

While the demand for raw materials that are not recycled (e.g. fossil fuels) or only partly recycled (e.g. metals) is likely to remain over time, even without economic growth in the world system, the social and environmental costs of extraction are increasing as the grade of metallic minerals and the EROI decreases. This is the case as oil or gas is extracted from distant places, as also happens with timber, soy or palm oil. At the same time, even if in the long term the demands remain, prices can fall sharply due to variations in the business cycles. Overall, reprimarization is a risky economic strategy. Therefore, it is not surprising that new Latin American voices call for different economic policies. For them, the local complaints against extractive industries (including biomass extraction) should not be seen as instances of NIMBY ("not in my backyard") or as attacks on the state, but instead as useful contributions towards a change in environmental governance.

Therefore the criticism of South American post-extractivist scholars (Maristella Svampa, Eduardo Gudynas, Alberto Acosta) not only has a social and environmental basis but also has economic and democratic foundations. The export of raw materials depletes natural resources and causes pollution and conflicts with local populations. Governments use repression as a method to facilitate raw-material extraction. On the other hand, the prices of these major exports are cheap in comparison to imports, hence a new march along the route to debt. These tendencies point to the need for a change in policies. In fact, there have been some attempts to curb the export of raw materials through public policies such as the Yasuní-ITT initiative in Ecuador from 2007 to 2013, aimed at leaving oil in the ground under zones with exceptionally high biodiversity in the Ecuadorian Amazon. Popular resistance is also expressed in many existing protests, often arguing in terms of indigenous land rights. And new institutions arise as referenda or local consultations (see Chapter 11). These local protests and initiatives for environmental justice are a response to the power of corporations and governments, a power that leads to a deficit in local democracy. In sum, next to physical and monetary trade deficits, the export of raw materials also produces a deficit in local democracy.

## Acknowledgement

We thank the journal *Ecología Política* journal and Julien Brun, Pedro Perez-Manrique and Ana C. Gonzalez-Martinez for granting us permission to use Figures 2.2 and 2.4.

## Notes

1. A recent industry study signals that, "With declining ore grades exacerbated by increasing energy and other costs, and significant deposits being found at greater depths or in more remote areas, the average capital costs for copper production capacity in new mines increased an average of 15% per year over the past 20 years, with much of the increase evident since 2008" (SNL Metals Economics Group, 2013).
2. Mining-related chemical pollution can be generated by the release into the environment of reagents added during mineral processing, such as the sulphuric acid that is used for the leaching of copper oxides, or the mercury or cyanide used to process gold. Pollution is also caused by the oxidation that naturally occurs in minerals that are present in the ore as a result of exposure to air, water and/or bacteria. Many metal ores, such as nickel, copper and lead, occur in the rock as sulphides. The contact with oxygen and water triggers an oxidation process that forms sulphuric acid. This process can result in the formation of acid rock drainages. This process has been pointed out as one

of the main environmental challenges of the mining industry (Bridge, 2004; Government of Australia, 2007; Giurco et al., 2010).

## References

Agarwal, B. (1994) *A Field of One's Own: Gender and Land Rights in South Asia* (Cambridge: Cambridge University Press).
Ali, S.H. (2006) "Gold Mining and the Golden Rule: A Challenge for Producers and Consumers in Developing Countries", *Journal of Cleaner Production* 14(3–4): 455–462.
Aranda, D. (2010) *Argentina Originaria: Genocidios, Saqueos y Resistencias* (Buenos Aires: La Vaca).
Asociación Civil Nodo Tau (2010) Asesinatos en Formosa: Indiferencia, Oídos Sordos y Represión, http://www.tau.org.ar/enredando2002-012/noticias_desarrollo.shtml?x=62518, date accessed 12 January 2015.
Bebbington, A. (2012a) *Social Conflict, Economic Development and the Extractive Industry: Evidence from South America* (London and New York: Routledge).
Bebbington, A. (2012b) "Underground Political Ecologies: The Second Annual Lecture of the Cultural and Political Ecology Specialty Group of the Association of American Geographers", *Geoforum* 43(6): 1152–1162.
Binimelis, R., Pengue, W. and Monterroso, I. (2009) "Transgenic Treadmill: Responses to the Emergence and Spread of Glyphosate-Resistant Johnsongrass in Argentina", *Geoforum* 40(4): 623–633.
Blacksmith Institute (2006) *The World's Worst Polluted Places. The Top Ten* (New York: Blacksmith Institute).
Bridge, G. (2004) "Contested Terrain: Mining and the Environment", *Annual Review of Environment and Resources* 29(1): 205–259.
Bullard, R. (1990) *Dumping in Dixie: Race, Class, and Environmental Quality* (Boulder, CO: Westview Press).
Bunker, S. (2007) "The Poverty of Resource Extraction", in A. Hornborg, McNeil, J.R. and Martinez-Alier, J. (eds), *Rethinking Environmental History: Worldsystem History and Global Environmental Change* (Lanham, MD: Altamira Press), 239–258.
Conde, M. and Walter, M. (2014) "Commodity Frontiers", in G. D'Alisa, F. Demaria and G. Kallis (eds), *Degrowth. A Vocabulary for a New Era* (New York and London: Routledge), 71–74.
Douguet, J.M., O'Connor, M. and Noel, F. (2008) *Systèmes de Valeur et Modes de Regulation: Vers Une Économie Politique Écologique* (Paris: Cahiers du C3ED, UVSQ).
EJOLT (2014) Environmental Justice Atlas, http://elatlas.org, date accessed 16 January 2015.
Ericsson, M. and Larsson, V. (2013) "E&MJ's Annual Survey of Global Mining Investment Project Survey 2013", *E&MJ Engineering and Mining Journal*. http://pure.ltu.se/portal/files/100685420/EMJ_2013.pdf
EUROSTAT (2001) "Economy-Wide Material Flow Accounts and Derived Indicators. A Methodological Guide", *Luxembourg, Office for Official Publication of the European Communities*. (Luxemburg: European Commission).

Finer, M. and Orta-Martínez, M. (2010) "A Second Hydrocarbon Boom Threatens the Peruvian Amazon: Trends, Projections, and Policy Implications", *Environmental Research Letters* 5(1): 014012.

Fischer-Kowalski, M. (1997) "Society's Metabolism: On the Childhood and Adolescence of a Rising Conceptual Star", in M. Redcliff and G. Woodgate (eds), *The International Handbook of Environmental Sociology* (Cheltenham: Edgard Elgar), 119–137.

García-López, G.A. and Arizpe, N. (2010) "Participatory Processes in the Soy Conflicts in Paraguay and Argentina", *Ecological Economics* 70(2): 196–206.

Gavaldà, M. (2013) *Gas Amazónico. Los Pueblos Indígenas frente al Avance de las Fronteras Extractivas en Perú* (Barcelona: Icaria).

Gerber, J.F. (2011) "Conflicts over Industrial Tree Plantations in the South: Who, How and Why?", *Global Environmental Change* 21(1): 165–176.

Gerber, J.F., Veuthey, S. and Martínez-Alier, J. (2009) "Linking Political Ecology with Ecological Economics in Tree Plantation Conflicts in Cameroon and Ecuador", *Ecological Economics* 68(12): 2885–2889.

Giurco, D., Prior, T., Mudd, G., Mason, L. and Behrisch, J. (2010) *Peak Minerals in Australia: A Review of Changing Impacts and Benefits* (Sydney: University of Technology and Monash University).

Gonzalez-Martinez, A.C. and Schandl, H. (2008) "The Biophysical Perspective of a Middle Income Economy: Material Flows in Mexico", *Ecological Economics* 68(1–2): 317–327.

Government of Australia (2007) *Managing Acid and Metalliferous Drainage*. Leading practice sustainable development program for the mining industry. Department of Industry, Tourism and Resources. Canberra, Australia, http://www.industry.gov.au/resource/Documents/LPSDP/LPSDP-AcidHandbook.pdf

GRR (2009) "Pueblos Fumigados: Informe sobre la Problemática del Uso de Plaguicidas en las Principales Provincias Sojeras de La Argentina", Grupo de Reflexión Rural, http://www.grr.org.ar/trabajos/Pueblos_Fumigados__GRR_.pdf, date accessed 3 December 2014.

Haberl, H. et al. (2007) "Quantifying and Mapping the Human Appropriation of Net Primary Production in Earth's Terrestrial Ecosystems", *Proceedings of the National Academy of Sciences of the United States of America* 104(31): 12942–12947.

Henriquez, V. (2012) "Latin America to Receive 50% of Global Mining Investments Up to 2020", *Business News Americas*. http://www.bnamericas.com/news/mining/latin-america-to-receive-half-of-global-mining-investment-until-2020-codelco

Hornborg, A. (1998) "Commentary: Towards an Ecological Theory of Unequal Exchange: Articulating World System Theory and Ecological Economics", *Ecological Economics* 25(1): 127–136.

Hornborg, A. (2006) "Footprints in the Cotton Fields: The Industrial Revolution as Time–Space Appropriation and Environmental Load Displacement", *Ecological Economics* 59(1): 74–81.

Hornborg, A. (2009) "Zero-Sum World: Challenges in Conceptualizing Environmental Load Displacement and Ecologically Unequal Exchange in the World-System", *International Journal of Comparative Sociology* 50(3–4): 237–262.

IUCN (2011) "Mining Threats on the Rise in World Heritage Sites", *International Union for Conservation of Nature*. http://www.iucn.org/knowledge/news/?7742/Mining-threats-on-the-rise-in-World-Heritage-sites

Jorgenson, A.K. (2009) "The Sociology of Unequal Exchange in Ecological Context: A Panel Study of Lower-Income Countries, 1975–2000", *Sociological Forum* 24(1): 22–46.

Kapp, K.W. (1950) *The Social Costs of Private Enterprise* (Cambridge: Harvard University Press).

Leff, E. (2003) "La Ecología Política en América Latina, un Campo en Construcción", *Polis* 1(5): 1–15.

M'Gonigle, R.M. (1999) "Ecological Economics and Political Ecology: Towards a Necessary Synthesis", *Ecological Economics* 28: 11–26.

MAGyP (2011) "Sistema Integrado de Información Agropecuaria", *Buenos Aires*.

Martínez-Alier, J. (1997) "Conflictos de Distribución Ecológica", *Revista Andina* 29(1): 41–66.

Martinez-Alier, J. (2001) "Mining Conflicts, Environmental Justice, and Valuation", *Journal of Hazardous Materials* 86(1–3): 153–170.

Martinez-Alier, J. (2002) *The Environmentalism of the Poor: A Study of Ecological Conflicts and Valuation* (Delhi: Edward Elgar; Cheltenham: Oxford University Press).

Martinez-Alier, J. (2009) "Social Metabolism, Ecological Distribution Conflicts, and Languages of Valuation", *Capitalism Nature Socialism* 20(1): 58–87.

Martinez-Alier, J., Healy, H., Temper, L., Walter, M., Rodriguez-Labajos, B., Gerber, J. F. and Conde, M. (2011) "Between Science and Activism: Learning and Teaching Ecological Economics with Environmental Justice Organisations", *Local Environment* 16(1): 17–36.

Martinez-Alier, J., Kallis, G., Veuthey, S., Walter, M. and Temper, L. (2010) "Social Metabolism, Ecological Distribution Conflicts, and Valuation Languages", *Ecological Economics* 70(2): 153–158.

Martinez-Alier, J. and O'Connor, M. (1996) "Ecological and Economic Distribution Conflicts", in R. Costanza (ed.), *Getting down the Earth: Practical Applications of Ecological Economics* (Washington DC: Island Press), 277–286.

Moore, J.W. (2000) "Sugar and the Expansion of the Early Modern World-Economy: Commodity Frontiers, Ecological Transformation, and Industrialization", *Review* 23(3): 409–433.

Mudd, G.M. (2007a) "Global Trends in Gold Mining: Towards Quantifying Environmental and Resource Sustainability", *Resources Policy* 32(1–2): 42–56.

Mudd, G.M. (2007b) "Gold Mining in Australia: Linking Historical Trends and Environmental and Resource Sustainability", *Environmental Science & Policy* 10: 629–644.

Mudd, G.M. (2010) "The Environmental Sustainability of Mining in Australia: Key Mega-Trends and Looming Constraints", *Resources Policy* 35(2): 98–115.

Muradian, R. and Martinez-Alier, J. (2001) "Trade and the Environment: From a 'Southern' Perspective", *Ecological Economics* 36(2): 281–297.

Muradian, R., Walter, M. and Martinez-Alier, J. (2012) "Hegemonic Transitions and Global Shifts in Social Metabolism: Implications for Resource-Rich Countries. Introduction to the Special Section", *Global Environmental Change* 22(3): 559–567.

Orta-Martínez, M. and Finer, M. (2010) "Oil Frontiers and Indigenous Resistance in the Peruvian Amazon", *Ecological Economics* 70(2): 207–218.

PEA (2010) "Plan Estratégico Agroalimentario y Agroindustrial Participativo y Federal 2010–2020".

Pengue, W. (2001) The Impact of Soybean Expansion in Argentina, Seedling, http://www.grain.org/es/article/entries/292-the-impact-of-soybean-expansion-in-argentina, date accessed 4 December 2014.

Pengue, W. (2004) "Producción Agroexportadora e (In)seguridad Alimentaria: El Caso de la Soja en Argentina", *Revista Iberoamericana de Economía Ecológica* 1: 46–55.

Perez-Manrique, P.L., Brun, J., González-Martínez, A.C., Walter, M. and Martínez-Alier, J. (2013) "The Biophysical Performance of Argentina (1970–2009)", *Journal of Industrial Ecology* 17(4): 590–604.

Pérez-Rincón, M.A. (2014) *Conflictos Ambientales en Colombia: Inventario, Caracterización y Análisis* (Cali: Universidad del Valle, CINARA, EJOLT).

Prior, T., Mudd, G., Mason, L. and Behrisch, J. (2012) "Resource Depletion, Peak Minerals and the Implications for Sustainable Resource Management", *Global Environmental Change* 22(3): 577–587.

Raikes, P., Friis Jensen, M. and Ponte, S. (2000) "Global Commodity Chain Analysis and the French Filière Approach: Comparison and Critique", *Economy and Society* 29(3): 390–417.

Rice, J. (2007) "Ecological Unequal Exchange: International Trade and Uneven Utilization of Environmental Space in the World System", *Social Forces* 85(3): 1369–1392.

Robbins, P. (2004) *Political Ecology: A Critical Introduction* (London: Blackwell Publishing).

Roberts, J.T. and Parks, B.C. (2009) "Ecologically Unequal Exchange, Ecological Debt, and Climate Justice: The History and Implications of Three Related Ideas for a New Social Movement", *International Journal of Comparative Sociology* 50(3–4): 385–409.

Russi, D., Gonzalez-Martinez, A.C., Silva-Macher, J.C., Giljum, S., Martinez-Alier, J. and Vallejo, M.C. (2008) "Material Flows in Latin America. A Comparative Analysis of Chile, Ecuador, Mexico, and Peru, 1980–2000", *Journal of Industrial Ecology* 12(5–6): 704–720.

Samaniego, P., Vallejo, M.C. and Martinez-Alier, J. (2014) "Déficit Comercial y Déficit Físico en Sudamérica", Documento de Trabajo, Proyectos CSO2010-21979 e ENGOV, Institut de Ciència i Tecnologia Ambientals (ICTA), Universidad Autónoma de Barcelona, FLACSO Sede Ecuador.

Santarcángelo, J. and Fal, J. (2009) "Production and Profitability in Livestock in Argentina. 1980–2006", *Mundo Agrario* 10(19).

Schoer, K., Giegrich, J., Kovanda, J., Lauwigi, C., Liebich, A. Buyny, S. and Matthias, J. (2012) *Conversion of European Product Flows into Raw Material Equivalents* (Heidelberg: Ifeu; Eurostat).

Singh, S. J., Krausmann, F., Gingrich, S., Haberl, H., Erb, K.H., Lanz, P., Martinez-Alier, J. and Temper, L. (2012) "India's Biophysical Economy, 1961–2008. Sustainability in a National and Global Context", *Ecological Economics* 76–341(100): 60–69.

Sneddon, C., Howarth, R.B. and Norgaard, R.B. (2006) "Sustainable Development in a Post-Brundtland World", *Ecological Economics* 57(2): 253–268.

SNL Metals Economics Group (2013) "SNL Metals Economics Group's Copper Study Reveals Lower Grades, Higher Costs for Copper Production in 2012", PRWeb.

Stern, D.I. (2004) "The Rise and Fall of the Environmental Kuznets Curve", *World Development* 32(8): 1419–1439.

Svampa, M. (2012) "Consenso de los Commodities, Giro Ecoterritorial y Pensamiento Crítico en América Latina", *Observatorio Social de América Latina* 32: 15–38.

Teubal, M. (2006) "Expansión del Modelo Sojero en La Argentina. De la Producción de Alimentos a los Commodities", *Realidad Económica* 220: 71–96.

Tsoskounoglou, M., Ayerides, G. and Tritopoulou, E. (2008) "The End of Cheap Oil: Current Status and Prospects", *Energy Policy* 36(10): 3797–3806.

UMSEF (2007) *Informe sobre Deforestación en Argentina* (Buenos Aires).

UNEP and CSIRO (2013) *Recent Trends in Material Flows and Resource Productivity in Latin America* (Nairobi). http://www.unep.org/dewa/portals/67/pdf/RecentTrendsLA.pdf

Urkidi, L. and Walter, M. (2011) "Dimensions of Environmental Justice in Anti-Gold Mining Movements in Latin America", *Geoforum* 42(6): 683–695.

Vallejo, M.C., Pérez-Rincón, M.A. and Martinez-Alier, J. (2011) "Metabolic Profile of the Colombian Economy from 1970 to 2007", *Journal of Industrial Ecology* 15(2): 245–267.

Wallerstein, I. (1974) *The Modern World-System, Vol I.: Capitalist Agriculture and the Origins of the European World-Economy in the Sixteenth Century* (New York and London: Academic Press).

Walter, M. (2014) *Political Ecology of Mining Conflicts in Latin America* (Barcelona: Autonomous University of Barcelona).

Walter, M., Brun, J., Perez-Manrique, P., Gonzalez-Martinez, A.C. and Martinez-Alier, J. (2013). Análisis de flujo de materiales de la economía Argentina (1970–2009). Tendencias y Conflictos extractivos. *Ecología Política* 45: 94–98.

West, J. and Schandl, H. (2013) "Material Use and Material Efficiency in Latin America and the Caribbean", *Ecological Economics* 94: 19–27.

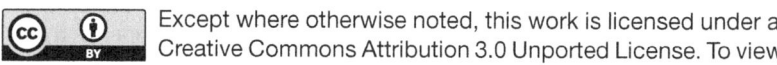 Except where otherwise noted, this work is licensed under a Creative Commons Attribution 3.0 Unported License. To view a copy of this license, visit http://creativecommons.org/licenses/by/3.0/

OPEN

# 3
# Indigenous Knowledge in Mexico: Between Environmentalism and Rural Development

*Mina Kleiche-Dray and Roland Waast*

## Introduction

Since the 1990s, several international agreements (Article 8J of the Biological Diversity Convention, 1992) and international protocols (Nagoya Protocol, 2010) have begun to assess the capacity of indigenous knowledge to contribute to socioeconomic progress as well as to environmental protection. In the course of this process, the knowledge and practices of peasants and natives have been called to the rescue to resolve a number of new problems. These include the loss of biodiversity, threats from carbon dioxide emissions and environmental conservation, with consequent debates about the property rights of local and autochthonous populations – such as that on "biopiracy" versus "bioprospection". However, the farming methods favoured by the indigenous populations often conflict with national development projects oriented towards the market economy. This discrepancy gives rise to tensions and to local, national and international conflicts that can be observed throughout Latin America. They are typified in a country such as Mexico, which will serve here as an example. Mexico has been the subject of a number of studies[1] and is often seen as a laboratory of both ideas and long-term development projects related to these issues. It has 12% of the biodiversity of the planet; natural vegetation occupies more than 71% of its territory, and its forest resources occupy 64.8 million Ha,[2] 70% of which belong to autochthonous and peasant communities (OCDE, 2013). Agriculture remains a highly important activity in the country, covering 24% of the territory (102 million Ha), of which half is *ejidataria* (communal land covered by extension services). Some

16 million of its 112 million inhabitants identify themselves as belonging to the *población indígena* and 7 million speak a native language. The population that lives in the areas of greatest biodiversity is generally classified as being one of the poorest. Some 88% of the 1,033 indigenous municipalities are classified as being in "great poverty". In fact, Mexico is the country that has the most revealing poverty rate in the OCDE.[3] Furthermore, its natural resources are deteriorating, under pressure from grazing, from slashing and burning brush in preparation for tillage, from excessive tillage and from intensive irrigation (OCDE, 2007). In this context, more and more social movements and proponents of environmental projects – such as the local branch of Vía Campesina – have emerged. They rely on autochthonous knowledge in the struggle against the rapid expansion of intensive agriculture, the monoculture of GM organisms, extensive ranching, biofuels, land-grabbing and extractive industries.

Of course, these social movements are by no means recent. However, everything indicates that they have gained a fresh impulse from the institutionalization of a national environmental policy, the boom of alternative rural development projects and the initiatives of new actors, such as movements of identity assertion and the national and international NGOs that support them (i.e. GRAIN).

These new actors favour decentralized management of natural resources, the setting up of local seed banks, the promotion of an agriculture free from chemical inputs, and the development of local markets. Family farming and small-scale agriculture – a political category that also covers the agricultural practices of the native and peasant populations – constitute the prime area targeted by their projects. In this complex context, "developmentalist" policies enter into competition with new projects classified as "socioenvironmental" (Léonard and Foyer, 2011).

New issues consist of the acknowledgement of indigenous and peasant knowledge, and its inclusion in the design, elaboration, implementation, execution and evaluation of projects that support family and small-scale agriculture.

Similarly, questions have arisen regarding ways of setting up a national environmental project that would involve native and peasant populations as well as new actors – NGOs, state and municipal authorities, and national and international private organizations (including large farmers and multinational firms) – in governance and decision-making. How can a sustainable and equitable use of natural resources be guaranteed? Is such an environmental project compatible with a particular development project?

This chapter focuses on the sociocognitive dynamics underlying the practical use of natural resources in family and small-scale agriculture. We shall first review the literature in social science studies and in Latin-American post-colonial studies on these dynamics. We shall then turn to the treatment of indigenous knowledge in mainstream social sciences and its promotion by certain policy-makers. Finally we will analyse the pragmatic combination of autochthonous and scientific knowledge in the process of governance, incorporating environmental matters by means of constant political, local and historical reconfiguration. These field perspectives are based on work in the Mixteca region (State of Oaxaca, Mexico).

## Decolonizing indigenous and peasant knowledge

The objective of this chapter is to understand how, on the one hand, indigenous and peasant knowledge penetrates technoscientific knowledge and how, on the other hand, it becomes part of rural-development projects and environmental issues. Of major help in this attempt are the general concepts of "translation" (Callon and Latour, 1981; Akrich et al., 2006), "boundary-object" (Leigh Star and Griesemer, 1989; Trompette and Vinck, 2009) and "transcodification" (Lascoumes, 1994). They have been forged in the field of social studies of science in order to deal with similar problems (Callon, 1986 on scientists, fishermen and the plan to breed sea shells). These concepts postulate a continuity between the logics of knowledge production and political logics, and a centrality of the dynamics of translation and hybridization in different epistemic spheres (Harding, 1997). Social studies of science examine the mediations between knowledge of differing types (and especially between scientific and profane knowledge), and between scientific knowledge and the political logics involved in action.[4]

Meanwhile, the anthropology of local knowledge has analysed the categories grouped under the term "traditional knowledge". Agrawal (1995, 2002) points out the context of their use (and the political dimensions involved in asymmetrical exploitation of this knowledge compared with that of "scientific knowledge"), particularly in development projects. In regard to environmental issues, several authors have stressed the embedding of different types of knowledge in their conditions of production, their historical, social and institutional settings, and the need to study the full context of practices and circulation when they are put into operation (Fairhead and Leach, 2003; Goldman et al., 2011).[5] All these aspects have to be analysed if one is to understand exchanges between types of knowledge and the construction of new hybridized forms in the processes of environmental governance. These various types of knowledge also have to be viewed in the asymmetrical perspective of North/South encounters (Gaillard et al., 1997; Escobar, 1995; Waast, 1996) and centre/periphery geopolitical relationships (Polanco, 1989; Raj, 2007).

Other useful perspectives have been developed over the last 15 years in Latin American post-colonial studies (Escobar, 2003; Boidin, 2010). Their Latin American proponents (e.g. in the Modernity/Coloniality/Decoloniality (M/C/D) programme, school of thought represented in Latin America) have catalysed a current of critical rethinking of "Eurocentric modernity". Using the notion of coloniality of power and of knowledge (Quijano, 1994; Lander, 2000; Mignolo, 2000; Dussel, 2007), the M/C/D programme describes colonization in a much more complex way, going beyond the conventional analysis in terms of political and economic oppression. A racial and ethnic classification of the world has given rise to a cultural oppression in which only one type of awareness and a single form of reason are taken into account. It is on this basis that geocultural identities have been attributed to the regions and populations of the world (Crespo, 2014).

The notion of "coloniality" reveals three parallel processes of "modernization": (1) the exclusion of other cultures or civilizations from participation in the construction of modernity; (2) the imposition of geocultural identities (Crespo, 2014); and (3) the exclusion of any forms of knowledge (other than the colonial) in the historical construction of the world. The M/C/D programme is an invitation to perform a "decolonial spin" (Castro Gomez and Grosfoguel, 2007) that involves taking into account the various places of enunciation and their critical or resistant approach towards colonial modernity. The programme uses the notion of "frontier epistemology" (Mignolo, 2007) to rewrite the narrative of modernity from alternative standpoints, re-evaluating dominated cultures and peoples and their histories of resistance. It aims, for instance, to retell the history of Latin America by taking into account relationships between society and nature.

The essential "coloniality of nature" in Latin America is linked to the disruption of indigenous ecosystems and methods of production, annulling the potential autonomy of these societies (Leff, 1986; Castro Herrera, 1996) and leading to a "subalternation" of the dominated bodies of both human beings and nature (Castro-Gómez, 2005). Arturo Escobar uses the concept of "nature regimes" to define the processes, articulating modes of perception and experience that determine the ways of using space. These processes are identified as "resistance", "compromise" and "hybridization".

Taken up by political ecology, along with the notion of "colonized nature" (Escobar, 2011), this sort of thinking enables us to understand that the categories of "traditional knowledge" and "local knowledge" can only be grasped in opposition to that of "scientific knowledge". All knowledge is produced within social, political and economic relationships of certain types. And the actors who promote one or another type of knowledge in modern society always do so through a binary classification: modernity/coloniality or universality/pluriversality.

"Decolonizing nature" involves understanding, first and foremost, how "subaltern knowledge" has been identified and characterized by science – that is, disqualified, and sometimes reappropriated in downgraded form as a mere resource – and also the ways in which all actors relate to nature. Nature is not merely seen as a resource but in a different framework altogether: as culture.

## From "traditional and local" to "indigenous" knowledge

This statement by A. Escobar leads us to examine the ways in which mainstream science has treated indigenous knowledge.

In the early 1980s, agronomists, in evaluating the technical component of farmers' agricultural practices, began to write about indigenous knowledge and know-how. The agronomists resumed observations and studies made by naturalists, ethnologists and linguists during and after the colonial period, focusing on instruments (tools), crop rotation, preparation of the land and so on. Within the social sciences, specialists in "development" subsequently took up the topic, accompanied by a few anthropologists.[6] This eventually muted into a craze, despite the fact that level-headed specialists stressed that local knowledge should not be made into a fetish.

In the 1990s the notion moved from agricultural questions to environmental studies, passing from issues of production and productivity to those of conservation and the management of natural resources. It came to the attention of experts, research centres and international organizations (Bell, 1979; Chambers, 1988). Many anthropologists climbed on the bandwagon. Their intervention opened up two distinct perspectives. On the one hand, the majority supported recognition of traditional knowledge, as it represented for them – at the very least – new fields of study, new sources of finance for applied anthropology, and access to a "specialist" status. On the other hand, the term "indigenous knowledge" began to develop as a more militant concept, highlighting the dependence and marginalization of "indigenous" peoples. This latter term differs from the previously predominant notions of "traditional" and "local knowledge", which have now come to be seen as condescending. The former term is linked to a modernizing project for society, and the latter to the universality of "scientific knowledge". These two notions enabled that of "indigenous knowledge" to emerge as a relatively open-minded alternative. Its promoters stressed that indigenous knowledge cannot be reduced to a recipe for development (Agrawal, 1995; Sillitoe, 1998). The notion of "indigenous knowledge" has been instrumental to the recognition of local knowledge in the legal field, in that of intellectual property rights and more generally in the right of peoples to their own culture.

Work on this subject continued to develop in the 2000s, massively appropriated in environmental studies and anthropology. In these circles, there has been passionate debate on the subject. The arguments deployed have often helped "indigenous" peoples and peasants to obtain the benefits brought about by development as well as greater political autonomy.[7] The journal *Human Ecology* has become a major vector of this environmental and anthropological work.[8] The notion of "traditional knowledge" has since followed its own developmental path, with a strong environmental focus. Many authors use the two concepts – traditional and indigenous – interchangeably (Godoy et al., 2005).

As for Latin America, the local history of all these notions is not very different. The term "indigenous knowledge" appeared very early on and spread primarily through Brazil, Mexico, Bolivia and Chile. Interestingly, it eventually deserted scientific literature and was linked mainly to social movements. At present there are few studies published on the topic in the social sciences and humanities. Possibly the recognition of intellectual property rights after the Rio Conference in 1992 put an end to debate in the region.[9]

Very few studies deal with the way in which companies avoid complex negotiations with local communities – buying, for example, medicinal plants on local markets, and hiring and training collectors and growers of plants required for natural cosmetics. Likewise, few authors now undertake studies of traditional knowledge in regard to medicinal plants, experiments with traditional knowledge in public health services, and discussions about climate change and other current issues.

While the term "indigenous knowledge" has been fading out, that of "agroecology" has grown in popularity, especially in Latin America. Agroecology as a scientific field valorizes native and peasant farming practices as a socioproductive alternative to modern agriculture (Altieri et al., 2006) that is also environmentally friendly. According to its protagonists, native and peasant practices can inspire the ecological scientific approach and at the same time become a sustainable way of farming.

Scientific and institutional interest in indigenous and peasant farming practices is not really new, however. In Mexico, a key figure in this intellectual tradition was Efraim Hernandez Xolocotzi (known as Efraim H.X.), an agronomist who was educated in the USA and taught at the University of Chapingo. He was called back to Mexico to support the Green Revolution at its very beginning but soon became critical of it (Jiménez Sánchez, 1984). He contributed to the creation of an agroecological movement in Mexico. Basically, his objective was to show how important it was to study traditional agrosystems, stressing the fact that resource scarcity drives man's creativity and encourages him to develop a set of cultural and productive practices to adapt to the environment and to the conditions of production (Díaz León and Cruz León, 1998). According to Efraim H.X. and his disciples, especially Victor Toledo (1992), "the indigenous model" of agriculture can serve as a basis for the development of agroecological knowledge and practices. In the 1980s a socialist current in Mexico – consisting primarily of biologists, ethnobotanists and agronomists – joined in social and environmental thinking and engaged directly with native and peasant communities.

Agroecology has been politicized in different ways for different purposes, depending on whether it is being promoted by academic activists, by peasants, by religious militants, by agronomy advisors or by officials. This can be said about projects ranging from the design of public policies to initiatives of an extremely local nature. This is what we will now discuss, tracing this shift in the political field and, in particular, in public rural development policies aimed at small-scale family farming.

## Indigenous knowledge as a lever for rural development and environmental policies

After a period of liberalization of structural adaptation plans following the financial crisis of 1982 – which resulted in the ratification of the North American Free Trade Agreement (NAFTA) and the political and financial crisis of 1994–2005 – Mexican agriculture had to face international competition in a context of market deregulation and trade liberalization. A policy of food security[10] replaced that of food self-sufficiency, which had been the credo of agrarian reform and the Green Revolution. By the 1990s the *ejidos* had been privatized and extension services reduced. As a result, foreign purchases of foodstuffs increased (Warman, 2001).[11]

Nevertheless, political discourse has continued to defend the importance of developing autonomous and efficient agrifood systems. In a country where only 6% of farmers are classified as "modern",[12] the Mexican Government has had to propose various programmes and measures to mitigate the impacts of rising food prices for the poorest strata of the population (Gravel, 2009). The main measures aimed at the poorest farmers were a distribution of grants according to cultivated acreage (such as the so-called Procampo Programme) and aid to the poorest women (Progresa/Oportunidad). The less marginalized categories were urged to adopt the Green Revolution technology package (hybrids, fertilizers, pesticides and mechanization) in programmes such as Object Income and Masagro.

Thus in 2007 the state designed a new national policy for rural development as a whole. With the programme *Nuevo Programa Especial Concurrente* (PEC), the government began to take an interest in the integration of the native and peasant population into national development. This PEC was launched in areas of great and very great marginalization, the population itself taking part, thanks to the organization of a forum (*Foro de Consulta Popular*), to which all stakeholders in the rural sector were invited.[13]

However, only 15.7% of all financial resources considered in the PEC were directed towards the support of agricultural food production (Gomez-Oliver, 2008). Furthermore, programmes that targeted small farmers – either by distributing a technology package or by granting subsidies – encouraged deforestation, and this gave rise to further intensification of farming.

This seems to be at odds with the aim of developing a national environmental policy. Yet ratification of the Convention on Biological Diversity and recognition of native struggles (in the San Andrés agreements of 1994) finally led to the creation of the Environment Ministry (the Secretariat of the Environment and Natural Resources/Secretaria de Medio Ambinete y Recursos Naturales (SEMARNAT)) in 1994. An environmental policy that attempts to integrate the international standards of Agenda 21[14] was established. In 2000 a National Plan for Sustainable Development was adopted. To top it off, in 2000 the Mexican Constitution was changed so as to acknowledge the cultural and ethnic diversity of Mexican society. This particular interest has been reinforced since 2007 in the sustainable development programmes in which ecological viability is treated as one of the five cornerstones of federal action. This functions in tandem with the Sector Programme for the Environment and Natural Resources, the objective of which is to "associate the conservation of natural capital with economic and social development" (OCDE, 2013: 40). The dual process involved in the recognition of indigenous knowledge has thus been made part of the development pattern for agricultural and environmental policies.

In this dual ministerial context, the Mexican Government undertook the task of integrating the participation of the native and peasant population into its agricultural policy and also into its political agenda, thereby institutionalizing national environmental policy.

The Ministry of the Environment has confirmed that "the native populations that maintain a very close link with natural resources and biodiversity actively support sustainable development through on-site conservation of ecosystems and natural habitats, and the maintenance and recuperation of viable populations of species in their natural surroundings".[15]

In 1997 the Ministry of the Environment initiated the Conservation and Restoration Programme of soils. In 1998 it launched the National Reforestation Programme and other programmes that sought to combine economic and social development with environmental conservation. The objective was to devote economic resources to National Protected Areas and to the restoration of regions identified as priorities from an environmental perspective.

The main tools that the government has used have always been aimed at the conservation of biodiversity and of forests, in accordance with the National Strategy for Biodiversity (2000), complemented by the Mexican Strategy for the Conservation of Plants (which has existed since 2008 and was revised in 2012) and subsequently enhanced by the National Strategy combating invasive species. Major programmes within this framework have been specifically dedicated to native and peasant populations.

A twist was introduced, however, when the Ministry of the Environment developed its Regional Sustainable Development Programme (*Programa de Desarrollo Regional Sustentable* (PRODERS)) in an attempt to link the environmentalist vision to a developmentalist one. The programme was presented as a comprehensive initiative by means of which SEMARNAT contributed to the support of sustainable development in poor rural regions. These regions often include native and peasant populations who live where the major biological and environmental riches are located, far from the rural nodes. The management of this programme was supposed to be decentralized and participative, based on a long-term vision (Toledo and Bartra, 2000).

Thus it would seem that – despite almost ten years of government efforts to institutionalize an environmental policy linked to the development of sustainable agriculture in the most disadvantaged areas of the country – most observers agree that the main thrust of agricultural policy has been, and remains, the pursuit of greater productivity (OCDE, 2013). The bulk of financial resources are still being oriented towards commercial agriculture and "modern farmers": the most important subjects in the sector. This conclusion is congruent with the criticism emanating from the post-colonial school, which interprets from these policies a vision based on denial of all rationality and veritable knowledge in other forms of culture. This attitude does not leave room for any concepts other than those of a modernizing society and its links to high-productivity projects. Ultimately, it leaves no space for plurality or, in the words of Arturo Escobar, "pluriversity" (Escobar, 2011). Although this trend presents itself globally, the fact remains that conflicting logics – even at a government level – mean that heterogeneous projects are now being implemented for merely practical reasons. Several studies have attempted to bring visibility to the success of various local experiences that overcome this contradiction between developmental and environmental concerns. The government – notably two ministries (Environment and Agriculture: SEMARNAT and SAGARPA (*Secretaria de Agricultura, Ganadería, Desarrollo Rural, Pesca y Alimentación*)) – gave direct or indirect support to these local experiences, particularly (a recent development) various civil society groups that had made progress in the conservation of soil and water, the protection of biodiversity and wildlife, and the autonomy of their food systems. We shall now describe a case of this sort that illustrates the importance of practical reason in action.

## Towards an institutionalization of native and peasant knowledge

We will now deal with a case study that needs to be contextualized. Its whole story takes place in the Mixteca Region of Oaxaca, Mexico. To begin, we will discuss traditional knowledge and its evolution over the course of time.

### Construction of agricultural knowledge and practices, and their exchange over the course of time

The Mixteca region of south-east Mexico covers the eastern part of Oaxaca state. It extends over an area of 4 million Ha, in which there are

221 municipalities, 155 of them located in the state of Oaxaca. It is in the Mixteca region that the largest "indigenous population" of Mexico is concentrated, with more than 1 million people (34% of the Oaxaca population)[16] (INEGI, 2010). *Mixtec* inhabitants belong, however, to a diversity of peoples: Chocholteca, Tlapaneca, Nahuatl, Triqui, Zapotec and Amuzgo (Rivas Guevara et al., 2009). Their history is traversed by episodes of expropriation and reappropriation of their land.

The Aztecs and later the Spanish colonized the region and divided local political entities into small communities, grabbing the best land. Since the Mexican independence, the Agricultural Reform has redistributed the *haciendas* (large farming units) into *ejidos*, the privatization of which has been authorized by federal law since the 1990s. The result has been a broad diversity of land use and tenure in the native and peasant communities of the *Mixteca* – *bienes comunales* (commons), *ejidos* (public lands with extension services), *tierras de uso común* (collective lands managed by means of community meetings) and *tierras privadas* (private lands). Control is highly concentrated: 1.7% of the *ejidos* and communities control 70% of the land, and 0.41% of the private properties cover 20% of the total of privatized lands. Thus more than 85% of private units and *ejidos* are smaller than 5 Ha (Sanchez Lopez, 2013). This inequality has generated agrarian conflicts that continue to this very day.

However, despite this conflict-ridden history, periods of tranquillity have made it possible to introduce new plants, and new techniques of cultivation and food preparation, since colonization. This has been due to exchanges among communities during religious festivals and at markets, and migration to other regions (Katz, 1994, 2002). During the colonial period, the cultivation of wheat and sugarcane, extensive ranching, and the breeding of silkworms and cochineal progressed, gaining economic importance (Long and Attolini, 2009; Lazos, 2012). With the decline of the silk industry and cochineal at the end of the nineteenth century, artisanal palm weaving gained importance, driven primarily by the Spaniards, who managed to establish an international market. On the other hand, deforestation and the erosion of soils worsened when goats were introduced and lime was exploited (Velásquez, 2002).

Subsequently the Mexican Government's "developmentalist" project also had an impact on these dynamics, by influencing local agricultural practices. From 1935 to 1988, the Mexican Government implemented more than 19 "developmentalist" programmes (Altieri et al., 2006) dedicated to crops ranging from cochineal, fruit trees, coffee, hybrid corn, and vegetables to livestock and the improvement of agricultural infrastructure. During the 1970s the government also tried to promote a Green Revolution technology package (improved seeds, mechanization, the use of fertilizers and chemical pesticides) by means of aids and extension services within the framework of its Integrated Rural Development Programme (*Programa Integral de Desarrollo Rural* (PIDER)). Though PIDER achieved a significant volume of production, it led to the loss of native varieties of maize, beans and squash; the contamination of soil and water; the overexploitation of aquifers; deforestation; and soil erosion (Altieri et al., 2006).

The government saw the main problems of the Mixteca as matters of water and soil. By the 1970s, it tried to recover the Mixtecan technique of terraced agriculture that the inhabitants had lost (Mendoza García, 2002, 2004). This had been used in small valleys and heavy rainfall areas. The federal government attempted to restore the ancient terraces using heavy machinery. Facing poor results, it decreed that the Mixteca was unable to sustain the development of an alimentary agriculture. The main replacement project was to plant palm trees to supply a craft industry. As of 1973, weavers were organized into cooperatives (Velasco Rodriguez, 1994) supported by the Palm Trust (Fideicomiso de la Palma (FIDEPAL)). Unfortunately, the government neither managed to consolidate this cottage industry nor to diversify the uses of woven palm fibre. Marketing, support for cultivation, the development and exploitation of palm plantations, and the industrialization and export of goods made from natural fibres all disappeared during the 1990s.

Despite these setbacks, today in the Mixteca, small-scale and family farming cover areas larger than in other Mexican regions. Some 30 years ago, most of the Mixtec population was involved in agriculture. But migration has had a profound impact – especially since the 1990s, which saw extensive migration to the United States. Emigration now accounts for more than 30% of the population (Lazos, 2012). This has weakened local institutions considerably, including mutual aid, collective work (such as the *guetza* and *tequio*)[17] and social networks. The *milpa* – the food and agriculture system, associating representations and rituals with the cultivation of maize, beans and squash – seems to subsist only in homes that need fresh maize for the festivities of the Day of the Dead, which maintain a symbolic link with the land (Lazos, 2012). All the varieties of maize that needed a lot of work and a lot of space have gradually been abandoned and replaced by commercial crops, such as passion fruit and new varieties of tomatoes that are grown in gardens and greenhouses[18] (Katz, 1994). Today the farming system has to be complemented by additional income from welfare programmes and remittances sent by emigrant relatives. Thus the native and peasant population tends, on the one hand, to diversify their diet by buying more meat and industrial food and beverages; on the other hand, they consume fewer of the wild greens (*quelites*) that were always seen as "poor people's food" (Katz, 1992).

### The "farmer to farmer" model in the Mixteca region (Oaxaca state)

Life is difficult, and modernity, cash crops and intensive technology are attractive; but there are alternatives. Our case study accounts for a civil society group (*Centro de Estudios de Tecnologías Alternativas para México*/Center for the Study of Appropriate Technologies for Mexico (CETAMEX)) and the institution that was finally built by its efforts. Institutions of this sort were set up with the support of the government, although sometimes the support was indirect, as in the case of the Center for Comprehensive Peasant Development in the Mixteca (*Centro de Desarrollo Integral Campesino de la Mixteca* (CEDICAM)).

The CETAMEX group has roots in the vast experience of the team that worked with civil society in the Mixteca Alta from 1983 to 1997. CETAMEX (headquartered in Mexico, DF) is financed by the World Neighbors organization (*Vecinos Mundiales*), whose objective is to resolve internal community conflicts by means of collective work performed for the benefits of the community (Blauert, 1990). World Neighbors is a Protestant religious organization that comes from Oklahoma. It formed links with a Catholic movement, *Pastoral de la Tierra*, which emerged in indigenous and peasant communities in the Mixteca region of the state of Oaxaca in the 1980s, with the help of Guatemalan peasants who were there on missionary service for World Neighbors organization.

Thanks to the advice and support of these Guatemalan peasants, catechists of *Pastoral de la Tierra* as agricultural development promoters (Holz-Giménez, 2006) – who also gave agricultural advice derived from their own peasant experience – and the technology support of people from CETAMEX, a project was launched in Santiago Tilantongo (a Mixtecan municipality) by Jesús León Santos, a local farmer.[19] This was in the early 1980s, and Santos and his colleagues received some funding from World Neighbors (Blauert and Quintanar, 2000). They decided to adopt the strategy of the World Neighbors movement (i.e. to work only with local authorities and to avoid direct dealings with federal government agencies (Bunch, 1985)) and to build up farmer-to-farmer networks (*campesino a campesino*), which focused on improving native and peasant farming practices (Boege and Carranza, 2009; Holt-Giménez, 2010).

Initially the "parent group" of CETAMEX provided services that were instrumental to promoting the use of organic fertilizers, reforestation, and the construction of tree nurseries in the municipalities of Yodocono and Tilantongo by 1982 (Altieri et al., 2006). Jesús León Santos and his colleagues subsequently worked in different municipalities and in nine communities (Nochtixltan and neighbouring communities) of Mixteca Alta. They restored the fertility of the soil when the surface layer was exposed to the effects of agents of erosion (air, water and anthropogenic activity). They made fundamental contributions to the recovery of the *tequio* (*yeta* or *guetza*), to mutual aid and to collaborative organization of work. They also recovered several techniques such as *barbecho* (long-fallow land), *recorte* (delumping), *rayada* (planting in rows), *cajeteada* (planting corn or cornfields in pits or bowls), *coa* (plowing), *yunta* (the yoke) and other local devices that retained moisture and prevented soil compaction. Subsequently, to improve the soil, they used green manures (*bocashi*) and selected their own seeds. They dug trenches on field borders and on slopes of land, forming terraces to prevent erosion, to maintain moisture and to revive springs (Rivas Guevara, 2008; Rivas Guevara et al., 2009). As a first step they undertook reforestation, using local tree species that could generate firewood, timber and wood for crafts, and they created a new organization of community nurseries.

Their second step was to restore the cultivation of *maices de cajete* by accumulating in ravines a water supply and the *limon* that had been swept away by landslides. This system (known as *jollas*) makes it possible to use residual soil moisture at the end of the rainy season to plant *maices* and thus avoid a hunger gap by guaranteeing a full year's harvest of maize. The *jollas* system was created by the Mixtecs between the pre-classical and the post-classical ages in response to demographic pressure (Romero Frizzi, 1990); until the 1980s it functioned in the sub-region of the Mixteca Alta (in the Nochixtlan, Tiaxiaco, Teposcolula and Coixtlehuaca districts). At the time, this crop system was the second most important in the Oaxacan Mixteca (Romero Penaloza et al., 1986).

It is worth noting that, in a region where *tequio* and/or the *guetza* had often been abandoned, the conservation and restoration of soil and water required intensive labour.

Fortunately, the Ministry of the Environment became concerned with soil erosion, and subsequently the government launched a national programme for soil conservation (PRODERS). This included a specific project (ProArbol) that benefited CETAMEX. Free, adapted trees were distributed, enabling the CETAMEX members to save time and labour and to concentrate on agrifood systems.

Major institutions, such as the General Directorate of Regional Programmes, were established to harmonize the programmes of different ministries. This was notably instrumental in bringing together the three ministries of the Environment, Agriculture and Social Development in support of the Sustainable Productive Development in Marginal Rural Areas (*Programa de Desarrollo Productivo Sustentable en Zonas Marginadas Rurales* (PDPSZRM)) programme. In the late 1990s, this programme, supervised by eight secretariats, implemented about 50 regional projects. The community was considered to be the basic territorial unit within Regional Development Councils (which brought together institutional and civil-society actors in prioritized microregions). These councils had to design and implement development plans whenever involvement of the community was needed. PRODERS also organized local workshops for training and for developing new skills in communities.

In 1989 a new institution was created in the Mixtec region itself: CEDICAM. This brings us back to the beginning of our story: that of a peasant movement (CETAMEX, see above). CEDICAM (*Hita Nuni* in the Mixtec language) is based in Asuncion Nochixtlan. Its role is to promote the "farmer to farmer" relationships by means of workshops and educational demonstrations. It consists of 12 Mixtec farmers who have qualified as demonstrators in the 14 Tilantongo communities. Jesús León Santos is one of the founding members of CEDICAM. He is also in charge of networking with support agencies, including Mexican governmental programmes. Santos argues that care for water and soil are essential for sustainable agriculture (Velásquez Hernández and Santos, 2006).

Finally, the pioneers who were involved in the beginnings of the CETAMEX farmers' group have recovered both their agricultural practices and a balanced diet. Others are following the same path, but this is not why they have been praised throughout the world. What is admired, above all, is their contribution to ecology (conservation of soil and water, and reforestation). In 2008 their main representative, Jesús León Santos, was awarded the annual international Goldman Environmental Prize in recognition of their efforts. Santos embodies the success of traditional peasant and indigenous agricultural practices in combating desertification. He has spread the word to all arenas in which the environment is an issue of concern.

Several experiments of this type (see Chapter 10) have shown that the initiatives of communities themselves, supported by civil-society associations, constitute a warning call to governments. The governments, in turn, rely on these institutions to design and implement appropriate programmes. One of the most important actions in this programme, which has made Mexico an international model, is the National Programme for Payment of Environmental Services (PES), which covers 3.25 million Ha of forest. The ProArbol Programme establishes the principle of financial compensation for all actions that retard deforestation and promote the recovery of forest soils.

The teachings that have brought the Mexican experience into the limelight concern potentials and limits of projects that are "truly alternative", and which at some point need to rely on the state's capacity for action. In a way, this shows that nothing can be done without the state, but that with only state support nothing can be done at all.

## Conclusion

In Mexico the issue of environmental governance is linked to that of social and economic development by its explicit objective: "food sovereignty". We have examined this relationship at different levels – national, regional and local – and we have found that effective environmental governance calls for a simultaneous analysis of Mexican agricultural policy as a whole, including the "traditional" practices of the native and peasant world. Moreover, our analyses have been diachronic as well as synchronic, and historical as well as structural. Their aim is to clarify, identify and characterize economic trends and the ways in which different sorts of knowledge contribute to this aim, by their interplay in the process of constructing environmental standards.

We have described the construction of environmental governance in the Mixteca region in Mexico, which is home to numerous native and small-scale peasant communities, known for both its food requirements and its exemplary efforts in reforestation over the past 30 years. The environmental governance process has been worked out here in terms of participation. In practice this implies the integration, accommodation and hybridization of traditional native and peasant knowledge. How do these different sorts of knowledge fit in with knowledge of the modern technoscientific sort? We have attempted to unpack the intellectual framework involved and the steps through which the process passes. We have relied on a theoretical framework that involves both science and technology studies (STS) and post-colonial studies (with its Latin American version, the M/C/D Programme). We have explained that a historical trend has assigned a subordinate place to indigenous knowledge; but also that, for practical reasons, it is translated and exchanged when it is acted upon. Exchanges can be structural (e.g. in the Green Revolution) or merely circumstantial (e.g. in the course of colonization). They can also take place between different agricultural communities that have different types of knowledge. Since ancient times, market places have been the locus of an exchange of plants (and the ways to grow them) – that is, for an exchange, adaptation and transposition of knowledge brought in from abroad. Native and peasant knowledge is not fixed; it evolves, just as technoscientific knowledge does. "Pre-modern" knowledge has now come to inspire a number of academic works, and has also influenced technical and ecological thinking. Attention has been drawn to it, and it itself has become an object of knowledge. We have shown that this upsurge of scientists' interest has been aligned with the policy debates of the day, in such matters as technology and agriculture, ecology and the environment, and cultural and social issues. There is now in Mexico an agroecological approach that is recognized by the academy and that is used by technical operators.

All of this has drawn attention to the weight of practical considerations in the evolution and reception of different sorts of knowledge, including scientific knowledge. Practical reasons not only spur a few dissident approaches but also orient the action of farmers and governments. We have dealt, to some extent, with the case of a local initiative promoted by native and small-holding farmers. They began by resisting the options and programmes designed for them by the Ministry of Agriculture, but subsequently attempted to gain self-sufficiency by restoring their traditional collaboration and recovering discontinued agricultural techniques. In doing so they have contributed to the conservation of soils and wooded areas, and this in turn has brought them recognition and help from the Ministry of the Environment. Action can change perspectives, with some actors learning to see others in new ways and opening up opportunities to build alternative projects through interaction with partners who had not originally been envisaged.

To what extent can autochthonous and peasant populations seize such opportunities, which are generally based on "secondary contradictions"? The answer to this question is less clear. There are many contradictions between environmental and agricultural policy. There is, however, a dominant trend. In Mexico it would seem that (intensive) agriculture has gained the upper hand. But this does not prevent other concerns (social and environmental) from being asserted. There has been a focus on environmental protection through reforestation. Reforestation programmes have fostered the creation of opportunities for participation at a microregional level, complementing policies dedicated to nature reserves and support for community initiatives. Simultaneously, however, the "productivist" agricultural programme designed for marginal areas (*Procampo*) has been repeated (at least for 2007–2012), despite the fact that is has accentuated deforestation. History shows also that state support, direct or indirect, is necessary for small-scale initiatives to blossom, if not during their take-off period then at least for their subsequent development and replication in other regions. Unfortunately, today there is a downward trend in budgets dedicated to environmental protection and rural development.[20] This makes new local initiatives even more precarious.

Therefore, despite the number of programmes that have been devoted to marginalized populations over the last ten years, the National Strategy seems to lack an overall plan of action. What direction will this policy take? How will it take into account the multiple experiments that have been carried out in the more vulnerable and marginalized regions?

Similar contradictions exist at an international level. The Biodiversity Convention made a breakthrough when it obtained the FAO's agreement on phytogenetic resources, recognizing that autochthonous peoples owned pro parte biodiversity and its uses. But its implementation is still in question. It is true, furthermore, that recognition of the important part played by peasant and indigenous family agriculture (providing 70% of the global food production; the FAO dedicated the year 2014 to this sector) could have a leveraging effect in promoting an operational recognition of native and peasant knowledge. However, few people argue that it would be enough to feed the planet, to alleviate dramatic famines throughout the world and to supply large cities. This is what accounts for the dual system that exists today, and what legitimates the pursuit of other avenues of (scientific) research. For example, another Mexican citizen, Dr Sanjaya Rajaram, won a World Food Prize in 2014 for his work on the genetic improvement of maize, thanks to biotechnology.[21]

At the preparatory meetings of the international climate conference (COP21), held in Paris at the end of 2015, a wish was expressed: to combine concern for family and peasant farming with thinking about climate change. It is yet to be seen whether the international conference will provide native and peasant knowledge with a real opportunity to contribute to the construction of policies dealing with climate issues.

## Notes

1. Notably, studies of WP5 "Building and Exchanging Knowledges on Natural Resources in Latin America" within the ENGOV EU Programme.
2. Forests occupy 33% of the territory with 200,000 different species, which puts it in 12th place internationally, 2nd place in terms of variety of ecosystems and 4th in terms for species (OCDE, 2013).
3. It also appears in 12th place of the countries with the greatest inequality in terms of income.
4. Until now they have dealt little with specific mediations in agriculture projects, especially between scientific knowledge and native or peasant knowledge.
5. This analysis is detailed in Foyer et al. (2014).
6. The works in this field are abundant. We primarily cite Howes and Chambers (1979); Howes (1979); and O'Keefe and Howes (1979).
7. See the Waast and Rossi report (2014). The most cited works are Davis and Wagner (2003); Woods (2002); Greene (2004); and Turner, Davidson-Hunt and O'Flaherty (2003), cited in Waast and Rossi (2014).
8. See Hassink (2005); Berkes and Turner (2006); Godoy et al. (2005); Greene (2004); Aswani and Lauer (2006); and Kirsch (2001), cited in Waast and Rossi (2014).
9. The issue was resumed in Mexico after the controversy surrounding the International Cooperative Biodiversity Group-Maya (ICBG-MAYA) project in 2000: on the one hand it was denounced as "biopiracy" and on the other hand it was advocated as a development project respectful of local communities. See Alarcón Lavín (2011); see also Barreda (2001).
10. Food security is related to the healthy diet of a maximum of persons all over the world. Perhaps the social and indigenous movement forged the food sovereignty movement, which means that each group of people should design its own agriculture policy according to its needs and culture.
11. It has been observed that imports increased from 74% to 84% for oil, from 22% to 40% for cereals, from 18% to 27% for meat, and from 15% to 24% for milk. Despite the great proportion of the population linked to agriculture, Mexico has become one of the main import countries of agricultural products (in third place after the EU and Japan).
12. In other words, with sufficient capacity to integrate into the market. See Gravel (2009).
13. Seven regional forums of public consultation – coordinated by the Interministerial Commission for Sustainable Rural Development (*Comisión Intersecretarial para el Desarrollo Rural Sustentable* (CIDRS)) – were created with the objective of collecting the proposals and viewpoints of the rural population on five topics, among which were nutrition, welfare and care for the environment.
14. It was initially created as the Ministry of the Environment, Natural Resources and Fishing (Semamap) in 1994, but it later became the Ministry of the Environment and Natural Resources. Today, climate change is included within the transformation of the agency, changing the National Institute of Ecology (INE) into the National Institute of Ecology and Climate Change (INECC). The National Commission on Biodiversity (CONABIO) and the

reformulation and strengthening of the General Law of Ecological Equilibrium and Environmental Protection (LGEEPA, 1996) are also included. See Léonard and Foyer (2011).
15. The fundamental initiative in this regard is the Indigenous Peoples and Environmental Programme 2007–2012. See SEMARNAT, México, 2009, http://www.semarnat.gob.mx/apoyossubsidios/programmeasparalospueblosindigenas/Documents/programprogrammemea%20de%20pueblos%20indigenas%20y%20medio%20ambiente.pdf, date accessed 15 September 2014.
16. Population in Oaxaca State, 3.8 million (INEGI, 2010).
17. Flores Quintero, G. (2005) has clarified what differentiates *guetza* from *tequio*. In effect, despite what had been written, it has been shown that *guetza* is the collective work that was institutionalized during the colonial era. *Tequio* is a náhuatl word that designates the community service of the adult members of the community, whose origin dates back to colonial times.
18. Esther Katz has observed how, in the last 30 years, the variety of cultivated species has diminished considerably. This is the case for the maize of the humid highlands. See Katz and Kleiche (2013).
19. Olga Elena Lara, interview with Jesus Santos León, http://ssheltonimages.com/play/ptk9uDK0XuU/Part_1 (date accessed 15 September 2014).
20. By 2011 the budget of SAGARPA was 73 billion Mexican pesos, while the budget for the environment fell to 51.2 billion Mexican pesos (out of which 12.6% was for marginalized areas: 0.99 billion Mexican pesos went to the Comisión Nacional de Áreas Naturales Protegidas (CONANP) versus 3.35 billion Mexican pesos in 2002) and 6.42 billion Mexican pesos to the Comisión Nacional Forestal (CONAFOR) (OCDE, 2013).
21. Dr Sanjaya Rajaram belongs to Centro Internacional de Mejoramiento de Maíz y Trigo (CIMMYT), an organization that played a key role in the Green Revolution of the 1960s.

## References

Agrawal, A. (1995) "Dismantling the Divide Between Indigenous and Scientific Knowledge", *Development and Change* 26(3): 413–439.
Agrawal, A. (2002) "Indigenous Knowledge and the Politics of Classification", *International Social Science Journal* 54(173): 325–336.
Akrich, M., Callon, M. and Latour, B. (2006) *Sociologie de la traduction. Textes fondateurs* (Paris: Presses des Mines de Paris).
Alarcón Lavín, R.R. (2011) "La Biopiratería de los Recursos de la Medicina Indígena Tradicional en el Estado Chiapas. El Caso ICBG-Maya", *Revista Pueblos y Fronteras* 6(10): 151–180.
Altieri, M.A., Fonseca, S.A., Caballero, J.J. and Hernández, J.J. (2006) *Manejo del Agua y Restauración Productiva en la Región Indígena Mixteca de Puebla y Oaxaca* (México D.F.: CEDEC).
Aswani, S. and Lauer, M. (2006) "Incorporating Fishermen's Local Knowledge and Behavior into Geographical Information Systems (GIS) for Designing Marine Protected Areas in Oceania", *Human Organisation* 65(1): 81–102.
Barreda, A. (2001) "Biopiratería y Resistencia en México", *El Cotidiano* 18(110): 21–39.

Bell, M. (1979) "The Exploitation of Indigenous Knowledge or the Indigenous Exploitation of Knowledge: Whose Use of What for What?", *The IDS Bulletin* 10(2): 44–50.

Berkes, F. and Turner, N.J. (2006) "Knowledge, Learning and the Evolution of Conservation Practice for Social-Ecological System Resilience", *Human Ecology* 34(4): 479–494.

Blanc, J. and Georges, I. (2012) "L'Émergence de l'Agriculture Biologique au Brésil: Une Aubaine pour l'Agriculture Familiale? Le Cas de Producteurs de la Ceinture Verte de la Ville de São Paulo", *Autrepart* 64: 121–138.

Blauert, J.K. (1990) *Autonomous Approaches to Rural Environment Problems: The Mixteca Alta, Oaxaca, Mexico* (University of London, Wye College and Institute of Development studies, University of Sussex Brighton).

Blauert, J. and Quintanar, E. (2000) "Seeking Local Indicators: Participatory Stakeholder Evaluation of Farmer-to-Farmer Projects, Mexico", in M. Estrella (ed), *Learning from Change: Issues and Experiences in Participatory Monitoring and Evaluation* (Londres: Intermediate Technology Publications), 32–49.

Boege, E. and Carranza, T. (2009) *Agricultura Sostenible Campesina-Indigena, Soberania Alimentaria y Equidad de Género. Seis Experiencias de Organizaciones Indígenas y Campesinas en México* (México, Publicación realizada con el apoyo de Pan para el Mundo).

Boidin, C. (2010) "Études Décoloniales et Postcoloniales dans les Débats Français", *Cahiers des Amériques Latines* 62: 129–140.

Bunch, R. (1985) *Dos Mazorcas de Maiz: Una Guia para el Mejoramiento Agrícola Orientado hacia la Gente* (Oklahoma City: Vecinos Mundiales).

Callon, M. and Latour, B. (1981) "Unscrewing the Big Leviathan: How Actors Macro-Structure Reality and How Sociologists Help Them to Do So", in K. Knorr-Cetina (ed), *Advances in Social Theory and Methodology: Toward an Integration of Micro and Macro-Sociologies* (London: Routledge and Kegan Paul), 277–303.

Castro-Gómez, S. (2005) *La Poscolonialidad Explicada a Los Niños* (Popayan: Universidad de Cauca).

Castro-Gomez, S. and Grosfoguel, R. (eds) (2007) *El Giro Decolonial. Reflexiones para una Diversidad Epistémica más allá del Capitalismo Global* (Bogota: Universidad Javeriana/Universidad Central/Siglo del Hombre).

Castro Herrera, G. (1996) *Naturaleza y Sociedad en la Historia de América Latina* (Panama: Cedla).

Cervantes-Godoy, D. and Dewbre, J. (2010) *Importance Économique de l'Agriculture dans la Lutte contre la Pauvreté* (Paris: Éditions OCDE).

Chambers, R. (1988) *Sustainable Rural Livelihoods: A Strategy for People, Environment and Development* (Londres: Earthscan).

Crespo, J.M. (2014) "Propuesta de Políticas sobre Saberes y Conocimientos Ancestrales, Tradicionales y Populares en el Proyecto Buen Conocer/Flok", *Cumbre del Buen Conocer/FLOK Society* del 27 al 30 de mayo de 2014 en Quito, Ecuador, https://floksociety.co-ment.com/text/VpC768Jfmd6/view/

Davis, A. and Wagner, J.R. (2003) "Who Knows? On the Importance of Identifying 'Experts' When Researching Local Ecological Knowledge", *Human Ecology* 31(3): 463–489.

Díaz León, M.A. and Cruz León, A. (eds) (1998) *Nueve Mil Años de Agricultura en México. Homenaje a Efraím Hernández Xolocotzi* (Chapingo: Grupo de Estudios Ambientales, A.C. y Universidad Autónoma).

Dumont, R. (1969) *Réforme Agraire et Modernisation de l'Agriculture au Mexique* (Paris: PUF).
Dumont, R. (1981) *Le Mal-Développement en Amérique Latine* (Paris: Seuil).
Dumoulin, D (2003) "Les Savoirs Locaux dans le Filet des Réseaux Transnationaux d'ONGs: Perspectives Mexicaines", *Revue Internationale des Sciences Sociales* 4: 655–666.
Dussel, E. (2007) "Modernidad, Imperios Europeos, Colonialismo y Capitalismo (Para Entender el Proceso de la Transmodernidad)", in E. Dussel, *Materiales para una Política de la Liberación* (Madrid: Publidisa), 195–214.
Escobar, A. (1995) *Encountering Development: The Making and Unmaking of the Third World* (Princeton: Princeton University Press).
Escobar, A. (1999) "After Nature: Steps to an Anti-Essentialist Political Ecology", *Current Anthropology* 40(1): 1–30.
Escobar, A. (2003) "Mundos y Conocimientos de Otro Mundo. El Programa de Investigación de Modernidad/Colonialidad Latinoamericano", *Revista Tabula Rasa* 1: 51–86.
Escobar, A. (2011) "Ecología Política de la Globalidad y la Diferencia", in H. Alimonda (ed.) *La Naturaleza Colonizada. Ecología Política y Minería en América Latina* (Buenos Aires, Ciccus/Clacso).
Fairhead, J. and Leach, M. (2003) *Science, Society and Power: Environmental Knowledge and Policy in West Africa and the Caribbean* (Cambridge: Cambridge University Press).
Foyer, J., Jankowski, F., Blanc, J., Georges, I. and Kleiche-Dray, M. (2014) "Saberes Científicos y Saberes Tradicionales en la Gobernanza Ambiental: La Agroecología como Práctica Híbrida", ENGOV Working Paper Series, n°14.
Flores Quintero, G. (2005) "Tequio, Identidad y Comunicación entre Migrantes Oaxaqueños", *Amérique Latine Histoire et Mémoire. Les Cahiers ALHIM* (8) http://alhim.revues.org/423, date accessed 6 December 2014.
Gaillard, J., Krishna, V.V. and Waast, R. (eds) (1997) *Scientific Communities in the Developing World* (New Delhi: Sage Publications).
Godoy, R., Reyes-García, V., Byron, E. et al. (2005) "The Effect of Market Economies on the Well-Being of Indigenous Peoples and on Their Use of Renewable Natural Resources", *Annual Review of Anthropology* 34(1): 121–138.
Goldman, M.J., Nadasdy, P. and Turner, M.D. (2011) *Knowing Nature: Conversations at the Intersection of Political Ecology and Science Studies* (Chicago: University of Chicago Press).
Gomez-Oliver, L. (2008) "Crisis Alimentaria Mundial y México", *Agricultura Sociedad y Desarrollo* 5(2): 115–141.
Gravel, N. (2009) "La Gouvernance Rurale au Mexique en Reponse a la Vulnerabilité Paysanne Extreme", *Canadian Journal of Latin American and Caribbean Studies* 34(68): 111–145.
Greene, S. (2004) "Indigenous People Incorporated? Culture as Politics, Culture as Property in Pharmaceutical Bioprospecting", *Current Anthropology* 45(2): 211–237.
Harding, S. (1997) "Is Modern Science an Ethnoscience?", Shinn, T., Spaapen, J. and Krishna, V.V. (eds) *Yearbook of the Sociology of Sciences* 19 (Dordrecht: Kluwer).
Hassink, R. (2005) "How to Unlock Regional Economies from Path Dependency? From Learning Region to Learning Cluster", *European Planning Studies* 13(4): 521–535.

Holt-Giménez, E. (2006) *Campesino a Campesino: Voices from Latin America's Farmer to Farmer Movement for Sustainable Agriculture in Latin America* (Oakland: Food First).

Holt-Giménez, E. (ed.) (2010) "Linking Farmers' Movements for Advocacy and Practice", *Journal of Peasant Studies* 37(1): 203–236.

Howes, M. (1979) "The Uses of Indigenous Technical Knowledge in Development", *The IDS Bulletin* 10(2): 12–23.

Howes, M. and Chambers, R. (1979) "Indigenous Technical Knowledge: Analysis, Implications and Issues", *The IDS Bulletin* 10(2): 5–11.

INEGI (2010) Censos y Conteos de Población y Vivienda. Instituto Nacional de Estadística y Geografía. http://www.censo2010.org.mx/

Jankowski, F. (2012) *"Agro-Écologie, Gouvernance Environnementale et Dialogue des Savoirs dans l'État de Oaxaca (Mexique)"*, in M. Kleiche-Dray (ed.) ENGOV WP5 Report.

Jiménez Sánchez, L. (1984) "Entrevista a Efraim Hernández Xolocotzi", *Las Ciencias Agrícolas y Sus Protagonistas* 1.

Katz, E. (1992) "La Cueillette des Adventices Comestibles au Mexique", *Ecologie Humaine* 10(1): 25–41.

Katz, E. (1994) "Du Mûrier au Caféier: Histoire des Plantes Introduites en Pays Mixtèque (XVIe-XXe Siècle)", *Journal d'Agriculture Traditionnelle et de Botanique Appliquée (JATBA) N° Spécial Phytogéographie Tropicale* 36(1): 209–244.

Katz, E. (2002) Rites, Représentations et Météorologie dans la Terre de la Pluie (Mixteca, Mexique), in E. Katz, M. Goloubinoff and A. Lammel (ed.), *Entre Ciel et Terre: Climat et Sociétés* (Paris, Ibis Press/IRD Editions), 63–88.

Katz, E. and Kleiche-Dray, M. (2013) "Dynamic Processes in the Use of Natural Resources and Food Systems by Indigenous and Mestizo Communities in Mexico and Brazil", ENGOV Working Paper Series, n°3.

Kirsch, S. (2001) "Lost Worlds: Environmental Disaster, 'Culture Loss' and the Law", *Current Anthropology* 42(2): 167–197.

Lander, E. (2000) "Ciencias Sociales: Saberes Coloniales y Eurocentrismo", in E. Lander (ed.), *La Colonialidad del Saber: Eurocentrismo y Ciencias Sociales. Perspectivas Latinoamericanas* (Buenos Aires: CLACSO), chapter 13.

Lascoumes, P. (1994) *L'Éco-Pouvoir. Environnement et Politiques* (Paris: La Découverte).

Lazos, E. (2012) "Conocimiento, Poder y Alimentación en la Mixteca Oaxaqueña: Tareas para la Gobernanza Ambiental", in M. Kleiche-Dray (ed.), Engov WP5 Report.

Leff, E. (1986) *Ecología y Capital: Hacia una Perspectiva Ambiental del Desarrollo* (México: UNAM).

Léonard, E. and Foyer, J. (2011) *De la Integración Nacional al Desarrollo Sustentable: Trayectoria Nacional y Producción Local de la Política Rural en México* (Mexico: CEDRSSA).

Long, J. and Attolini, A. (2009) *Caminos y Mercados de México* (México, IIH-UNAM).

Mendoza García, E. (2002) "El Ganado Comunal en la Mixteca Alta. De la Epoca Colonial al Siglo XX. El Caso de Tepelmeme", *Historia Mexicana* 51(4): 749–785.

Mendoza García, E. (2004) *Los Bienes de la Comunidad y la Defensa de las Tierras en la Mixteca Oaxaqueña. Cohesión y Autonomía del Municipio de Santo Domingo Tepenene, 1856–1912* (México, D.F.: Senado de la República).

Mignolo, W. (2000) *Local Histories/Global Design: Coloniality, Subaltern Knowledge and Border Thinking* (Princeton: Princeton University Press).
Mignolo, W. (2007) *La Idea de América Latina: La Herida Colonial y La Opción Decolonial* (Barcelona: Gedisa Editorial).
OCDE (2007) *Política Agropecuaria y Pesquera en México: Logros Recientes y Continuación de las Reformas* (Paris: OCDE).
OCDE (2013) *Examens Environnementaux de l'OCDE* (Mexico: OCDE).
O'Keefe, L. and Howes, M. (1979) "A Select Annotated Bibliography: Indigenous Technical Knowledge in Development", *The IDS Bulletin* 10(2): 51–58.
Polanco, X. (ed.) (1989) *Naissance et Développement de la Science-Monde. Production et Reproduction des Communautés Scientifiques en Europe et en Amérique Latine* (Paris: La Découverte).
Quijano, A. (1994) "Colonialité du Pouvoir et Démocratie en Amérique Latine", in J. Cohen, L. Gómez and H. Hirata (eds), *Amérique Latine, Démocratie et Exclusion* (Paris: Harmattan), 93–101.
Raj, K. (2007) *Relocating Modern Science: Circulation and the Construction of Knowledge in South Asia and Europe, 1650–1900* (Houndmills/New York: Palgrave Macmillan).
Rivas Guevara, M. (2008) *Caracterización del Manejo de Suelo y Uso del Agua de Lluvia en la Mixteca Alta: Jollas y Maíces de Cajete Estudio de Caso: San Miguel Tulancingo, Oaxaca*. PhD Dissertation (Montecillo, Mexico: Colegio de Postgraduados).
Rivas Guevara, M., Rodriguez Haros, B. and Palerm Viqueira, J. (2009) "El Sistema de Jollas: Una Técnica de Riego no Convencional en la Mixteca", *Boletín del Archivo Histórico del Agua*. Número Especial Año 13: 6–16.
Rogé, P. et al. (2014) "Farmer Strategies for Dealing with Climatic Variability: A Case Study from the Mixteca Alta Region of Oaxaca, Mexico", *Agroecology and Sustainable Food Systems* 38(7): 786–811.
Romero Frizzi, M. (1990) *Economía y Vida de los Españoles en la Mixteca Alta:1519–1720* (Oaxaca: Instituto Nacional de Antropología e Historia/Gobierno del Estado de Oaxaca).
Romero Penaloza, J. et al. (1986) *Diagnóstico de la Producción Agrícola de las Mixtecas Oaxaqueñas Altas y Baja*, Tomo II y III, Centro Regional del Sur (México: UACH).
Sanchez Lopez, J. (2013) *Conflictos y Lucha Campesina en Oaxaca, 1970–80*. Master Thesis (Universidad de Sonora).
Sillitoe, P. (1998) "The Development of Indigenous Knowledge: A New Applied Anthropology", *Current Anthropology* 39(2): 232–252.
Leigh Star, S. and Griesemer, J.R. (1989) "Institutional Ecology, Translations and Boundary Objects: Amateurs and Professionals in Berkeley's Museum of Vertebrate Zoology, 1907–1939", *Social Studies of Science* 19(3): 387–420.
Toledo, V.M. (1992) "Utopía y Naturaleza. El Nuevo Movimiento Ecológico de los Campesinos e Indígenas de America Latina", *Nueva Sociedad* (Caracas) 122: 73–85.
Toledo, V.M. and Bartra, A. (2000) *Del Círculo Vicioso al Círculo Virtuoso: Cinco Miradas al Desarrollo Sustentable de las Regiones Marginadas* (México: Semarnap/Plaza y Valdés).
Trompette, P. and Vinck, D. (2009) "Retour sur la Notion d'Objet-Frontière", *Revue d'Anthropologie des Connaissances* 4(1): 11–15.

Turner, N.J., Davidson-Hunt, I.J. and O'Flaherty, M. (2003) "Living on the Edge: Ecological and Cultural Edges as Sources of Diversity for Social-Ecological Resilience", *Human Ecology* 31(3): 439–461.

Velasco Rodríguez, G.J. (1994) *La Artesanía de la Palma en la Mixteca Oaxaqueña* (Oaxaca: CIIDIR-IPN, Unidad Oaxaca).

Velásquez, J.C. (2002) "Sustainable Improvement of Agricultural Production Systems in the Mixteca Region of Mexico", NRG Paper 02–01 (Mexico, D.F.: CIMMYT).

Velásquez Hernández, J.C. and Santos, J.L. (2006) "CEDICAM: Una Organización de Campesinos para Campesinos en México", *LEISA* sept: 24–26.

Waast, R. (1996) *Les Sciences hors d'Occident au XXe siècle*, 7 tomes (Paris: ORSTOM), Editions: On-line sur la Base Horizon Pleins Textes, http://horizon.documentation.ird.fr

Waast, R. and Rossi, P.L. (2014) "Origins and Shifts in Meaning of ENGOV's Keywords. A Bibliometric Study", in M. Kleiche-Dray (ed.) Engov WP5 Report.

Warman, A. (2001) *El Campo Mexicano en el Siglo XX* (Mexico: Fondo de Cultura Económica).

Woods, C. (2002) "Life After Death', *The Professional Geographer* 54(1): 62–66.

Except where otherwise noted, this work is licensed under a Creative Commons Attribution 3.0 Unported License. To view a copy of this license, visit http://creativecommons.org/licenses/by/3.0/

# Part II
# New Politics of Natural Resources

OPEN

# 4
# The Government of Nature: Post-Neoliberal Environmental Governance in Bolivia and Ecuador

*Pablo Andrade A.*

## Introduction

In 2005 and 2006, anti-neoliberal coalitions won the elections in Bolivia and Ecuador, respectively. In both countries, this development put an end to the rules that had regulated the use of natural resources in hydrocarbon extraction during the latter part of the twentieth century (Hogenboom, 2014). The post-neoliberal governments constructed new institutions for the governance of extractive-industry activities. The new rules of the game have changed the way in which the Andean countries govern extractive industries. It has not put an end to their dependence on income generated from natural resources, but it has changed the way in which that income is distributed.

The process of change from neoliberalism to post-neoliberalism was fast, and fraught with confusion and abandoned experiments. This chapter describes that process. Two analytical objectives guide this description. First, I will identify the factors that guided the changes from neoliberalism to post-neoliberalism; and second, I will analyse the possibilities for the governance of mineral and hydrocarbon wealth and the creation of a "government of nature" that were opened up by the new regulatory framework.

## Natural resources, rentier states, development and post-neoliberalism

The contemporary debate about development based on natural resources has existed since the 1990s. Numerous academic studies conducted in that decade called attention to the relationship between income from natural resources and development, highlighting the negative impact of the former on the latter. In this century, however, the findings of those pioneering studies have been disputed by a growing body of literature primarily focused on political economy (Sachs and Warner, 1995; Karl, 2007; Whatchenkon, 1999; Auty and Gebb, 2001; Ross, 2001; Robinson, Tovik and Verdier, 2006; Acemoglu and Robinson, 2012).

The thesis of the "natural resources curse" questioned the policies advanced by international financial institutions and transnational companies. These stakeholders argued that the developing countries in the process of development could exploit their comparative advantages in the field of natural resources to accelerate their development (Bebbington et al., 2008). The neoliberal governments of the 1990s adopted this thesis. Critical studies developed in recent decades have examined the economic and social effects of those policies, stressing the effects of the rents from natural resources on the political and economic development of countries with an abundance of these resources.

The consequent debate failed to resolve the issue in the field of resource economics (Iimi, 2007; Collier, 2010), but not in the field of political institutions. In fact, political scientists and political economists who specialize in development have shown that an economy based on the extraction of natural resources actually has a negative impact on the development of political institutions that manage the appropriation and use of state income for these extractive activities (Bebbington et al., 2008; Collier, 2010). This adverse effect is mediated by a specifically political variable: the adoption of a rentier model of natural resource governance by the governmental decision-makers. The policy of the International Financial Institutions (IFIs) and transnational companies would instigate the governments of the developing countries to adopt some type of regulatory institution that would – in the medium and short term – guide the evolution towards a rentier state and very probably towards the creation of the conditions that produce an effect known as the "natural resource curse" (Bebbington et al., 2008).

Some Latin American scholars have criticized the idea of development based on natural resources in the thesis known as the "extractivist model": to the negative impacts of income from natural resources would have to be added two specifically Latin American effects. On the one hand, resource-based growth would have impeded the Latin American countries from earning great international autonomy. On the other hand, extractivist revenues would have induced the formation of a state that, in addition to being rentier, was also predatory by nature. This effect would be especially serious since that predation occurs in areas inhabited by indigenous peoples, thereby affecting particularly fragile ecosystems. Both effects thus imply a predatory and dependent capitalist social trajectory (Acosta, 2003; Acosta and Schuldt, 2009; Gudynas, 2012, 2009).

In recent years, various scholars have criticized the negative consensus on resource-based development. The criticisms have been focused on two major areas. First, the simple relationship between the abundance of natural resources and poor development does not hold. The evidence of countries rich in natural resources shows that – under certain conditions – they could achieve high income levels, relative equality, and a great degree of economic diversification, and that they are democracies. More importantly, these achievements have occurred among developed countries (Canada, the USA, the UK, Australia, Norway) as well as emerging countries (Brazil, Chile, South Africa, Indonesia) and developing countries (Botswana is typically the most cited example, but increasingly Bolivia and Ecuador are mentioned as well) (Dunning, 2008; Gylfason, 2012; Hujo, 2012; Thorp et al., 2012).

The second area of criticism has to do with the double directionality of the effects of rents from natural resources. A boom of natural resources can have a favourable effect on authoritarianism or on democracy; it can augment the interest of predatory elites who are in control of the state to preserve their control over the distribution of income (Acemoglu and Robinson, 2010); and it can simultaneously mitigate the redistribution of private income, thus increasing the appeal of democracy (Dunning, 2008). Similarly, it is possible that a natural resource boom would elevate the costs of economic diversification, but an active state could pay those costs from the tax revenue that it obtains from natural resource income (Bebbington, 2012; Thorp, 2012). By investing those fiscal resources in institutions that promote coordination between emerging economic sectors and the accumulation of human capital, the state would favour economic diversification (Dietsche, 2012; Ascher, 2012; Guajardo, 2012; Orihuela and Thorp, 2012).

This controversy can be resolved by distinguishing the rentier states from other types of state (Dunning, 2008). The key variable is not the abundance of resources but rather the abundance of rents that produces effects on the states. The exploitation of mineral resources, oil and gas generates revenues for the states and, given certain conditions, can transform them into rentier states. Why does this happen?

Rentier states support themselves on a set of regulations that govern the extractive industries. These rules determine the conditions of access to natural resources: how and how much of the profits obtained by extractive industries will be appropriated by the states; and who intervenes in the key decisions to authorize extractive activities and in the decisions corresponding to the distribution of income. This set of rules constitutes the core of natural resource governance.

Recent discussions have stressed the point that the distribution of income is the primary source of conflict and debate in rentier states. In particular, the literature asserts that such income may be used by governments in two ways. It can lead to a concentration of economic and political power in the hands of the elite. On the other hand, governments can also choose to use the revenues to reduce dependence on natural resources, diversify the economy, and provide benefit to the majority of its citizens. Bebbington (2012) has indicated that, in the study of development in the Andes, special consideration should be given to conflicts surrounding the extractive industries since they "have great significance for national and subnational political economic change". On the other hand, Gylfasson (2010) has argued that the investment of mineral incomes in social development is an integral strategy of economic growth. In particular, he states that "the level and composition of government expenditure should make a difference for growth".

Taking advantage of studies advanced by ecological economics and political ecology, social movements, environmental organizations, and intellectuals from Latin America as well as from outside the region have looked at the extraction of natural resources as something more than just development. The common element in these diverse perspectives is that they value the sustainability of ecosystems and society in a way that is entirely different from the utilitarianism inherent in mainstream economic thought (Nelson, 1995).

A second common element is the double criticism of neoliberal capitalism and the idea of development itself (e.g. Acosta, 2003; Gudynas, 2009; Alimonda, 2011; Escobar, 2011). The main thesis of this criticism is that the expansion of capitalism constantly requires new sources of natural resources, whose exploitation exclusively benefits industrialized countries, and in the short and medium term it generates an "illusion of development" in Latin American countries. This illusion is characterized by cycles of rapid economic growth, with partial and fragmented modernization of societies. These cycles are illusory to the extent that they have historically proved to be unsustainable over time. The cyclical behaviour produces great costs for societies, particularly the destruction of highly diverse ecosystems and the destruction of human populations whose way of life has been radically altered by the presence of extractive activities. These costs tend to crystallize in the political organization of Latin American societies, which aims to preserve and enhance social inequality and to keep the rural poor and indigenous populations out of political decisions.

The Latin American literature is very closely related to the arguments advanced by European and Anglo-Saxon ecological economists and ecological sociologists. The first have shown that the economic growth experienced by Latin American countries during natural resource booms has only been achieved on the basis of an unequal exchange of material flow (Vallejo, 2009; Martínez-Alier et al., 2010; Muradian et al., 2012). Similarly, Muradian et al. (2012) have noted that recent technological innovations in the extractive industries have made the exploitation of mineral and hydrocarbon deposits – located in remote areas inhabited by indigenous peoples (the Ecuadorian and Bolivian Amazon, for example) – economically profitable. The expansion of the "extractive frontier" implies the accelerated destruction of ecosystems that are essential for planetary survival, along with an increase in socioenvironmental conflicts that put the cohesion of Latin American, and especially Andean, societies at risk.

Environmentalist literature has made visible two innate elements of the rentier basis of the Bolivian and Ecuadorian states. First, the construction of rentier states represents a set of enormous environmental and social costs that are not only ignored by the literature of political economy and development economics but are also actively kept out of public discussion by academics, international financial institutions and the governments that have controlled these states. Second, the set of rules that govern the extractive industries in the rentier states is insufficient to achieve the objective of an environmental governance that ensures the sustainability of societies.

The set of debates that I have outlined allows me to present the central argument of this chapter in order to display and analyse in the next section the evidence offered by Bolivia and Ecuador on what I have called "post-neoliberal environmental governance". Analytically, post-neoliberal environmental governance in Bolivia and Ecuador – and possibly in other Latin American rentier states – can be understood as a system of three layers. In the centre would be the rules of natural resources governance. These are the rules that govern the extraction of resources and the production of revenues for the states. At this level the number of actors is minimal since it only includes governmental elites, certain state agencies and the companies (public and/or private) that conduct mining activities.

A second layer would consist of the rules that govern the distribution of income, particularly that which is intended to be some type of compensation for populations especially affected by extractive activities. It also includes rules that establish monitoring capabilities for the environmental damage caused by extractive activities and the organizational responsibility for such damage. This layer includes high-level policy-makers and specialized state agencies – just as in the previous level – but also other stakeholders such as organized citizen groups and professional experts who act as consultants for the assessment, monitoring and determination of environmental damage (van Dijck, 2014).

Finally, the third layer would contain the general way in which the relationships between the state, society and nature (or environment) are regulated. Besides being the least formalized of all the layers, it is also that which supports the greatest number of actors, and is especially open to the participation of citizens who, for whatever reason, have some interest in the decisions to be adopted about nature and the use of resources in their society. Therefore this is the level where organizations of environmental activists, specialized citizen groups (e.g. academic communities) and other groups are active.

## Bolivia and Ecuador: From the reconfiguration of rent-seeking to environmental governance

In order to function, the Bolivian and Ecuadorian states depend on the flow of rents to their treasuries. Both states capture this income directly from the activity of extractive industries of minerals and hydrocarbons, and these rents substitute other sources that are more politically expensive to obtain (e.g. taxes). Thanks to these rents, the states can carry out distributive policies that are less expensive than their alternatives (e.g. urban or rural property reforms). These characteristics interact to produce an overall effect of acceptance of the government in power and more generally of the state.

Beginning in the years from 2000 to 2002, approximately, Bolivia and Ecuador have regained significant economic growth rates; and this growth has been accompanied by significant reductions in poverty and inequality.[1] These trends are due to three main factors. First is the increase in world-market prices of the oil, gas and minerals exported by both countries.[2] Second, the Andean states have recovered their ability to capture the rents produced by the exploitation of natural resources. Third, the governments have invested in improving the state capacity to manage the rents, orienting them towards the broad distribution of the benefits of economic growth, and – to a lesser extent – trying to induce a change in the relationships between the rentier sector and the production of their economies. These trends are interdependent and mutually reinforcing.[3] The Bolivian and Ecuadorian states have improved their distributive capacities and therefore have contributed to improving the quality of life of their populations – especially the poorest – because they have the fiscal resources captured from extractive industry activities (Paredes, 2012). At the same time, the increased capacity of the Bolivian and Ecuadorian states to capture rents from natural resources has improved their tax bases.

The current situation in Bolivia and Ecuador contrasts sharply with that which dominated in the last decades of the twentieth century.[4] During that time, both states significantly reduced their capacities to provide social services to the poor populations, such as health, education and money transfers. Low international prices of natural resources and the inability of the governments to increase state revenues prevented states from implementing distributive policies. Therefore, in the 1980s and 1990s, Bolivia and Ecuador experienced a continued deterioration of the living conditions of the population, increased poverty – particularly in rural areas – and growing inequality (Lefeber, 2003).

The current natural resources boom is not, however, the cause of the formation of Bolivia and Ecuador as rentier states but rather only of its reactivation and reconfiguration. The Revolution of 1952, in Bolivia, and the oil boom of the 1970s – for both countries – were key events that shaped the current rentier states, as will be discussed below.

## Bolivia

During the boom period of tin (1910–1954) and before the nationalization of the mines in 1952, "the State's attempts to capture more rent...implied a substantial redistributive dynamic...any capture of rent by the State for purposes of greater public spending would tend to redistribute income from the tin oligarchy to...the rest of the population" (Dunning, 2008: 235). In simplified terms, the pressure of the social groups excluded from mining revenues – particularly tin workers and reformist intellectual groups – generated attempts by the governments to capture mining revenues, which were answered by the mining oligarchy with coups d'état and repression. The state wanted to be rentier, but the property ownership and the economic and political power of the mining elite would not allow it. The Bolivian administrations during those years had a single resource to expand its fiscal base: to increase taxes on the non-mining sector of the economy, which increased the discontent of the non-mining classes. Finally, this dynamic exploded with the Revolution of 1952.

The capture of the state by the Revolutionary Nationalist Movement (Movimiento Nacionalista Revolucionario (MNR)) and the Bolivian Workers' Union (Central Obrera Boliviana (COB)) in 1952 led to the nationalization of the mines in October of that same year and the formation of the state company Mining Corporation of Bolivia (Corporación Minera de Bolivia (COMIBOL)) (Paredes, 2012). Thanks to this direct control over mineral income, the mines became the main source of state income and the fuel for public spending in the rest of the economy. Between 1952 and 1964, when a military coup d'état put an end to the revolution, the Bolivian state used mining income to moderate the distributive conflict, to invest in the development of other sectors of the economy – particularly the manufacturing sector and the growth of the agricultural sector of eastern Bolivia – and to create a national citizenship (Klein, 2008; Soruco, 2010; Crabtree and Crabtree-Condor, 2012). However, domestic and international economic factors – primarily the prolonged and severe decline in the price of tin – conspired against this first attempt at the configuration of the Bolivian rentier state.

The decisive factor for the configuration of the current rentier state came with the oil boom of the 1970s. The administration of Hugo Bánzer approved a Hydrocarbon Law in 1972 that allowed for the opening of oil concessions, thus establishing new ways of capturing income. Oil exploitation throughout the 1970s expanded exponentially: in 1974, oil revenues allowed the state to balance its accounts, and in 1978, oil and natural gas exports represented 30% of Bolivian exports (Miranda, 2008). As Dunning notes, "by the end of the 1970s Bolivia had clearly witnessed an oil boom that...exerted a substantial impact on the coffers of the fisc" (Dunning, 2008: 244).

Although oil production and oil prices on the world market declined in the 1980s, oil revenues increased their share in the state treasury. In effect, the administration of Jaime Paz Zamora obligated the YPFB (Yacimientos Petroleros Fiscales de Bolivia) by law to transfer an increasing portion of its income to the central government, amounting to 60% of state revenues. In the 1990s the dependency of oil revenues tended to decline. This development initiated the neoliberal phase of the Bolivian state.

Confronted with serious macroeconomic imbalances, the government of Víctor Paz Estenssoro commissioned the minister of planning at the time – and future president – Gonzalo Sánchez de Losada to implement a reform of the oil sector. Inspired by neoliberal ideology, Sánchez de Losada pushed back the participation of the Bolivian state in oil revenues from 50% to 18% (Dunning, 2008). The idea behind these cuts was to attract foreign investment for the exploration of new oil fields and to develop the exploitation of newly discovered deposits of natural gas. Tax revenues from oil income dropped dramatically, reaching a low of only 7% of total tax revenues (Dunning, 2008). On the other hand, although foreign investment actually flowed into gas exploitation – especially from 1997 onwards – Sánchez de Losada's reforms prevented this development from contributing significantly to government revenues. Instead, Latin American companies (Petrobras, Pluspetrol) and transnational non-Latin American companies (Repsol, British Gas, Amoco-British Petroleum, Total ELF) benefited mainly from the exploitation of gas.

The growing opposition of popular sectors and of leftwing politicians to the effects of capitalization and the increased expectations of gas as the motor of a renewed national development finally exploded in 2003 in opposition to the government project of constructing a pipeline from the East to Chile. The Gas War put an end to the second administration of Sánchez de Losada. This led to an end of the political struggle for the capture of natural resource revenues by the Bolivian state, which caused a rapid turnover of governments between 2003 and 2005.

The neoliberal experiment of disarming the Bolivian rentier state came to an end with the election of Evo Morales as president. The Morales government nationalized the Bolivian oil and gas industry again in 2006, increasing the state's share in the income of the sector to 82%, although the effective participation of the state was stabilized at 50% of revenues after 2007 (Miranda, 2008). Finally, in 2009, the state secured its control over non-renewable natural resources in a way that was favourable to the central government, and to the detriment of the grievances of the Media Luna departments (Santa Cruz, Tarija, El Beni) and of the claims of the organized indigenous peoples in the Indigenous Native Peasant Territories, where the hydrocarbon deposits were located (Humphreys-Bebbington, 2012). The importance of these developments has been widely recognized and disseminated by the Bolivian Government, which in 2013 stated that the nationalization of hydrocarbons had "generated more than $5 million USD for redistribution", and that YPFB had become "the country's largest business corporation" (President of the Republic, 2013).

## Ecuador

More so than Bolivia, Ecuador benefited from the boom in oil prices in the 1970s. Along with the beginning of oil exploitation in the Ecuadorian Amazon, the military conducted a coup d'état and embraced a programme of oil nationalization and development guided by the state. The military government of General Rodríguez Lara (1972–1976) explicitly followed a policy of "planting oil". This consisted of the investment of fiscal oil revenue into infrastructure as well as industry loans and other policies that sought to diversify the country's industrial foundation and to improve its productivity – and that of the agricultural sector. While there is still debate about the achievements of the Rodríguez Lara government (North, 1985; Conaghan, 1988), there is a consensus that this administration actually succeeded in institutionalizing a path of development that linked the country's economic growth, maintenance and expansion of infrastructure and government capabilities with the provision of comprehensive tax revenue from oil exports.

The development towards a rentier state was completed in two phases. In the first phase (1972–1976) a progressive fraction of the military controlled the state and maintained nationalist and inclusive development policies, although without much support from weak popular sectors. The second phase (1976–1979) actually halted some of those policies and instead used oil revenues as collateral to obtain international loans that were used to pay a bloated state sector, and as a source of cheap loans channelled into a dominant rentier class (Acosta, 2003; Larrea, 2009; Oleas, 2013). In both instances, tax collection – except those obtained in customs – practically stopped to the point that, according to Acosta (2003), "the dictator himself, Guillermo Rodríguez Lara, boasted decades later that in his government taxes were not collected. Any fiscal emergency, when oil revenues were insufficient or declining due to economic reasons, was covered by foreign loans."

In 1979 the military gave back the state government to elected civilian governments. The first civilian government (1979–1984) partially resumed the project of the progressive military government, using oil revenues to postpone adjustments to the economy and to increase social investment (Oleas, 2013). However, the impact of the international debt crisis in 1982 and the deterioration of international oil prices tested the ability of these civilian governments to handle the problems that they had inherited from the rentier state: a mostly inefficient, oligopolistic and slow-growing industry, rising urban and rural poverty, and so forth.

The institutions that made the capture of oil revenues possible in the 1970s remained practically unchanged in the 1980s. Only at the end of the decade, as a result of a sharp drop in oil prices, did the Ecuadorian Government make efforts to reduce direct state control over some elements of the oil industry and to attract foreign investment.

During the government of Sixto Durán Ballén (1992–1996), a politician of clearly neoliberal orientation, the state ceded a large part of its regulatory capacity and economic participation to private companies, and simultaneously reduced its oversight of mining activities. In an attempt to attract private transnational companies, state participation – in the form of royalties – decreased in favour of the creation of income taxes. In this period there was a systematic increase in socioenvironmental conflicts with indigenous peoples residing in the Amazon.

Oil revenues improved from 2002 onwards with the opening of new oil fields and the construction of a pipeline complementary to that which was constructed in the 1970s. Acosta described the situation in 2003: "Ecuador will be what it has always been, a primary producer country. And oil looms as the source of income that will alleviate pressures... The wager is how to produce and transport the greatest quantity of crude oil." This was a situation that, according to the author, was not beneficial to the state because the developments of the 1980s and 1990s had reduced the production capacity of the state oil company. The capture of oil rents by the state had decreased significantly (from 80% in the late 1970s to 18% at the beginning of the 2000s).

This bleak picture changed dramatically with the election of the current president, Rafael Correa, in 2006 (re-elected in 2009 and 2013). Armed with overwhelming electoral support, the new administration resuscitated the 1970s scheme of controlling oil revenues: he cancelled existing contracts, returned most of the concessions to the state, obliged companies to cede most of their income to the state, and strengthened the state oil company. All of these changes occurred just in time for the boom in international oil prices of recent years (Ray and Kozameh, 2012).

The reconfiguration of what I have named "the core of post-neoliberal environmental governance" in Bolivia and Ecuador happened within the institutional patterns established in the 1970s evolution towards rentier states. The current boom revives the countries' historical heritage, as shown in Table 4.1.

Endowed with abundant fiscal resources, the Bolivian and Ecuadorian governments have managed to distribute income by investing in social policies that seek to improve the living conditions of citizens, and to undertake ambitious programmes of industrialization and technological innovation (SENPLADES, 2013; Agenda Bolivia 2025, 2013). This aspect corresponds with the component of income distribution and it can be explained by two factors. First, in both countries the struggle for control

*Table 4.1* Income capture in Bolivia and Ecuador

| Mechanism of income capture | Bolivia | Ecuador |
| --- | --- | --- |
| Royalties | 18% | 13.5% |
| Profit and export taxes | 69.5% | 60% |
| Total share of income | 87.5% | 73.5% |
| Non-taxed mechanisms | YPFB | PETROECUADOR |

*Source*: UNASUR (2013), prepared by the author.

of the rentier state was resolved in the second half of 2000 in favour of rival political elites from the traditional oligarchies who had controlled their respective states during the 1980s and 1990s. Second, the pressure for a better distribution of wealth that developed in those years came from organized popular sectors, including rural groups affected by the exploitation of natural resources.

In short, political developments in previous years pushed for an income distribution different from that which predominated in the years of neoliberalism. However, since these developments incorporated new demands, they led to increased attention by the Bolivian and Ecuadorian governments to the themes relegated to the resource agenda that prevailed in the last quarter of the twentieth century, particularly the environmental costs of the extractive industries.

The current Bolivian and Ecuadorian governments originate from heterogeneous coalitions in middle-class and popular urban sectors, and – more in the case of Bolivia than Ecuador – rural sectors. Silva (2009) distinguishes two forms of inclusion of the popular sectors. On one hand, the ruling party in Bolivia – Movimiento Al Socialismo (MAS) – achieves the direct incorporation of popular sectors into the state government in the form of a classic party of the masses. Furthermore, the ruling party in Ecuador – Alianza País – is an electoral machine that had a strong mobilization, and participation of indigenous and peasant organizations, social movements with environmental roots, and NGOs from 2006 to 2009 (Becker, 2011; Andrade, 2012; Ortíz, 2013; Silva, 2013).

The difference between the origins and mechanisms of the incorporation of the governments is important. In Bolivia, the social support of the organized indigenous and peasants is key for the survival of the government. This factor has significantly influenced the discourse – strongly tinged by indigenous Bolivian ideology – and the way in which the project of *Vivir Bien/Buen Vivir* is configured. In Ecuador, the indigenous have maintained a tense relationship with the government of President

Correa as well as a progressive distancing from environmental organizations since 2010. This item is also reflected in the discourse of *Buen Vivir* (Dominguez and Caria, 2013).

One would expect, given these differences, that the policies of the two governments with respect to the economy–society–nature relationship would also be distinct. A government with high indigenous participation should have a policy that is more pro-environment than one with low participation; however, this is not the case. In fact, if a difference exists between Bolivia and Ecuador, it is in the degree of translation of environmental concerns into specialized state agencies. The strange thing is that, contrary to the prediction by indigenous theorists, the degree of incorporation of the environmental issue in Ecuador is higher than in Bolivia.

## Environmental compensations and claims

Political sociological studies of the state administration (or management) of the environment have shown that it is composed of the following elements: a network of actors who operate – within and outside the state – around problems defined as "environmental"; certain professionals who define the situation and develop solutions to problems; institutional rules of the political process of decision-making; and the cultural ideas that legitimate these decisions (Lahusen and Münch, 2001). I have suggested that in Bolivia and Ecuador the core of resource governance consists of a strict set of governmental actors, namely, specialized ministers and state companies. Institutional rules in this core are highly formalized in their respective constitutions (state ownership of oil, gas and minerals being the basic rule). The relevant professions are basically administration, geology and – to a lesser extent – a diverse set of "environmental consultants". Finally, the cultural ideas that legitimate decisions are fairly simply: oil, gas and minerals are resources to be exploited for the benefit of national development (SENPLADES, 2013; Framework Law of Mother Earth and Integral Development for Living Well O431 Official Gazette, 2012; Agenda Bolivia 2025, 2013).

Outside this nucleus, both Bolivia and Ecuador have ministries of the environment (the Ministry of Environment of Ecuador (MAE) and the Ministry of Environment and Water in Bolivia (MAyA)), departments and other state agencies that integrate a diverse network of professionals. Also, in both cases, final decisions are taken by the government. The principles that structure the cultural ideas of this sector are precaution, the need to restore environmental damage; the prevention of such damages; and the concern for ensuring sustainability. The diagnosis of environmental problems includes, in both cases – and even more clearly in Ecuador – checking for damages caused by oil activities, such as deforestation, soil and water pollution, and loss of biodiversity and cultural diversity.

The solution to the detected problems is also common. In Ecuador, environmental governance is defined as the realization of the "citizen's right to live in a healthy environment, free of pollution and sustainable, and the guarantee of the rights of nature through comprehensive planning to manage habitats, to manage resources efficiently, to holistically repair and return life systems to real harmony with nature" (SENPLADES, 2013: 222). The Bolivian Government affirms that it has an obligation to "create the conditions to ensure the sustainability of the State itself in all its territorial areas in order to attain the standards of Living Well... to incorporate integral development in harmony and balance with Mother Earth in order to Live Well in the policies, rules, strategies, plans, programmes and projects at the central level of the State and of the autonomous territorial entities... to formulate, implement, monitor and evaluate policies, standards, strategies, plans, programmes and projects for the compliance of the objects, targets and indicators of Living Well, through integral development..." (Gaceta Oficial, 2012: 12).

In both countries, and as the culmination of long historical evolutions of the twentieth century (Baud and Ospina, 2013), the respective ministries of the environment administer "systems of environmental management". Key components of these systems are national parks and ecological reserves. In Ecuador the National System of Protected Areas (Sistema Nacional de Áreas Protegidas (SNAP)) comprises the State Heritage of National Areas (Patrimonio de Áreas Naturales del Estado (PANE)) – managed by the central government – and three other "subsystems" to make room for the participation of subnational governments, organized local communities and the private sector: "the Autonomous Decentralised Governments, the Subsystem of Protected Community Areas and the Subsystem of Private Protected Areas". Together these areas of conservation and protection comprise nearly 8 million Ha of the country.

The Bolivian Government, meanwhile, has organized a complex institutional framework that grants powers to the Public Ministry, the Ombudsman of Mother Earth, the Agro-environmental Court, the Ministry of Environment, and the Plurinational Council for Living Well in Harmony and Balance with Mother Earth. It integrates the ministry of Developmental Planning (the Bolivian equivalent to the Secretaría Nacional de Planificación y Desarrollo/National Secretary for Planning and Development (SENPLADES)), the Autonomous Departmental Governments and so forth. This organization multiplies the actors and entry points in environmental issues. As in Ecuador, the basic component of this system is the National System of Protected Areas (Sistema Nacional de Áreas Protegidas (SERNAP)). The Plurinational Council is directly hinged to the presidency of the Republic.

Another important environmental agenda of the two countries is climate change. The respective ministries and other state agencies have created plans for adaptation to and mitigation of climate change. The development of this theme, and of the environmental agencies overall, has relied heavily on international cooperation. Prominent international actors, who are common to both countries, are the World Bank, UNEP and the official German cooperation.

Finally, the Bolivian and Ecuadorian governments agree that the rich biodiversity of the two countries provides opportunities for some kind of "green" development, and they have advanced policies in this direction. Since 2001, Ecuador has been developing a National Program of Bio-knowledge, whose management depends on the ministries of environment and agriculture under the National Biosafety Framework (MAE, 2013; Andrade and Zenteno, 2014). In Bolivia the "Framework Law..." and the "Bolivia Agenda, 2025" contemplate a similar development, but the government has not made progress in the implementation of these policies.

As indicated above, the Ecuadorian environmental policy differs from that of Bolivia in the importance that it gives to the environmental damage caused by oil exploitation. Since 2008 the Ecuador's government has promoted an active policy of environmental remediation, executed by the Reparation Program of Environmental and Social Liabilities (Programa de Reparación de Pasivos Ambientales y Sociales (PRAS)).

The notion of "shared responsibility" – between the state and local communities in the management of environmental problems that prevail in institutional environmental designs – opens up opportunities for the participation of local communities and municipal, provincial and (in Bolivia) departmental governments. The role of scientific knowledge in this layer of environmental governance is important. Agencies generate and require scientific knowledge for the installation of environmental indicator systems, environmental accounts, early detection of environmental damage and so on. This necessity has created state organizations populated by local experts – specialized in public

128  *The Government of Nature*

administration and in certain branches of knowledge, such as biology and geography – and scientists mostly of international origin or trained in first-world universities (Andrade and Zenteno, 2014, http://www.conocimiento.gob.ec).

Although the Bolivian Government shares this point of view to a great extent, it gives senior ranking to the generation of knowledge and technology to add value to "food processing, lithium, gas and hydrocarbons..." (Agenda Bolivia 2025, 2013). In fact, the sixth objective of the development of the Bolivian Government's agenda indicates that such technological advances will be accompanied by an increase in hydrocarbon and metallic and non-metallic minerals. The incorporation of technology refers not only to processes of industrialization but also to minimizing environmental damage.

In summary, this level of post-neoliberal environmental governance – summarized in Table 4.2 – incorporates not only a range of actors but also well-established international actors and issues of the global environmental agenda (deforestation, environmental remediation, environmental services, climate change, etc.). The latter should not be surprising given that the state agencies that organize the sector originated precisely from pressures and institutional global designs, or they at least count on international cooperation for their operation. Environmental administration is focused on environmental management, and its basic attention is devoted to widely accepted global issues – deforestation, the preservation and administration of water resources, the remediation of various forms of environmental pollution, and increasingly climate change – and its function is to produce public policies on these issues. Its fundamental political component is the administration of national and international resources for the reproduction of environmental management.

There remains to be examined the third layer. Unlike the previous two layers, which are directly hinged to the state, this last one is the domain of civil society. Even when it resorts to formal rules, it is mainly informal and is open to a number of state and non-state actors. This level is important because, on the one hand, it has provided some of the discursive resources that comprise the environmental rhetoric of the Bolivian and Ecuadorian governments and, on the other hand, civil actors use this rhetoric as a resource of political action.

A cursory examination of the rhetoric of "living well" and "good living", in Bolivia and Ecuador, respectively, indicates the constant appeal to three ideas: harmony with nature, the sacredness of nature (revealed in the frequent use of names such as Mother Earth and *Pachamama*), and

*Table 4.2* Environmental administration in Bolivia and Ecuador

|  | Bolivia | Ecuador |
|---|---|---|
| Formal rules | Framework Law<br>Agenda Patriótica 2025<br>Specific laws | Constitution<br>National Plan for Living Well<br>Specific laws |
| State actors | Ministry of the Environment and Water<br>Plurinational Council for Living Well | Ministry of the Environment<br>Various ministries and departments |
| Other actors | Subnational governments<br>International cooperation | Subnational governments<br>International cooperation |
| Scientific knowledge | Integrated into the identification of problems and solutions<br>Dependence on standard scientific knowledge | Integrated into the identification of problems and solutions<br>Dependence on standard scientific knowledge |
| Issues | Administration of national parks<br>Policies of conservation and environmental reparation<br>Climate change | Administration of national parks<br>Policies of conservation and environmental reparation<br>Climate change |

the rights of this entity. The Ecuadorian Constitution, both in its preamble and in its Chapter 4, recognizes the right of Ecuadorians to live in a healthy and balanced environment, in harmony with nature. A similar phrase appears in Chapter 1, Article 1 of the Bolivian "Framework Law..." in the form of "comprehensive and balanced development" and as a guarantee of the "continuity of the regenerative capacity of the systems and components of Mother Earth". The "living well" and "good living" discourses also agree on two other points. First, this state of harmony does not exist at the moment, but it will be obtained in the more or less distant future as a result of social efforts led by the state. Second, a key component of this company is the respect and use of "ancestral knowledge" ("originating" in the Bolivian rhetoric) (SENPLADES, 2009, 2013; Domínguez and Caria, 2013).

Regardless of the ideological value that these discourses may have to legitimate governmental actions, "living well" and "good living" have encouraged complaints, protests and demands of indigenous and environmentalist actors as much in Ecuador as in Bolivia. In effect, the anti-mining protests in Ecuador in 2012, the staging of anti-mining referenda in that country (see Chapter 11, this volume) and the failed Yasuní-ITT initiative articulated the idea that the achievement of "good living" depended on at least three conditions. These comprised the preservation of ecological balance, the need for governments to take into account the voice of those who are possibly affected (van Teijlingen, 2013), and, in the case of Yasuní, the obligation of the Ecuadorian state to preserve cultures whose ancestral knowledge preserves the rights of nature (Rival, 2012). In Bolivia the conflict over TIPNIS national park was also articulated and could be processed through the resource of the "living well" and rights of nature rhetorics (Ortiz, 2013).

Both the Yasuní-ITT initiative and the TIPNIS conflicts show some of the processes, mechanisms, actors, potentials and limits of the "living well" and "good living" rhetorics. In both cases, policies initiated by their respective governments tried to protect the rights of the indigenous peoples who lived in areas of the Amazon. Similarly, in both cases these policies implied that the state would abstain from exploiting oil resources in those territories. Finally, when both governments changed their policies, they incited intense conflicts between the executives and national indigenous and environmentalist groups that had international support.

In summary, the third layer provides discursive and legal resources for stakeholders to advance their environmental demands. These actors are, in principle, any group of citizens; and even those citizens are not limited to national boundaries as they may be international organizations. In special circumstances – such as the temporary control of the state by "green" coalitions – actors, issues and modes of operation that arise in this sphere can become national and international public policies (Sodërbaum, 2000), as happened in Ecuador between 2007 and 2010. In Bolivia this position was occupied by indigenous movement organizations (Hogenboom, 2014). However, when that careful step contradicts the preservation of the core of natural resource governance in a rentier state, these same actors and themes are again expelled to the periphery, as indeed happened with the Yasuní-ITT initiative and the Bolivian TIPNIS conflict. The expulsion depends on how the decisive power is organized in the Bolivian and Ecuadorian states. In both cases the standard decisive power falls on the president and state agencies that are nuclear to the rentier states; this group can veto policies that would infringe on their reproduction.

## Conclusion

The Bolivian and Ecuadorian experiences show that although new forms of regulating the exploitation and use (income) of natural resources can be created, they have prioritized the preservation of the states' access to income and, by implication, of the extractivist activities themselves. This burden differentiates environmental government at various levels, as long as their existence does not compromise the reproduction circuit of the rentier state (extraction cycle, income and distribution). Bolivia and Ecuador have abundant natural resources, both in the narrow sense of mineral resources – oil and gas – and in the extended sense of ecosystem diversity. Additionally, in both countries the long-term historical development has been towards the installation and consolidation of rentier states. The current commodity boom created room so that governments that might have followed a different path opted to recreate the rentier states of the 1970s. The policy option resulted in the differentiated post-neoliberal mode of environmental governance that is currently being consolidated in the two countries.

In both countries the original formation of the rentier states depended both on internal political struggles and the existence of high international prices for hydrocarbons – and in the case of tin in post-war Bolivia, the collapse of these international markets. The current reactivation of the rentier states reflects factors similar to those of the past: the boom in mineral exports enabled the Bolivian and Ecuadorian governments to reconfigure the rent-seeking mechanisms that ensure their access to the abundant returns produced by extraction and export to international markets. This development, in turn, increased the ability of states to provide basic services, and consequently legitimized the extractive activities supported (and to some extent controlled) by the states.

The explanation is not only economic. Politics has also played a role in creating post-neoliberal environmental governance. The Bolivian and Ecuadorian governments are the result of processes of dispute over the use of natural resources. The arrival of new players to the control of the state, and the means by which they attained that power, would seem to explain the construction of a sort of macroideology with strong environmental tones: "living well" and "good living". This element completes the set of environmental governance and gives it ideological coherence. The regulation of natural resources, including the use of income from exploitation, makes sense only to the extent that it serves a greater purpose: to achieve a new relationship between Bolivian and Ecuadorian societies and their natural surroundings.

The dynamics of post-neoliberal environmental governance are complex. On the one hand, the rentier status of the Bolivian and Ecuadorian states promotes the social and biological reproduction of societies and new attempts at industrialization. On the other hand, rentier states have an interest in promoting the expansion of resource frontiers, which compromise fragile ecosystems and the survival of rural societies, thereby increasing political conflicts. However, it is still incipient and relatively exclusive, and its mechanisms are insufficient to solve the operation/preservation dilemma. Finally, the open possibilities in the ideological or cultural layer provide symbolic and material resources for the expression of socioenvironmental conflicts, and some mechanisms for its processing. However, its implementation depends on the strength of the democratic regime.

It is reasonable to assume that the tensions, conflicts and dynamics that gave rise to the current mode of environmental governance will continue to influence future developments. At the moment, however, it is difficult to say if at some point in this development it will organize itself in a more pluralistic and open way than it is at present, or whether – as in periods of decline in international prices – it will be reconfigured in an increasingly exclusive and unstable direction.

## Notes

1. In Bolivia, poverty improved more rapidly than inequality, which actually seems to be increasing, while in Ecuador the two indicators have decreased simultaneously and at accelerated rates. A report from the Central Bank of that country indicates that the accelerated rate is due to two factors: "the improvement of the international environment" and the degree of destruction provoked by the crisis of 1998–2002. See Dirección General de Estudios, Banco Central del Ecuador, *La Economía Ecuatoriana luego de 10 años de Dolarización* (*The Ecuadorian Economy After 10 Years of Dolarisation*) (Quito: Banco Central del Ecuador).
2. The exportation of minerals is not important for Ecuador, but the high prices of mineral ores have stimulated the government to promote the development, albeit still incipient, of metal mining in Ecuador, for which Chinese investments have flowed into the country.
3. For an overview of the financing of social policy from mineral (or hydrocarbon) resources, see Hujo, K. (2012).
4. See CEPAL (2013).

## References

Acemoglu, D. and Robinson, J.A. (2012) *Why Nations Fail: The Origins of Power, Prosperity, and Poverty* (New York: Random House).

Acosta, A. (2003) "Ecuador: Entre la Ilusión y la Maldición del Petróleo", *Ecuador Debate* 58: 77–100.

Acosta, A. and Schuldt, J. (2009) "Petróleo, Rentismo y Subdesarrollo", in CAAP-CLAES, *Extractivismo, Política y Sociedad* (Quito: CAAP-CLAES).

Andrade, P. (2012) "El Reino (de lo) Imaginario: Los Intelectuales Políticos Ecuatorianos en la Construcción de la Constitución de 2008", *Ecuador Debate* 85: 35–47.

Andrade, P. and Zenteno, J. (2015) "Ecuador: Changing Biosafety Frames and New Political Forces in Correa's Government", in B. Bull and M.C. Aguilar-Stoen (eds), *Environmental Politics in Latin America: Elite Dynamics, the Left Tide and Sustainable Development* (London: Routledge), 92–112.

Alimonda, H. (2011) "La Colonialidad de la Naturaleza. Una Aproximación a la Ecología Política Latinoamericana", in H. Alimonda (ed.), *La Naturaleza Colonizada. Ecología Política y Minería en América Latina* (Buenos Aires: CLACSO).

Ascher, W. (2012) "Mineral Wealth, Development and Social Policy in Indonesia", in K. Hujo (ed.), *Mineral Rents and the Financing of Social Policy, Opportunities and Challenges* (New York: UNRISD), 223–256.

Auty, R. and Gebb, S. (2001) "Political Economy of Resource-Abundance States", in R. Auty (ed.), *Resource Abundance and Economic Development* (Oxford: Oxford University Press), 126–144.

Baud, M. and Ospina, P. (2013) *The Emergence of New Modes of Governance of Natural Resources Use and Distribution in Latin America and Ecuador*, http://www.engov.eu/documentos/working_paper/Working_Paper_ENGOV_4_BaudandOspina.pdf, date accessed 14 January 2015.

Bebbington, A. (2012) "Extractive Industries, Socio-Environmental Conflicts and Political Economic Transformations in Andean America", in A. Bebbington (ed.), *Social Conflict, Economic Development and Extractive Industry: Evidence from South America* (London: Routledge), 3–26.

Bebbington, A., Hinojosa, L., Humphreys Bebbington, D., Burneo, M.L. and Warnaars, X. (2008) "Contention and Ambiguity: Mining and the Possibilities of Development", *Development and Change* 39(6): 887–914.

Becker, M. (2011) "Correa, Indigenous Movements, and the Writing of a New Constitution in Ecuador", *Latin American Perspectives* 38–1(176): 47–62.

Crabtree, J. (2008) "Introduction: A Story of Unresolved Tensions", in J. Crabtree and L. Whitehead (eds), *Unresolved Tensions: Boliva Past and Present* (Pittsburgh, PA.: University of Pittsburgh Press), 1–7.

Crabtree, J. and Crabtree-Condor, I. (2012) "The Politics of Extractive Industries in the Central Andes", in A. Bebbington (ed.), *Social Conflict, Economic Development and Extractive Industry: Evidence from South America* (Londres: Routledge).

CEPAL (2013) *Recursos Naturales en UNASUR. Situación y Tendencias para una Agenda Regional* (Santiago de Chile: Naciones Unidas), 46–64.

Collier, P. (2010) *The Plundered Planet* (Oxford: Oxford University Press).

Conaghan, C. (1988) *Restructuring Domination: Industrialists and the State in Ecuador* (Pittsburgh: University of Pittsburgh Press).

Dietsche, E. (2012) "Institutional Change and State Capacity in Mineral-Rich Countries", in K. Hujo (ed.), *Mineral Rents and the Financing of Social Policy, Opportunities and Challenges* (New York: UNRISD), 122–152.

Dirección General de Estudios, Banco Central del Ecuador (2010) *La Economía Ecuatoriana luego de 10 años de Dolarización* (Quito: Banco Central del Ecuador).
Domínguez, R. and Caria, S. (2013) "La ideología del Buen Vivir: La Metamorfosis de una 'Alternativa al Desarrollo' en Desarrollo de Toda la Vida", *Pre-Textos para el Debate* 2: 53.
Dunning, T. (2008) *Crude Democracy. Natural Resource Wealth and Political Regimes* (New York: Cambridge University Press).
Escobar, A. (2011) "Ecología Política de la Globalidad y la Diferencia", in H. Alimonda (ed.), *La Naturaleza Colonizada. Ecología Política y Minería en América Latina* (Buenos Aires: CLACSO).
Guajardo, J.C. (2012) "Mineral Rents and Social Development in Chile", in K. Hujo (ed.), *Mineral Rents and the Financing of Social Policy, Opportunities and Challenges* (New York: UNRISD), 185–222.
Gudynas, E. (2009) "Diez Tesis Urgentes sobre el Nuevo Extractivismo. Contextos y Demandas bajo el Progresismo Sudamericano Actual", in CAAP/CLAES (ed.), *Extractivismo, Política y Sociedad* (Quito: CAAP/CLAES).
Gudynas, E. (2012) "Estado Compensador y Nuevos Extractivismos. Las Ambivalencias del Progresismo Sudamericano", *Nueva Sociedad* 237: 128–146.
Gylfason, T. (2012) "Development and Growth in Resource-Dependent Countries", in K. Hujo (ed.), *Mineral Rents and the Financing of Social Policy, Opportunities and Challenges* (New York: UNRISD), 26–61.
Hogenboom, B. (2014) "South American Minerals at the Crossroads of Global Markets, National Politics, and Local Needs", in F. de Castro, P. van Dijck and B. Hogenboom (eds), *The Extraction and Conservation of Natural Resources in South America. Recent Trends and Challenges.* Cuaderno Series 27 (Amsterdam: CEDLA), 1–22.
Hujo, K. (2012) "Introduction and Overview: Blessing or Curse? Financing Social Policies in Mineral-Rich Countries", in K. Hujo (ed.), *Mineral Rents and the Financing of Social Policy, Opportunities and Challenges* (New York: UNRISD), 3–25.
Humphreys Bebbington, D. (2012) "State-Indigenous Tensions over Hydrocarbon Expansion in the Bolivian Chaco", in A. Bebbington (ed.), *Social Conflict, Economic Development and Extractive Industry: Evidence from South America* (London: Routledge), 134–152.
Iimi, A. (2007) "Escaping from the Resource Curse: Evidence from Botswana and the Rest of the World", *IMF Staff Papers* 54(4): 663–699.
Karl, T.-L. (2007) "Ensuring Fairness: The Case for a Transparent Fiscal Contract", in M. Humphreys, J. Sachs and J. Stiglitz (eds), *Escaping the Resource Curse* (New York: Columbia University Press).
Klein, H.S. (2008) *Historia de Bolivia* (La Paz: G.U.M.).
Lahusen, C. (2001) "Political Regulation in Sociological Perspective: A Multidimensional Framework of Analysis", in R. Münch, C. Lahusen, M. Kurth, C. Borgards, C. Stark and C. Jaus (eds), *Democracy at Work. A Comparative Sociology of Environmental Regulation in the United Kingdom, France, Germany, and the United States* (Westport: Praeger), 19–46.
Larrea, C. (2009) *Hacia una Historia Ecológica del Ecuador. Propuestas para el debate* (Quito: Universidad Andina Simón Bolívar, Sede Ecuador – Corporación Editora Nacional).

Ley Marco de la Madre Tierra y Desarrollo Integral para Vivir Bien, Gaceta Oficial O431, 2012.

Lefeber, L. (2003) "Problems of Contemporary Development. Neoliberalism and Its Consequences", in L.L. North and J.D. Cameron (eds), *Rural Progress, Rural Decay. Neoliberal Adjustment Policies and Local Initiatives* (Bloomfield, CT: Kumarian Press), 25–45.

Leftwich, A. (2010 [1983]) *Redefining Politics, People, Resources and Power* (London: Routledge).

Martínez-Alier, J., Kallis, G., Veuthey, S., Walter, M. and Temper, L. (2010) "Social Metabolism, Ecological Distribution Conflicts and Valuation Languages", *Ecological Economics* 30: 1–6.

Miranda, C. (2008) "Gas and Its Importance to the Bolivian Economy", in J. Crabtree and L. Whitehead (eds), *Unresolved Tensions: Bolivia past and Present* (Pittsburgh, PA: University of Pittsburgh Press), 177–193.

Muradian, R., Walter, M. and Martinez-Alier, J. (2012) "Hegemonic Transitions and Global Shifts in Social Metabolism: Implications for Resource-Rich Countries. Introduction to the Special Section", *Global Environmental Change* 22: 559–567.

Nelson, R. (1995) "Sustainability, Efficiency, and God: Economic Values and the Sustainability Debate", *Annual Review of Ecology and Systematics* 26: 135–154.

North, L. (1985) "Implementación de la Política Económica y la Estructura del Poder Político en el Ecuador", in L. Lefeber (ed.), *Economía Política del Ecuador. Campo, Región, Nación* (Quito: Corporación Editora Nacional – FLACSO – York University).

Oleas, J. (2013) *Ecuador 1972–1999: Del Desarrollismo Petrolero al Ajuste Neoliberal* (Quito: Universidad Andina Simón Bolívar), Doctoral Thesis.

Orihuela, J.C. and Thorp, R. (2012) "The Political Economy of Managing Extractives in Bolivia, Ecuador and Peru", in A. Bebbington (ed.), *Social Conflict, Economic Development and Extractive Industry. Evidence from South America* (London: Routledge), 27–45.

Ortíz, C. (2013) *La Sociedad Civil Ecuatoriana en el Laberinto de la Revolución Ciudadana* (Quito: FLACSO Ecuador), Doctoral Thesis.

Ortíz, P. (2013) "Gobernanza Territorial y Conflictos entre Estado y Pueblos Indígenas. Una Perspectiva Comparada de Ecuador y Bolivia", *ENGOV Project*, WP2 Final Report.

Paredes, M. (2012) "Extractive Dependence in Bolivia and the Persistence of Poor State Capacity", in R. Thorp, S. Battistelli, Y. Guichaoua, J.C. Orihuela and M. Paredes (eds), *The Developmental Challenges of Mining and Oil. Lessons from Africa and Latin America* (Londres: Palgrave Macmillan).

Presidencia de la República (2013) *Evo 2006–2009, 100 Logros* (La Paz: Presidencia de la República Estado Plurinacional de Bolivia).

Ray, R. and Kozameh, S. (2012) *Ecuador's Economy since 2007* (Washington: Center for Economic and Policy Research).

Rival, L. (2012) "Planning Development Futures in the Ecuadorian Amazon: The Expanding Oil Frontier and the Yasuní-ITT Initiative", in A. Bebbington (ed.), *Social Conflict, Economic Development and Extractive Industry: Evidence from South America* (Londres: Routledge), 153–171.

Robinson, J., Tovik, R. and Verdier, T. (2006) "Political Foundations of the Resource Curse", *Journal of Development Economics* 79: 447–468.

Ross, M. (2001) "Does Oil Hinder Democracy?", *World Politics* 53(3): 325–361.
Sachs, J.D. and Warner, A.M. (1995) "Natural Resource Abundance and Economic Growth", *NBER Working Papers* 5: 398.
Silva, E. (2009) *Challenging Neoliberalism in Latin America* (New York: Cambridge University Press).
SENPLADES (2013) *Plan Nacional para el Buen Vivir 2013–2017* (Quito: SENPLADES).
Soruco, X. (2010) "De la Goma a la Soya: El Proyecto Histórico de la Elite Cruceña", in X. Soruco, W. Plata and G. Medeiros (eds), *Los Barones del Oriente. El Poder en Santa Cruz ayer y hoy* (Santa Cruz: Fundación Tierra).
Sodërbaum, P. (2000) *Ecological Economics. A Political Economics Approach to Environment and Development* (London: Earthscan).
Fundación Tierra (2012) *Marcha Indígena por el TIPNIS: La Lucha en Defensa de los Territorios* (La Paz: Fundación Tierra).
Thorp, R., Battistelli, S., Guichaoua, Y., Orihuela, J.C. and Paredes, M. (2012) *The Developmental Challenges of Mining and Oil. Lessons from Africa and Latin America* (Londres: Palgrave Macmillan).
Vallejo, M.C. (2009) "Estructura Biofísica de la Economía Ecuatoriana: Un Estudio de los Flujos Directos de Materiales", in F.M. Mayoral (ed.), *Deuda Externa y Economía Ecológica: Dos Visiones Críticas* (Quito: Ministerio de Cultura-Flacso Ecuador).
Van Dijck, P. (2014) "Linking Natural-Resource Exploitation with World Markets: Road Infrastructure and Its Impact on Land Use Conversion in Amazonia", in F. de Castro, P. van Dijck and B. Hogenboom (eds), *The Extraction and Conservation of Natural Resources in South America. Recent Trends and Challenges.* Cuaderno Series 27. (Amsterdam: CEDLA).
Van Teijlingen, K. (2013) *Negotiating Values and Development at the Mining Frontier: Private, Public, and Civil Interactions over El Mirador Mine in South-East Ecuador* (Amsterdam: CEDLA).

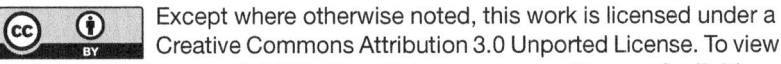

Except where otherwise noted, this work is licensed under a Creative Commons Attribution 3.0 Unported License. To view a copy of this license, visit http://creativecommons.org/licenses/by/3.0/

OPEN

# 5
# Changing Elites, Institutions and Environmental Governance

*Benedicte Bull and Mariel Aguilar-Støen*

## Introduction

The topic of elites has always been controversial in Latin American social sciences. Elites have been studied indirectly as landowners, capitalists, business-leaders or politicians, and have also been approached directly using concepts and theory from elite studies. Although there is a significant amount of literature on the role of elites in democratic transformations (see e.g. Higley and Gunther, 1992), elites have often been considered to be an obstacle to the formation of more democratic, prosperous and egalitarian societies (e.g. Paige, 1997; Cimoli and Rovira, 2008). This is also the case in the literature on environmental governance, in which elite groups are often considered to be an obstacle to sustainable development and an obstacle to establishing more equitable influence over the use and benefits of natural resources. Therefore, although an elitist conservation movement has long existed in Latin America, struggles to protect the environment from overexploitation and contamination have commonly been related to struggles against local, national and transnational elites by subaltern groups (Martínez-Alier, 2002; Carruthers, 2008; chapters 1 and 2 in this volume).

Over the last decade a number of changes, which might have an impact on the composition and attitudes of elites, have occurred in Latin America. Such changes could have consequences for environmental governance in the broad sense of the concept, as outlined in the Introduction of this book. Out of 49 presidential elections in the 2003–2013 period, 22 were won by centre-left candidates, and with the exception of Mexico and Colombia, centre-left governments were in power in all the large economies in Latin America for most of this period (Bull, 2014). Many of these governments represent groups that have previously been marginalized from politics and antagonized by the elites, including indigenous and environmental movements. With the changes in the global political economy, including the rise of China and a number of other emerging economies, Latin America has also seen the entry of a number of new economic actors, including new transnational companies and new lenders. Furthermore, in key sectors, new technologies have changed the production structure and therefore also the concentration of resources – and, in turn, possibly the composition of elites.

In spite of several such changes, the initial optimism regarding implications for environmental governance has subsided. In 2010, environmentalist Eduardo Gudynas (2010) rhetorically asked the new governments: If you are so progressive, why do you destroy the environment? In the aftermath, several other questions have been posed about why governments that publicly rejected genetically modified agriculture later promoted it; why they accelerated the issuing of mining concessions in spite of protests from their former constituencies; and why they expanded logging and oil exploitation in vulnerable areas in spite of pledging to protect them.

The aim of this chapter is to provide new insight into the elite dynamics that may provide answers to some of the questions outlined above. The chapter empirically interrogates elite shifts based on six case studies, outlining how new elites have emerged, how old elites have continued to influence politics and the economy, and how the relationship between new and old elites has affected environmental governance in the region.

For our analysis we use a "resource-based" definition of elites in environmental governance: "Groups of individuals that due to their economic resources, expertise/knowledge, social networks, or positions in political or other organizations stand in a privileged position to influence in a formal or informal way decisions and practices with key social and environmental implications" (Bull, 2015: 18). This is a multifaceted definition of elites that allows for the existence of both parallel and competing elites. Nevertheless, our analysis places particular emphasis on elites that control economic resources, including business elites and landowners. Therefore below we discuss the relationship between the concept of elite and that of class, and we discuss how elites and classes are considered to contribute to or hinder democracy and development, how they might change, and how they might be thought to impact environmental governance.

The rest of the chapter is structured as follows. The second part presents the puzzles motivating our study. In most cases these refer to environmental practices or environmental policies that were less sustainable than what was expected. Yet there are also cases in which surprising progress has been made. The third part summarizes the different problems of elites as discussed in the literature. The discussion includes the structural limitations to the transformative potentials related to a shift in the command of liberal political institutions, the predominance of "elite circulation", and what we call the "state/development" imperative based on a Weberian understanding of the need for state construction. The fourth part discusses different ways in which our case studies illustrate and confirm the problems discussed in the elite literature: how entrenched elites have hindered structural transformations towards an environmental governance that ensures more sustainable and equitable production; the conflicts over land use and how they have their roots in institutions that are kept weak due to historical control by elites; and how new governments accommodate their politics to the demands of the elites. However, some of the findings also challenge the rather pessimistic outlook of elite theory. In the fifth part we concentrate on the role of global economic transitions and technology, and elite shifts. The sixth part discusses the possibilities for change due to the emergence of new elites with different attitudes towards environmental governance. These include both new political elites and new knowledge elites. Finally, the conclusions are presented.

## The puzzles: Progress and setback in environmental governance under leftist governments

In 2009, when El Salvador got its first president supported by a leftwing and former guerrilla party (Frente Farabundo Martí para la Liberación Nacional (FMLN)), it was a major break with the past. Before Mauricio Funes' electoral victory, El Salvador had been ruled for 20 years by the same business elite-led party (Alianza Republicana Nacional (ARENA)). During this period the country had undergone a major economic transformation from an agroexport country to one dependent on remittances and the service sector (Segovia, 2002). This conversion brought temporary relief to the environmental impact of agroexport-oriented agricultural production, of which the most damaging products were cotton and sugar (Hecht et al., 2006). Nevertheless, when Funes took power, the country faced a triple crisis – economic, social and environmental – which enhanced the vulnerability of the population to natural catastrophes. Yet the country was also characterized by significant rural as well as urban political mobilization, and by the existence of various organizations that had developed sophisticated alternatives to the conventional agricultural development model. In spite of this, the Funes government struggled hard to set El Salvador on a different path, and, in the end, new ideas related to alternative agricultural development became marginalized in the national agenda, while there was no consensus on why no new development model was allowed to emerge while the old one continued to perform badly.

In Bolivia, Evo Morales – who came to power in 2006 with his Proceso de Cambio (process of change) – gave priority to family and small-scale agriculture over industrialized agriculture. This also implied a rejection of all GM organisms due to their environmental and health implications. This position echoed the viewpoints of a broad array of social movements upon which the governing party, MAS, was based. Since coming into power, MAS has been in deep conflicts with the country's old elite, with their stronghold in the "half-moon" states (Santa Cruz, Beni, Pando y Tarija) and controlling most of the country's economic sectors, including agriculture. Paradoxically, during the government of Evo Morales, the amount of genetically produced soy in Bolivia more than doubled (Zeballos, 2012), and the question of how the government could make this compatible with the official anti-GM discourse became more and more pressing.

Argentina has also been marked by deep conflicts between the governments of Nestor Kirchner and Cristina Fernández de Kirchner of the Peronist party (Partido Justicialista) and agricultural elites. Yet, at the same time, Argentina has become one of the major producers of genetically modified soy in the world, with major implications for the structure of the agricultural sector as well as for the environment (Trigo, 2011). Why was there so little public debate about it?

In Ecuador a major puzzle regarding biotechnology also arose. A major shift in the governance of the country occurred when Rafael Correa came to power, leading a broad coalition (Alianza PAIS) with strong participation from both indigenous and environmental movements. The platform for the coalition strongly rejected GM organisms and other uses of biotechnology in agriculture. However, once in government, Correa strongly promoted their use.

The mining sector is probably the most controversial in Latin America today, with its notable expansion, and its obvious environmental and social impacts, along with the large number of conflicts mining has generated across the region (see also chapters 2 and 11). In Guatemala, in the silver mining project El Escobal in the south-eastern part of the country, a transnational company (Tahoe Resources Inc.) and national elites face the protests of the indigenous Xinka people and their organizations. However, in spite of stated good intentions by the government as well as the companies, repressive practices against the protesters have continued and there has been virtually no room for dialogue. We attempt to explain why it has proved so difficult to mediate between the conflicting parties.

However, there are also positive cases. In the Brazilian states of Acre and Amazonas, there have been significant improvements in forest policy and forest protection over recent years. This stands in stark contrast with the rather disappointing record regarding forest protection of the federal government during the three consecutive governments led by the Worker's Party (Partido dos Trabalhadores (PT)), those of the Luiz Inácio Lula da Silva (2003–2007 and 2007–2011) and Dilma Rousseff (2011–2014) administrations. Moreover, advances in forest governance have occurred in states governed by different parties. What can explain Acre and Amazona's success?

Environmental policies and practices are not only influenced by domestic and local politics. There are also questions to be asked of international initiatives. One of the international initiatives with the most far-reaching consequences for forest governance in Latin America at the moment is the project known as Reducing Emissions from Deforestation and Forest Degradation (REDD). In spite of its rhetoric of inclusion, not all involved parties seem to have found their views expressed in the initiative. Instead, REDD is generating its own "elite" and its own discourse, and the question is how this can really address the pressing issues in Latin America's environmental governance.

Our approach to answering these questions has been to focus on elites. Thus in the following section we will discuss what elite theory might say about the questions above.

## The "elite problem" in theories of development, democracy and environmental governance

The recent increase in interest about elites and development in academic literature is closely connected to an increasing consensus about the importance of institutions for development, and the role of elites in shaping those institutions (Acemoglu and Robinson, 2012; Amsden, Di Caprio and Robinson, 2012). A focus on elites and institutions is in no way new in development theory. It has been a central element in Weberian-inspired development theory, from Gunnar Myrdal to the literature on the "developmental state"[1] (Myrdal, 1968; Woo-Cumings, 1999). Elites have sometimes been considered a hindrance to the emergence of such a state. As argued by Myrdal, "In fact, [elites] are best defined as people who are in a position to hinder reforms or to manipulate them, and, in the end, to obstruct their implementation" (Myrdal, 2010 [1979]: 335). However, others have considered elites to be capable of the efficient and productive channelling of resources, although they have frequently also acted as rent-seekers and have directed resources towards favoured and inefficient social groups (Amsden, Di Caprio and Robinson, 2012: 5).

Much less discussed is the relationship between elites, institutions and sustainable development, a dynamic that also necessitates the analysis of environmental governance. The literature referred to above is almost exclusively focused on economic growth and industrial upgrading. Furthermore, the term "institutions" is largely equated with "state institutions", and "development" is understood as economic growth at the national level. This literature has, to a very limited degree, problematized the environmental sustainability of development, and its distributional implications are only considered to the extent that they have consequences for long-term national economic growth. In other words, distribution of the benefits of growth and development across social groups and geographical areas is only considered a problem if it leads to decelerated growth – for example, if the majority are too poor to constitute a market or lack the health and education to provide the necessary human resources.

This view of development is often rejected by the literature on political ecology that takes "Capitalism and its historical transformations [as] a starting point for any account of the destruction of nature" (Peet, Robbins and Watts, 2010: 23). What was characterized above as "development" is, in political ecology literature, considered to be the privileging of certain exploitative productive systems over others, causing intertwined distributive and ecological conflicts and the degradation of the environment (Alimonda, 2011). In the political ecology literature, however, elites are largely "black boxed". Elites appear as the perpetrators: they are the capital owners, the business and knowledge elites, and the groups controlling the state, thereby contributing to the marginalization of people inhabiting rural landscapes and to the overexploitation and pollution of natural resources (Carruthers, 2008). However, elites in the political ecology literature are rarely the object of direct scrutiny. Their interests are considered to be dependent on their location in the structural relations of domination, and their privileges to be derived from their positions in the structures that configure Latin America as a subaltern region open to exploitation according to the needs of a globally integrated capitalism. The double exploitation of people and nature also forms the basis of the construction of the modern states, dominated by national elites, but it is also often based on the control of natural resources and local groups in different parts of the territory (Alimonda, 2011).

Against the backdrop presented above, we should not only ask under what conditions we can expect elite objectives to be aligned with national development goals. We also have to discuss how to make those goals aligned with the interests, needs and aspirations of all population groups across social classes and territories, as well as with those of future generations. A common answer to the question of how to achieve that has been to emphasize pluralism and democracy; in other words, to ensure that there are good mechanisms of representation, participation and accountability, which can lead to the establishment of institutions of environmental governance with the potential of less elitist and more sustainable development outcomes. This has been what many hoped would occur in Latin America in recent decades after the return to formal democracies and the historical rise of previously marginalized groups to power in the government.

Elite theory has nevertheless never been convinced of the merits of pluralist democracies to make societies more egalitarian. To the contrary, elite theories of all kinds have had a quite dismal view of the potential of democracy to transform society, a matter that is partially rooted in their view of the state. Marxist elite theory, which defines elites based on their relationship to capital and means of production, is generally sceptical of the possibility for changes in the state without underlying changes in the mode of production upholding it (see e.g. Jessop, 1990). As a democratic government depends on public support, it will suffer if it presides over a serious drop in the level of economic activity as a result of conflicts with capitalists (Block, 1977). Therefore, in spite of the establishment of pluralist institutions, the state cannot really be democratized within a capitalist economy.

The other major classical political-economy theory of elites and democracy, developed by Schumpeter, was highly critical of the Marxist equation of true democracy with socialism, although not discharging the possibility that they could coexist.[2] He does not have much more faith in pluralism either. Schumpeter's main point is that democracy is inherently elitist: "democracy does not mean and cannot mean that the people actually rule... Democracy means only that the people have the opportunity of accepting or refusing the men (sic) who are to rule them" (Schumpeter, 1976: 285). However, this should not lead one to be pessimistic about the decisions made in democratic institutions. The functioning of democracy would depend on the degree to which a government is restrained by autonomous state powers (most importantly, the judiciary), the self-restraint used by such powers (also parliamentary) and the existence of an independent bureaucracy.

The so-called "Italian school" of elite theory was also sceptical of the virtues of pluralist democracies. Originating in the writings of Mosca, Mitchells and Pareto (see Pareto 1997 [1935]; Mosca, 1939; Michels, 1962), it defines the elite as a distinct group of society that enjoys a privileged status and exercises decisive control over the organization of society (Wolf, 2012: 120). Mosca regarded universal suffrage and parliamentarism as unable to dissolve the principle that, in any society, an "organized minority" is able to "impose its will on a disorganized majority" (Mosca, 1939: 154), while Vilfredo Pareto argued that elites would slowly be replaced by ascending families and groups without changing the elitist structures of society (Pareto, 1935). Yet it is this elite circulation, not the revolutions led by the dispossessed classes, that would lead to change (Pareto, 1916, cited in Hartmann, 2007).

For this study we adopted a "resource-based definition" as outlined above, which combines some of the elements of the Italian approach with that of the Marxist approach. The definition we adopted here considers elites to potentially emerge from their control of various and possibly overlapping resources, including organizational (control over organizations, including the state), political (public support), symbolic (knowledge and ability to manipulate symbols and discourses) and personal (such as charisma, time, motivation and energy) (Etzioni-Halevy, 1997: xxv). Yet we also include a focus on the actual influence that these groups have on the environmental outcomes of changing policies and practices.

Also, our view on how elites shift is eclectic. In the Marxist view, rather than through a democratic shift of government, change would emanate from below, based on the construction of political subjects among the dispossessed classes. However, Marxism has also envisioned changes emerging from the space opened for the "relative autonomy of the state" in situations of weak or split class forces (Jessop, 1990). The capitalist classes were considered to be unable to establish a "political hegemony" by themselves, thus ensuring the dominance of the lower classes. This is rather the role of the state, which in the process assumes a relative autonomy from the capitalist classes (Poulantzas, 1978). This makes room for the emergence of a state elite that is functionally set apart from the capitalist class.

This is an issue that is also essential to Weber, who regards the state bureaucracy not only as a by-product of capitalism but as the most effective form of legitimate power and the source of the emergence of an entirely new class (Weber, 1978). The structure and power of the bureaucracy is much more important than the electoral institutions since the demos itself is "a shapeless mass [that] never 'governs' larger associations, but rather is governed" (Weber, 1978). The dilemma presented to new political forces gaining formal power over a state apparatus is that, while the bureaucracy may hinder a shift in policies and practices, it may take decades to construct. Irrespective of how much popular support a ruler may enjoy, without the instrument of a modern bureaucracy, his or her ability to enact, implement and enforce his or her will is severely limited.

In sum, with the exception of the Marxists, elite theorists doubt the possibility of elite-free societies. Moreover, they all have reservations against the belief that a shift in government will automatically result in a shift in elites. Nevertheless, there are venues open for change. We focus on shifts in the elites' control of resources that result in changes in their ability to exert influence over decisions and practices with environmental implications. In Latin America recently, we identified four such shifts, which will be discussed below.

## Leftwing governments, elite circulation and limitations to environmental governance shifts

The first such process of change is the shift in control of political resources related to the entering of power of centre-left governments, many of which represented groups that had previously been excluded from political power, including indigenous movements, labour movements, environmental movements and diverse social movements constituted by dispossessed groups. In spite of getting electoral support from these groups, many of the governments have later disqualified or consciously attempted to co-opt some of them (Zibechi, 2010; Bowen, 2011), while new elites emerge. Thus we may observe a process of "circulation of the elites", controlling political resources with a possible impact on environmental governance.

One example of that is found in Bolivia, where groups associated with the governing party MAS have started to gain political resources and power (Ayo Saucedo, 2012), but also economic resources through, for example, the processes of nationalization of enterprises (Ayo Saucedo et al., 2013). The soy sector has long been dominated by a landed elite, with diverse origins (including large groups of immigrants from Brazil), but with a common discourse on the use of GM, the benefits of industrial agriculture and the desire to be independent from the Morales government (Plata, 2008; Soruco, 2008). This traditional elite still control important economic resources (particularly through their control of land). Nevertheless, a new group of people, with significantly fewer economic resources than the traditional economic and political elite, has entered the political arena and is exerting influence over the way in which the environmental consequences of GM agricultural production are addressed (Høiby and Zenteno Hopp, 2015). This new group is composed of soy farmers who have accessed their productive capacity due to contacts in the MAS party, and political groups. While standing quite far apart from the old soy elite on several matters of economic policy and so forth, they coincide with them on the issue of the desirability of expansion of GM soy. Soy production contributes substantially to government revenues and perhaps, therefore, the expansion of GM soy production into forested areas is not rejected by the government.

In El Salvador, the entering of a centre-left government had quite different consequences. El Salvador is a country that has historically been dominated by a closely knit agroexport-based elite that have had political power for most of the country's history, historically in conjunction with the military (Paige, 1997). Between 1989 and 2009 they ruled through the rightwing ARENA party, led by some of the country's richest families. Thus they awaited the coming of a government supported by the FMLN with significant fear and contempt, and the old elite put up both political and economic resistance. However, the right wing was already split when the Funes government came into power, partially due to the prior transformation of El Salvador from an agroexport- to a service-based economy dependent on remittances from migrants in the USA. Although the old elite families diversified their portfolios to benefit from the new economy (Bull, 2013), the economic transformation also produced the ascendance of new economic elites that eventually challenged the old elite dominating the ARENA party. That resulted in the breakout of the GANA party (Gran Alianza por la Unidad Nacional) soon after the Funes government took power. The Funes government attempted to include broad groups of the society in a multistakeholder dialogue to establish new forms of governance of agricultural and other productive activities. The purpose was to confront the grave environmental crisis in which El Salvador was submerged. The groups that advocated a different agricultural model more focused on small farms and ecological production included both members of the new government, particularly linked to the Ministry of the Environment and Natural Resources (MARN), and a broad set of civil society organizations working locally to create economies based on principles of ecology and solidarity.

However, the government could not ignore the economic crisis, with low or negative growth for many consecutive years. As predicted by Block and other Marxists, the government's dependence on the economic elites for investment limited strongly its freedom of action. The domestic economic elite also represented the political opposition, although it was split between ARENA and GANA. Although ARENA, GANA and the private sector peak association ANEP (Asociación Nacional de la Empresa Privada) initially participated in different forums of dialogue to reach solutions to pressing problems (including the Social and Economic Council established on the model of a similar institution in the EU), the relationship soon soured. The government was required to re-establish a relationship with the private sector in the context of the US-funded Alliance for Growth program, but then chose to deal directly with a narrow group of the country's most powerful businessmen in order to attempt to entice them to invest in El Salvador. In the process, however, the development plans became more and more aligned with the business elite's priorities and less and less to the groups proposing alternative models within the government (Bull, Cuéllar and Kandel, 2014). There was also an incipient economic elite emerging as a result of the policies of the new government. This had links to the governing party, but benefited from its role in the companies established with funds from the Venezuela-led Bolivarian Alliance for the Peoples of Our America (ALBA) (Lemus, 2014). However, this elite showed little inclination to support the groups within a ruling party that advocated a shift towards a more sustainable development model.

The case of Ecuador is illustrative of a different solution to similar structural constraints. When Rafael Correa came to power in 2008, it was as head of a broad coalition with support from grassroots organizations, and with a strong environmentalist faction within the government. While new groups entered the governmental corridors, these were not considered to be a new elite but rather a counterweight to the traditional elites in Ecuador that had previously – and simultaneously – incorporated and marginalized grassroots organizations (Bowen, 2011). The environmentalists in the government were able to influence how environmental issues were framed in the official discourse, and important changes in the status of the environment and its relation to human activity were introduced to the constitution of the country (Andrade, 2012; Basabe, Pachano and Acosta, 2012). One of the changes made was that the government openly rejected GM organisms.

However, Correa's government was equally challenged by old elites that, although lacking a recent past of 20 years of relatively stable rule that ARENA in El Salvador had enjoyed, were equally enmeshed in the international economy (both countries converted to US dollars in 2000) and had enjoyed strong privileges in association with multinational companies in the past. Yet Correa managed to challenge the old elites to a quite different extent than his El Salvadoran counterpart by ensuring income from the oil industry, strengthening the incipient mining industry and engaging in a process of strengthening the Ecuadorean state.

During Correa's second term (2009–2013) his political project was increasingly formulated as that of a developmentalist project, resting on the parallel strengthening of technology and industrial upgrading and the intensification of resource extraction. This resulted in the weakening of the environmentalist faction of the government and in the emergence and gradual strengthening of a young technocratic elite. This elite not only supported the industrialization efforts but also had a positive view on GM organisms. These young professionals, owing their influence to specialized knowledge of biotechnology, are becoming key players in defining strategies to achieve the diversification of agricultural production in Ecuador. Their view fits well with the developmentalist ideas pursued by Correa, seeking rapid diversification of the Ecuadorian economy led by experts and guided by scientific knowledge (SENPLADES, 2013). While these ideas made room for the influence of this new technocratic elite, it is also the case that the emergence of the technocratic elite reinforces and supports the plan.

Thus in the cases above we have seen the entry of new political groups in government that have struggled against old elites in their pursuit of political and economic projects. However, in the process, new elites have formed based on access to economic- and knowledge-based resources in addition to political ones. Yet the elite circulation we have seen in Ecuador and Bolivia has had a limited positive environmental impact, as requirement for funding for social projects, the strengthening of the state, and the continued struggle against old elites have often weighed stronger than environmental concerns. Moreover, emerging new elites have had equally strong economic interests in the continuation of an extractivist model, while political elites (particularly in the case of Ecuador) have sought support from groups controlling a technical knowledge and ideology of continued industrialization and the conquering of nature.

## The role of global economic transitions and technology

The second process of elite change is a shift in the control of economic resources due to changes in the global economy. Parallel to the so-called "left tide" in Latin America, three major interrelated trends have occurred in the global economy: a rise in the demand and prices of commodities; the strongly related rise of China as a major economic power, lender and investor in Latin America (Durán Lima and Pellandra, 2013); and the strengthening of regional integration schemes such as ALBA, MERCOSUR and UNASUR, which have favoured the emergence of new economic elites associated with, for example, state-controlled or supported companies. These processes have enabled new groups to control significant shares of the economy.

The rise of China and booming commodity prices have allowed the South American countries to speed up debt repayment to international institutions, and to form new economic alliances. This has resulted in a decrease in importance of elites that have traditionally been very influential in the region, among them those related to Western multinational companies, the World Bank, the IDB and the International Monetary Fund. As a result we are currently observing new relationships and arrangements between national states on the one hand and, on the other hand, diverse international elites of various origins, including North American, European, Chinese and Latin American.

In all of the cases discussed here there has been, to a certain degree, an interplay with commodity prices, particularly the booming of the soy market and the opportunities that new elites have had to emerge. One case in particular, Guatemala, suggests that when rising commodity prices have resulted in the entrance of new transnational elites to the country, the scope of possibilities to influence environmental governance and outcomes of these new elites is limited not only by the features of the industry (i.e. mining) but also by the dynamics found in the relationship between new elites controlling access to markets and technology, and old elites controlling political resources and land. New transnational elites have opted to operate within a status quo determined by the power that the traditional elite holds over major knobs of the economy and the government, and a series of corrupt practices and relationships between the old, entrenched elite and the government

(Aguilar-Støen, 2015). Guatemalan business elites have been successful in keeping transnational elites, including transnational companies from Canada, Australia, the USA and Russia, in a subordinate position. This is explained by the control of different but complementary resources. Domestic elites control important political resources, networks and information; transnational companies could not operate without such resources (Schneider, 2012; Bull, Castellacci and Kasahara, 2014). Local elites have also established different types of partnership with transnational mining companies. In many cases, local elites have interests in junior mining firms that are subsidiaries of transnational mining companies. The drafting of mining legislation in Guatemala involved the participation of Canadian and Guatemalan businessmen, and the resulting mining law disproportionately favours mining companies over the interests of local populations, including their environmental concerns (Dougherty, 2011). A mix of local and foreign capital finances mining operations in Guatemala. Canadian groups in association with Guatemalan capital dominate metallic mineral exploitation. The largest non-metallic mining company is the Guatemalan company Cementos Progreso, which makes the second largest contribution to mining investments in Guatemala (Lee and Bonilla de Anzueto, 2009). Mining contributed 2% to the GDP in 2013 but it is estimated that, with the development of planned exploitation, mining could contribute approximately 4% in the future (Lee and Bonilla de Anzueto, 2009). This growth, however, is expected to occur in a context where 51% of the population of the country (15 million) live in rural areas and rely on agriculture for their livelihoods.

Mining operations have caused massive protests and discontent among local populations in Guatemala. One of the main reasons is that the law does not require companies to inform communities about mining operations before applying for licences. In this context, local communities have felt that their opinion has not been considered before mining operations have started, something to which they are entitled by law. Another source of discontent is that mining royalties were reduced from 6% to 1% by a new mining bill (Decree 48–97) and this is perceived among the general Guatemalan population as extremely unfair. Another source of conflict is that mining projects are often established in areas with longstanding conflicts related to access to land and land tenure, before the conflicts have been resolved. In most cases the government has responded to the demands of participation from the local population and to the protests with violence and repression. Also, as a response to the complaints regarding royalties, the Chamber of Industry unilaterally decided to propose a voluntary agreement by way of which mining royalties could, based solely on the decision of mining companies, be increased from 1% to 3% for gold and from 1% to 4% for silver, whereas for other activities – such as cement production controlled by a Guatemalan family – royalties were kept at 1%. The government is then supposed to launch agreements with local authorities regarding royalties in their communities. This has been strongly rejected by local populations.

In Argentina there has been quite a different process of elite shift dependent on a combination of technological shifts, a changing world market and political changes. Soy production in the Pampa region in Argentina started to expand after GM soy was legalized in 1996, but it soon expanded in magnitude in other parts of the country as well, currently occupying approximately 22 million Ha, which is between 50% and 60% of all the cultivated land in Argentina (USDA, 2013). However, rather than being predicated on the entry of a new governing elite, it has generated a shift in economic elites. As its leftwing government has drawn its main leaders from the ranks of the Peronist party, it can hardly be considered a new political elite in Argentina. However, soy production has generated shifts in the power relations among agricultural producer groups. Although not completely displacing the traditional landowning elite, new groups related to agribusiness have gained significant influence in the governance of agricultural production. This group is composed of agricultural producers, utilizing a management model in which several individuals or companies have different roles in the system, from renting land from landowners to administering external investments and managing the total production (Benchimol, 2008). They run what is commonly called "agroenterprises", in which landowners, contractors and investment brokers are involved. Such agribusinesses agreements can take the form of investment funds, agroassociations (*pools de siembra*), financial trust coalitions and simple contract alliances, among others. The most recent attempt to quantify it argued that agroenterprises are responsible for about 70% of total grain production in Argentina (Barri and Wahren, 2010). Today the figure is probably higher.

At the same time as the soy expansion generated a new agricultural (but not necessarily rural) elite, the strained relations between the four governments of the Kirchners (two of Néstor Kirchner and two of Cristina Fernández de Kirchner) and the traditional agriculturalists contributed to the speed of the soy expansion. The main reason behind the conflicts was the increase in export taxes on agricultural products, particularly during the first government of Cristina Fernández de Kirchner. However, conflicts also arose due to the perceived lack of governmental support for and interest in agriculture in general, due to lack of both predictability in "framework conditions" (including adjustments in export taxes) and technical support. This contributed to a weakening of the influence of the old rural elite (Zenteno Hopp, Hanche-Olsen and Sejenovich, 2015). Moreover, in a context characterized by high levels of uncertainty for many farmers, many of them either leased the land to agrientreprises for soy production or turned to soy production since its profitability was considered almost guaranteed over time (Calvo et al., 2011). While depending on transnational companies, first and foremost Monsanto for seed and fertilizer, there has also been a prolonged conflict between Argentinean farmers and the agricultural giant. Argentinean farmers first objected to paying royalties for the fertilizer Roundup Ready as Monsanto had failed to obtain a valid patent for it in Argentina, and later farmers opposed the payment of new royalties for the new soy seed BTRR2.

Initially it was also argued that GM soy would result in less environmental impact than conventional soy. It was argued that soy production would minimize soil cover loss due to the no-till method, and that the use of the herbicide glyphosate would prevent the use of other, and more toxic, agrochemicals applied in conventional production (Bindraban et al., 2009). GM soy soon acquired a privileged position among the nation's exports and also became a main source of governmental income. Currently the production of GM soy generates approximately one-tenth of the GDP and one-quarter of the nation's export value (Loman, 2013). The conversion to the GM soy model generated a net value of US$65,435 million for Argentina between 1996 and 2010, due to savings in costs and higher profitability (Trigo, 2011). This source of funding has been of key importance for the government's ambitious programmes of social redistribution. Added to this, the economic interest by national and international agribusiness companies explains the government's unwillingness to impose more ambitious environmental guidelines on GM soy production. Only very recently has there been a broader public debate due to increasing opposition against and conflicts related to soy production, exposing the severe soil degradation resulting from soy production and glyphosate's negative impact on human health, among other issues (Skill and Grinberg, 2013).

## The role of knowledge and the contours of elite reorientation

However, we also see the contours of a third process: "elite reorientation", or, in other words, the shift in the dominating ideas of an elite.

Both Schumpeter and Weber emphasized the orientation and capacity of elites as a major factor in understanding the role of the state in development, rejecting that this could be directly inferred from their position in the capitalist economy (as the Marxists would argue). In recent Latin American history we have two major examples of such elite reorientation: the process of democratization of the 1980s and 1990s, and the neoliberal transformation in the same period. Neither processes of elite reorientation happened out of the blue. Rather, the new ideas achieved influence due to a crisis and exhaustion of prior models and a gradual shift in interests. Currently the seriousness of the environmental crisis, and the climate crisis more specifically, could open up space for new ideas brought about by new elite groups, the reorientation of old groups, or a new dynamic interplay between different elite groups.

Despite the numerous contradictions evident in the environmental policies pursued by Brazil's three leftwing governments (two under Luis Inácio Lula da Silva and one under Dilma Rousseff), in the Amazonian states of Acre and Amazonas a shift in elites and in the environmental policies pursued in these states occurred at the state level. Despite the differences (in size among other things), around 2009/2010, Acre and Amazonas were the least deforested states in Brazil, with small Acre having lost 14% of its original forest and Amazonas only 3% (Lemos and Silva, 2011). Our research found that this was closely related to a shift in elites occurring in different ways. The turning point in Acre was the coming to power of the PT candidate in 1998, whereas in Amazonas it occurred as a candidate linked to the old elites shifted towards a more environmentalist and less developmentalist strategy to distance himself from the old ruling elite in order to gain votes in the local elections of 1992 (Toni, Villarroel and Taitson Bueno, 2015). Thus the process at the local level has been very different from that at the federal level. At the federal government level an "elite settlement" between economic elites and rightwing parties, on the one hand, and elites of the leftwing parties, on the other hand, has led to the favouring of developmental goals over the environment (Arretche, 2013). In contrast, at the state level there has been some room for elite shifts through elite reorientation. The autonomy given to lower politicoadministrative levels in the federal model has thus been crucially important for the latter process.

Global initiatives, such as REDD, are also fostering a possible "elite reorientation" through the emergence of a new knowledge-based elite that is organized in wide and often transnational networks. These networks have been able to influence the attitudes and strategies of certain elites, although this has not implied a complete reorientation of old elites (i.e. those linked to agroexport activities) or of government elites, particularly because of the centrality of resource extraction in economic growth in the region (Aguilar-Støen and Hirsch, 2015). The global REDD initiative was launched at the climate negotiations in 2005 but only gained political traction in 2007, when donor governments agreed to commit substantial economic resources to establish a fund that would pay developing countries not to deforest. The principle of REDD is relatively simple: it is based on the idea that it is possible to pay countries and communities for not cutting down their forests. However, the implementation of REDD is not so simple. Latin America is endowed with vast amounts of forested land but as a whole the region has the world's highest rate of deforestation (Hall, 2012). Because of that, much attention and efforts have been invested in trying to successfully develop REDD projects in the region. These projects are, to date, only demonstration activities that will allow implementers to understand how REDD would work on the ground. That means understanding how payments are to be implemented and to whom, how to monitor that the area covered by forest is effectively not deforested, and how to ensure that economic benefits are distributed in a fair manner among those that contribute to forest conservation and constitute a legitimate beneficiary of REDD. Since forests are valuable for a range of different actors, from forest-dwellers to drug cartels, control of forested land is a contested issue and thus establishing national or local REDD projects is a complex task. In addition, many valuable non-renewable resources, such as minerals and oil, are often located in forested areas and several governments in Latin America have declared extractive activities as being key to national economic development. REDD has attracted the attention of various and disparate actors, including environmental NGOs, research centres, extractive industry companies, indigenous peoples' organizations and international development agencies.

REDD is a broad and vague enough idea to allow different interpretations of it that can fit the goals of different actors (Angelsen and McNeil, 2012). This has allowed these actors to distinctly define the actions necessary to implement REDD at local levels. In the process, certain narratives, values and visions gain prominence and those promoting such ideas gain power to define how REDD should look in specific contexts. Controlling the production of knowledge seems to be a prominent strategy of different actors to position themselves in the REDD debate, particularly in the countries in the Amazon basin (Aguilar-Støen and Hirsch, 2015). The knowledge required to participate in the REDD debates is not just any type of knowledge. It has to be maintained and strengthened through particular networks, in which different concepts and arguments are socially constructed and legitimated through complex processes that have produced new dominant forms of expertise and consultancy (Fairhead and Leach, 2003; Bumpus and Liverman, 2011). These networks that are coalitions of actors who share values, interests and practices can be conceptualized as elites insofar as they control key resources: the production and promotion of specific knowledge or forms to generate knowledge and access to policy-making forums. Ideas, values and resources circulate within networks, and as such the networks may set the limits or boundaries of how reality is to be understood or to set apart what constitutes expert and non-expert knowledge. A range of different private actors and companies support REDD activities, forming alliances and promoting certain models, particularly those that are positive to carbon markets. In this way, REDD is offering a new regime of profit-making possibilities in the trade of carbon offsets, but also in fostering the development of new forms of consultancy and expertise. REDD science-policy networks are influencing, although not necessarily reorienting, the position of other elite actors. For example, various transnational and national companies, such as mining and energy-producing companies, plantation companies, forestry companies and carbon-market companies, engage in REDD demonstration activities by funding specific projects. Since dominant REDD science-policy networks have ideological positions that do not conflict with the ideological position of corporations, it has been possible to establish alliances between them. But since resource extraction continues to be central to the economies of most Amazon countries (Bebbington and Bebbington, 2012), often at the expense of forests, the degree to which REDD elites can influence other elites is limited. Mining, gas and oil extraction are the most important activities to generate economic revenues for most of the countries in the Amazon basin. The development of infrastructure such as hydropower and road-building are also priorities for these countries. All these activities are, in most cases, planned to occur in forest areas. In addition, the agricultural frontier is expanding in many Latin American countries. Therefore we cannot affirm that REDD elites have a strong influence in the Amazon countries' broader development policy-making or in the national visions of development, but REDD elites have indeed been successful in engaging actors from the agricultural and industrial sectors in the funding of demonstration activities.

Taken together, the cases of Brazil and of REDD show that a shift in elites sometimes leads to more ambitious environmental goals and regulations. Whether or not this happens depends on the degree to which new elites are able to influence the positions and views of old elites. Chapter 6 suggests that the views, aspirations and environmental orientation of elites are not homogeneous. It is conceivable that we will see the ascendance of elites in the future with aspirations of a more sustainable development policy and environmental governance. It is also necessary to remember that centre-left governments in Latin America won the elections thanks to the support of wide segments of the population, particularly the marginalized and subaltern ones. These governments depended on various types of alliance between different grassroots organizations and social movements. If these movements and grassroots organizations are able to exert some pressure on their governments to address environmental concerns in the future, we may see a shift towards more equitable and sustainable models of environmental governance. If popular mobilization continues to be crucial for maintaining leftist governments in power, at some point the environmental concerns of the population need to be addressed.

## Conclusion

Back in 1977, Marxist scholar Fred Block rejected the possibility that a leftwing government in power could make a significant change to the productive structure of a country, as any government presiding over a capitalist economy inevitably has to care about the creation of employment and economic growth, and therefore would never counter the interests of the capitalists. Over the last decade we have seen a multitude of strategies applied by leftwing Latin American governments to overcome the constraints presented by old elites that are often also political adversaries. Although, judging from media reports, the relationship between the centre-left government and the old economic elites is strained, under the surface they are more often than not characterized by accommodation and consent than outright conflict. However, in the process there has been a gradual elite shift where groups that have benefited from the centre-left governments policies gradually gain influence at the expense of old rural and business elites. This has occurred in Argentina with the strengthening of agrienterprises; in Bolivia with the emerging soy elite; in Ecuador with the new cadres of technocrats in the ministries; and in a more incipient form in El Salvador with new elites related to ALBA investments.[3]

In addition to new governmental policies, we have found two factors to be of key importance to the emergence of new elites: knowledge and technology. Controlling capital or politics without also controlling knowledge and technology has shown to be insufficient to dominate the development agenda and the environmental governance of it. Knowledge and technology can be "bought" by those who control capital, but this is only partly true because it is necessary to have the sufficient knowledge, relevant technology and appropriate attitude towards innovation to know where to invest in it. Also, obtaining and making use of these resources are long-term processes. The corollary to that is that groups that control knowledge and technology may also influence environmental governance to an extent disproportionate to their political position or economic resources, as we have seen in the cases of the REDD networks, and in a different way in the Ecuadorean Ministry of Agriculture.

This may have positive and negative implications for the environment. The control of knowledge can be an obstacle to better environmental governance, such as when it is used by a technocracy to pursue an agenda that pays little attention to environmental or distributional concerns, or when it is controlled by a transnational company as a means to strengthen its own profit generation. However, it can also be used to influence the agenda in a more sustainable way, such as has been observed in the case of, for example, El Salvador, where groups of environmentalists with high levels of technical education were included in the government. In spite of not having achieved the influence that they had hoped for, they did influence parts of the governmental agenda to become more directed towards adapting to climate change and avoiding new environmental catastrophes induced by intensive export agriculture. The emergence of what could be called a "new, environmental technocratic elite" was also observed in other countries, including Chile and Bolivia (Reyes, 2012; Høiby and Zenteno Hopp, 2015). This new technocratic elite differs from other historical groups of technocrats, not only by being unified by a different body of knowledge from, for example, the neoliberal economists that constitute the technocrats supporting the neo-liberal conversion. They also show a different attitude towards relating to non-elite groups. Many have been involved in environmental movements at local, national and transnational levels, and many stay in touch with communities through everything from frequent visits to membership of Facebook groups. Although their actual influence varies, their strengthening may lead to stronger environmental governance over time. Moreover, where the government favours party cadres over technically competent officials in important positions, the likelihood that such "new technocracies" emerge diminishes as, for example, in the case of Argentina (Hanche-Olsen, 2013).

Yet it is impossible to ignore at least three "constants" in environmental governance in Latin America. One is the importance of global markets. During the last decade, Latin America as a region has made significant progress in a number of social indicators, but it has also reinforced its dependency on natural resource export, and therefore its vulnerability to changes in the global markets for a limited set of export goods. This is less so in Mexico and Central America than in South America, but across the region there is little in the way of a "structural transformation" towards a production structure dependant more on knowledge and innovation and less on cheap labour and natural resources. As noted by CEPAL (2014), without such a conversion, it will be difficult to sustain incipient processes towards a more just resource distribution, or to counteract the serious processes of environmental degradation.

The second "constant" is limitation in resources. For leftist governments with little support from, and often in conflict with, the economic elite, to stay in power and to implement ambitious programmes for societal transformation has required both to employ policies to strengthen the state and to confront the opposition from old elites. State-building has been an unavoidable priority for the centre-left governments in Latin America to be able to deliver strong programmes of resource redistribution to address historical inequalities, and in this way to lift millions out of poverty. Several strategies have been employed to face opposition from old elites: grooming new elites, confronting competing elites or allying with outside elites. Given that the international context has been very favourable for resource extraction, focusing on these sectors (particularly mining and agriculture) has allowed centre-left governments to increase their revenues and deliver their promises of resource redistribution. At the same time, larger revenues have permitted governments in Latin America to transform their relationships with traditional international elites (weakening their influence in domestic politics) and to enter into relationships with new international elites. In this context it can be said that leftist governments in Latin America have taken a pragmatic approach to be able to secure their position; this approach implies that, in development policy, economic revenues take precedence over environmental concerns. We can then affirm that the effects of the elite shift on environmental governance in Latin America have been limited thus far.

The third "constant" is the abyss between the traditional elite and non-elite groups in terms of the meaning given to nature and what constitutes a just governance of it, in terms of both processes and outcome. Although, as we have shown, the elites go through processes of change that lead to episodes of "elite circulation" as well as "elite conversion", we still find elite groups across the region with a very limited understanding of the local environmental impact of developmental projects, the importance and meaning of resources such as land and water to rural communities, and what it takes to actually reach understandings across cultural and class divides. Without this, reaching more sustainable and just environmental governance in Latin America may still be far away.

## Notes

1. This approach focused on the conditions for – and evolution of – a state with a monopoly on legitimate violence, and an institutional bureaucracy capable of implementing policies and controlling the masses (e.g. Migdal et al., 1994; Evans, 1995). Such a state, in which a given set of institutions' right to tax and demand loyalty in return for protection and the extension of benefit are no longer questioned, is, for example, considered to be a precondition for the high-growth policies and business–state relationship of the East Asian developmental states (Amsden, 2001) as well as the more historical examples of development, such as that of Europe (Tilly, 1992).
2. He argued rather that "Between socialism as we defined it and democracy as we defined it there is no necessary relation" (Schumpeter, 1976: 284).
3. The tendency observed in El Salvador would probably have been more pronounced had we included Nicaragua and Venezuela in the study.

## References

Acemoglu, D. and Robinson, J.A. (2012) *Why Nations Fail? The Origins of Power, Prosperity and Poverty* (London: Profile Books).

Aguilar-Støen, M. (2015) "Staying the Same: Transnational Elites, Mining and Environmental Governance in Guatemala", in B. Bull and M.C. Aguilar-Støen (eds), *Environmental Politics in Latin America: Elite Dynamics, the Left Tide and Sustainable Development* (London: Routledge Studies in Sustainable Development), 131–149.

Aguilar-Støen, M. and Hirsch, C. (2015) "REDD and Forest Governance in Latin America: The Role of Science-Policy Networks", in B. Bull and M.C. Aguilar-Støen (eds), *Environmental Politics in Latin America: Elite Dynamics, the Left Tide and Sustainable Development* (London: Routledge Studies in Sustainable Development), 171–189.

Alimonda, H. (2011) "La Colonialidad de la Naturaleza: Una Aproximación a la Ecología Política Latinoamericana", in H. Alimonda (ed.), *La Naturaleza Colonizada: Ecología Política y Minería en América Latina* (Buenos Aires: CLACSO).

Amsden, A. (2001) *The Rise of "The Rest" Challenges to the West from Late-Industrializing Economies* (Oxford: Oxford University Press).

Amsden, A., Di Caprio, H.A. and Robinson, J.A. (2012) *The Role of Elites in Economic Development UNI-Wider Studies in Development Economics* (Oxford: Oxford University Press).
Andrade, P. (2012) "En el Reino (de lo) Imaginario: Los Intelectuales Ecuatorianos en la Creación de la Constitución de 2008", *Ecuador Debate* 85: 35–48.
Angelsen, A. and McNeil, D. (2012) "The Evolution of REDD+", in A. Angelsen, M. Brockhaus, W.D. Sunderlin and L.V. Verchot (eds), *Analysing REDD+ Challenges and Choices* (Bogor Indonesia: CIFOR).
Arretche, M. (2013) "Quando Instituições Federativas fortalecem o Governo Central?", *Novos Estudos-CEBRAP* 95: 39–57.
Ayo Saucedo, D. (2012) *Nuevas Élites Económicas y su Inserción en la Política*, Working Paper N°01 (La Paz: Foro de Desarrollo Friedrich Ebert Stiftung/Fundación Boliviana para la Democracia Multipartidaria).
Ayo Saucedo, D., Fernández Morales, M. and Kudelka Zalles, A. (2013) *Municipalismo de Base Estrecha. La Guardia, Viacha, Quillacollo: La Difícil Emergencia de las Nuevas Élites* (La Paz: PIEB).
Barri, F. and Wahren, J. (2010) "El Modelo Sojero de Desarrollo en la Argentina: Tensiones y Conflictos en la Era del Neocolonialismo de los Agronegocios y el Cientificismo-Tecnológico", *Realidad Económica* 255: 43–65.
Basabe, S., Pachano, S. and Acosta, A. (2012) "Ecuador: Democracia Inconclusa", in M. Cameron and J.P. Luna (eds), *Democracia en la Región Andina* (Lima, Peru: IEP).
Bebbington, A. and Bebbington, D. (2012) "Post-What? Extractive Industries Narratives of Development and Socio-Environmental Disputes Across the (Ostensible Changing) Andean Region", in H. Haarstad (ed.), *New Political Spaces in Latin American Natural Resource Governance* (London: Palgrave Macmillan), 17–38.
Benchimol, P. (2008) La Concentración de la Tierra, Latifundios y Agro-Societies, http://www.pagina12.com.ar/diario/suplementos/cash/17-3460-2008-04-20. html, date accessed 16 November 2013.
Bindraban, P.S., Franke, A.C., Ferraro, D.O., Ghersa, C.M., Lotz, L.A.P., Nepomuceno, A., Smulders, M.J.M. and Van de Wiel, C.C.M. (2009) *GM-Related Sustainability: Agro-Ecological Impacts, Risks and Opportunities of Soy Production in Argentina and Brazil*. Report 259 (Wageningen: Wageningen University – Plant Research International).
Block, F. (1977) "The Ruling Class Does Not Rule: Notes on the Marxist Theory of the State", *Socialist Review* 33: 6–27.
Bowen, J.D. (2011) "Multicultural Market Democracy: Elites and Indigenous Movements in Contemporary Ecuador", *Journal of Latin American Studies* 43(3): 451–483.
Bull, B. (2013) "Diversified Business Groups and the Transnationalisation of the El Salvadoran Economy", *Journal of Latin American Studies* 45(2): 265–295.
Bull, B. (2014) "Latin America's Decade of Growth: Progress and Challenges for a Sustainable Development", in A. Hansen and U. Wethal (eds), *Emerging Economies and Challenges to Sustainability* (London: Routledge), 123–134.
Bull, B. (2015) "Elites, Classes and Environmental Governance: Conceptual and Theoretical Challenges", in B. Bull and M.C. Aguilar-Støen (eds), *Environmental Politics in Latin America: Elite Dynamics, the Left Tide and Sustainable Development* (London: Routledge Studies in Sustainable Development), 15–32.

Bull, B., Castellacci, F. and Kasahara, Y. (2014) *Business Groups and Transnational Capitalism in Central America: Economic and Political Strategies* (Houndmills, Basingstoke, Hampshire: Palgrave Macmillan).

Bull, B., Cuéllar, N. and Kandel, S. (2014) "El Salvador: The Challenge to Entrenched Elites and the Difficult Road to a Sustainable Development Model", in B. Bull and M.C. Aguilar-Støen (eds), *Environmental Politics in Latin America: Elite Dynamics, the Left Tide and Sustainable Development* (London: Routledge Studies in Sustainable Development), 33–50.

Bumpus, A.G. and Liverman, D. (2011) "Carbon Colonialism? Offsets Greenhouse Gas Reduction and Sustainable Development", in R. Peet, P. Robbins and M.J. Watts (eds), *Global Political Ecology* (New York: Routledge), 203–224.

Calvo, S., Salvador, M.L., Giancola, S., Iturrioz, G., Covacevich, M. and Iglesias, D. (2011) "Causes and Consequences of the Expansion of Soybean in Argentina", in H. El-Shemy (ed.), *Soybean Physiology and Biochemistry* (open access: InTech), 365–388.

Carruthers, D.V. (2008) *Environmental Justice in Latin America: Problems, Promise and Practice* (Cambridge MA: MIT Press).

CEPAL (2014) *Pactos para la Igualdad: Hacía un Futuro Sostenible* (Santiago, Chile: Comisión Económica para América Latina y el Caribe).

Cimoli, M. and Rovira, S. (2008) "Elites and Structural Inertia in Latin America: An Introductory Note on the Political Economy of Development", *Journal of Economic Issues* XLII(2): 327–347.

Dougherty, M.L. (2011) "The Global Gold Mining Industry, Junior Firms, and Civil Society Resistance in Guatemala", *Bulletin of Latin American Research* 30(4): 403–418.

Durán Lima, J. and Pellandra, A. (2013) "El Efecto de la Emergencia de China sobre la Producción y el Comercio en América Latina y el Caribe", in E. Dussel Peters (ed.), *América Latina y El Caribe – China Economía Comercio e Inversiones* (Mexico: UNAM).

Etzioni-Halevy, E. (1997) "Introduction", in E. Etzioni-Halevy (ed.), *Classes and Elites in Democracy and Democratization* (New York and London: Garland Publishing).

Evans, P. (1995) *Embedded Autonomy States and Industrial Transformation* (Princeton: University Press).

Fairhead, J. and Leach, M. (2003) *Science Society and Power: Environmental Knowledge and Policy in West Africa and the Caribbean* (Cambridge: Cambridge University Press).

Gudynas, E. (2010) "Si eres tan Progresista ¿Por qué destruyes la Naturaleza? Neoextractivismo, Izquierda y Alternativas", *Ecuador Debate* 79: 61–81.

Hall, A. (2012) *Forests and Climate Change: The Social Dimensions of REDD in Latin America* (London: Edward Elgar Publishing Ltd).

Hanche-Olsen, E. (2013) *Argentine Farmers Search for Collective Action. A Case Study of Rural Cooperatives Within the Genetically Modified Agricultural Production in Argentina*, unpublished master thesis Latin America Studies (Oslo: University of Oslo).

Hartmann, M. (2007) *The Sociology of Elites* (London: Routledge).

Hecht, S.B., Kandel, S., Gomes, I., Cuellar, N. and Rosa, H. (2006) "Globalization, Forest Resurgence and Environmental Politics in El Salvador", *World Development* 34: 308–323.

Higley, J. and Gunther, R. (1992) *Elites and Democratic Consolidation in Latin America and Southern Europe* (Cambridge: Cambridge University Press).

Høiby, M. and Zenteno Hopp, J. (2015) "Bolivia: Emerging and Traditional Elites and the Governance of the Soy Sector", in B. Bull and M.C. Aguilar-Støen (eds), *Environmental Politics in Latin America: Elite Dynamics, the Left Tide and Sustainable Development* (London: Routledge Studies in Sustainable Development), 51–70.

Jessop, B. (1990) *State Theory Putting Capitalist States in Their Place* (Pennsylvania: Penn State University Press).

Lee, S.L. and Bonilla de Anzueto, M.A. (2009) *Contribución de la Industria Minera al Desarrollo de Guatemala* (Guatemala: Centro de Investigaciones Económicas Nacionales).

Lemos, A.L.F. and Arimatea Silva J. (2011) "Desmatamento na Amazônia Legal: Evolução, Causas, Monitoramento e Possibilidades de Mitigação Através do Fundo Amazônia", *Floresta e Ambiente* 18(1): 98–108.

Lemus, E. (2014) "La Milionaria Revolución de ALBA", *El Faro* 19 January 2014, http://www elfaro net/es/201401/noticias/14423/, date accessed 17 November 2014.

Loman, H. (2013) Country Report Argentina Rabobank, https://economics. rabobank.com/publications/2013/october/country/report/argentina/, date accessed 21 January 2014.

Martinez-Alier, J. (2002) *The Environmentalism of the Poor: A Study of Ecological Conflicts and Valuation* (Cheltenham: Edward Elgar Publishers).

Michels, R. (1962) *Political Parties* (New York: Free Press).

Migdal, J.S., Kohli, A. and Shue, V. (1994) *State Power and Social Forces. Domination and Transformation in the Third World* (Cambridge: Cambridge University Press).

Mosca, G. (1939) *The Ruling Class* (New York: McGraw Hill).

Myrdal, G. (1968) *Asian Drama: An Inquiry into the Poverty of Nations* (New York: Pantheon).

Myrdal, G. (2010 [1979]) "Underdevelopment and the Evolutionary Imperative", *Third World Quarterly* 1(2), reprinted in B. Bull and M. Bøås (2010), *International Development*, Volume II (London: Sage).

Paige, J. M. (1997) *Coffee and Power Revolution and the Rise of Democracy in Central America* (Cambridge, MA: Harvard University Press).

Pareto, W. (1997 [1935]) "The Governing Elite in Present-Day Democracy", in E. Eztioni-Halevy (ed.), *Classes and Elites in Democracy and Democratization* (New York and London: Garland Publishing).

Peet, R., Robbins, P. and Watts M. (eds) (2010) *Global Political Ecology* (London: Routledge).

Plata, W. (2008) "El Discurso Autonomista de las Élites de Santa Cruz", in X. Soruco, W. Plata and G. Medeiros (eds), *Los Barones del Oriente. El Poder en Santa Cruz Ayer y Hoy* (Santa Cruz, Bolivia: Fundación Tierra).

Poulantzas, N. (1978) *State, Power, Socialism* (New York: Verso).

Reyes, C. (2012) *Las Elites y los Nadies: Caso de Estudio sobre la Influencia de la Elite Chilena en un Conflicto Ambiental*, unpublished master thesis Latin America Studies (Oslo: University of Oslo).

Schneider, A. (2012) *State-Building and Tax Regimes in Central America* (New York: Cambridge University Press).

Schumpeter, J.A. (1976) *Capitalism, Socialism and Democracy* (London and New York: Routledge).

Segovia, A. (2002) *Transformación Estructural y Reforma Económica en El Salvador. El Funcionamiento Económico de los Noventa y sus Efectos sobre el Crecimiento la Pobreza y la Distribución de Ingreso* (Guatemala: FandG Editores).

SENPLADES (2013) *"Plan Nacional del Buen Vivir"*, República del Ecuador Consejo Nacional de Planificación (Quito, Ecuador: Secretaria Nacional de Planificación y Desarrollo).

Skill, K. and Grinberg, E. (2013) "Controversias Sociotécnicas en Torno a las Fumigaciones con Glifosato en Argentina. Una Mirada desde la Construcción Social del Riesgo", in G. Merlinsky (ed.), *Cartografías del Conflicto Ambiental en Argentina* (Buenos Aires: CLACSO and Fundación CICCUS).

Soruco, X. (2008) "De la Goma a la Soya: El Proyecto Histórico de la Élite Cruceña", in X. Soruco, W. Plata and G. Medeiros (eds), *Los Barones del Oriente. El Poder en Santa Cruz Ayer y Hoy* (Santa Cruz, Bolivia: Fundación Tierra).

Tilly, C. (1992) *Coercion Capital and European States AD 990 – 1992* (Cambridge MA and Oxford UK: Blackwell Publishers).

Toni, F., Villarroel, L. and Taitson Bueno, B. (2015) "State Governments and Forest Policy: A New Elite in the Brazilian Amazon?", in B. Bull and M.C. Aguilar-Støen (eds), *Environmental Politics in Latin America: Elite Dynamics, the Left Tide and Sustainable Development* (London: Routledge Studies in Sustainable Development), 190–205.

Trigo, E. (2011) *Fifteen Years of Genetically Modified Crops in Argentine Agriculture* (Buenos Aires: ArgenBio).

USDA (2013) *USDA Agricultural Projections to 2022*, Long-Term Projections Report OCE-2013–1 (Washington: Office of the Chief Economist World Agricultural Outlook Board US Department of Agriculture/Interagency Agricultural Projections Committee).

Weber, M. (1978) *Economy and Society* (Berkeley: Berkeley University Press).

Wolf, A. (2012) "Two for the Price of One? The Contribution to Development of the New Female Elites", in A. Amsden, A. DiCaprio and J.A. Robinson (eds), *The Role of Elites in Economic Development. UNI-Wider Studies in Development Economics* (Oxford: Oxford University Press), 120–139.

Woo-Cumings, M. (ed.) (1999) *The Developmental State* (Ithaca, NY: Cornell University Press).

Zeballos, H. (2012) *Bolivia: Desarrollo del Sector Oleaginoso 1980–2010* (Santa Cruz: Instituto Boliviano de Comercio Exterior – IBCE).

Zenteno Hopp, J., Hanche-Olsen, E. and Sejenovich, H. (2015) "Argentina: Government-Agribusiness Elite Dynamics and Its Consequences for Environmental Governance", in B. Bull and M.C. Aguilar-Støen (eds), *Environmental Politics in Latin America: Elite Dynamics, the Left Tide and Sustainable Development* (London: Routledge Studies in Sustainable Development), 71–91.

Zibechi, R. (2010) "América Latina: Nuevos Conflictos Viejos Actores", *América Latina en Movimiento,* http://alainet.org/active/39400, date accessed 10 November 2014.

 Except where otherwise noted, this work is licensed under a Creative Commons Attribution 3.0 Unported License. To view a copy of this license, visit http://creativecommons.org/licenses/by/3.0/

OPEN

# 6
# Water-Energy-Mining and Sustainable Consumption: Views of South American Strategic Actors

Cristián Parker, Gloria Baigorrotegui and Fernando Estenssoro

## Introduction

Mining activity has undeniable environmental impacts due to the nature of its operations, processing plants and foundries. Mining companies proclaim their environmental responsibility by implementing policies that limit environmental risk and impact, while also applying new technologies and production processes that are more respectful of the environment. The degree of efficacy of these sustainability measures and the degree to which companies voluntarily ensure environmental care cannot belie the fact that – no matter what – mining activity has and will always have environmental impacts. There are two major points of view about the subject, according to Whitmore (2006). On the one hand, there are the views of companies – that is to say, the actors who control the mining bulldozers and claim to ensure that everything goes well and that mining is, or can be, sustainable. On the other hand, there are the views of those who are affected by mining activity, such as the communities, peasants and indigenous peoples who are displaced without proper consultation, who suffer illnesses, and whose lifestyles, health and environment are impacted.

This chapter will not address the mining problem from the conventional perspective of whether or not mining is sustainable. The majority of the socioenvironmental conflicts that arise around mining are focused on this problem.[1] We refer to the fact that mining consumes large quantities of water and energy and is one of the most widespread productive activities. As AngloGold Ashanti's sustainability report declares,[2] mining activity has a direct impact on the environment because it requires access to land, water and energy, scarce resources that should be shared with the communities in which it operates. Mining processes also require "considerable amounts of water" and "significant quantities of energy" in order to function.

The sustainable consumption of strategic natural resources such as water and energy in South American mining is a key theme that challenges environmental governance, but it is rarely studied by the social sciences. This is especially true in the case of the mining sector. Since the 2000s, the mining boom has resulted in expanded investment in all of the countries in the region, in many cases generating socioenvironmental conflicts (Svampa and Antonelli, 2009; Teijlingen, 2012). And this trend is likely to continue in the coming years.[3]

The research that we present here looks into the different social representations[4] of strategic actors with respect to the sustainable consumption[5] of energy and water in the mining sector. These social representations of environmental issues are fundamental to understanding the social and institutional practices aimed towards sustainable consumption and environmental governance (Hajer and Versteeg, 2005). With strategic actors we refer to members of elites who have the capacity for long-term influence, and who may come from the private sector or the public sector as well as from organized civil society. We include strategic actors who are linked to a few paradigmatic mining cases in four South American countries: Argentina, Chile, Colombia and Ecuador.

## Problem under study: The water-energy-mining complex

The main questions of the study are related to the configuration of the social representations – of an institutional nature and pertaining to strategic actors – of water and energy, and actors' views of nature and development. In order to understand the viability of forms of governance for the sustainable and equitable consumption of water and energy in the cases studied, we want to see which different representational models can be observed and on which points they coincide. As a result of climate change (PNUMA, SEMARNAT, 2006), nature – and in particular water and energy – is increasingly understood to be of strategic significance (Bruzzone, 2010; Sunkel, 2011). The aspiration to capitalist economic growth makes these sources highly sought by both Latin American countries and emerging powers.

In the economic interpretation of development, energy and water are vital resources for human life and production, and they cannot be separated from the environment. From this perspective, the strategic character of water and energy is linked to their availability for use in productive processes. However, mining is an economic activity that proportionally uses more water and energy and, for that reason, it is more controversial in environmental terms (Norgate and Haque, 2010; Superneau, 2012). Another conceptualization of water and energy comes from an ecological perspective, from which they are not – in a strict sense – economic "resources". They are rather "common goods", and their use has a greater value than simply their exchange value. In this chapter we will consider – from a holistic perspective – water, energy and mining as a complex of interrelated parts[6]; a complex that, in recent years, appears to have been critical to complying (or not) with ecological and environmental principles in Latin America.

Just as energy is required for the consumption of water, so is water for the production of energy (Wu et al., 2013). As both resources are indispensable for mining, it cannot function without the industrial consumption of water and energy (Mudd, 2008). For this reason the mining sector faces the huge challenge of resolving the problem of its high water demand without affecting the availability of water for agriculture and for the urban population, and without increasing pollution (Pizarro, 2012). As for its growing energy demand, the mining sector should seek to satisfy it with maximum efficiency and without increasingly relying on polluting energy sources (e.g. electricity generated by coal, gas or oil) (Zuñiga and Ana, 2009). Along these lines, contentious scenarios lay ahead for every strategic actor interested in defending their legitimacy. In other words, the water-energy-mining complex continues to form a Gordian knot of environmental governance in the mining sector in Latin America and beyond, throughout the socioeconomic structure.

## Studying the representations of strategic actors

We have sought to study the social representations of natural resources and their sustainable consumption among actors and institutions with the capacity for leadership and influence in long-term public policies related to environmental governance. Our main topic of concern is the sustainable consumption of water and energy in the mining sector in Argentina, Chile, Colombia and Ecuador.[7] In these countries, in distinct stages and with different emphases, metal mining has become one of the pillars of their development policies. Here we focus on the network of actors (Bebbington, 2012) involved in cases of paradigmatic mining projects (some in the exploration phase, most in the operating phase) in these four countries, as shown in Table 6.1. In all countries,

*Table 6.1* Reference cases

| Country | Argentina | Chile | Ecuador | Colombia |
|---|---|---|---|---|
| Projects | Cerro Vanguardia[1] La Alumbrera[2] | Mantos Blancos; Manto Verde; Soldado; Chagres, Los Bronces[3] | Fruta del Norte[4] Mirador[5] | La Colosa[6] |
| Companies | AngloGold Ashanti (South African) and FormiCruz (Argentinean) Xstrata (Swiss), Goldcorp and Yamana Gold (Canadian) | Anglo American (British) | Kinross Gold (Canadian) Ecuacorriente (Chinese) | AngloGold Ashanti (South African) |

*Notes*
1. Cerro Vanguardia is a gold and silver mining project in Santa Cruz province.
2. Bajo de la Alumbrera, located in Catamarca, is one of the major metal deposits of copper and gold in the world and is being exploited by means of open pit mining.
3. Anglo American has several, mostly copper, open pit mines in Chile: Los Bronces in the Metropolitana region, Mantos Blancos in the Antofagasta region, El Soldado in the Valparaíso region, Mantoverde in the Atacama region and Collahuasi in the Tarapacá region.
4. The Fruta del Norte gold and silver deposit is a Kinross Gold project that quickly entered into conflict with the Shuar communities. It signed an initial agreement in 2011, but the resistance as well as the company's non-conformity with government regulations has caused Kinross to withdraw from the project.
5. The El Mirador Project in Zamora Chinchipe province, in the Cóndor mountain range, is a copper deposit that is in exploration and its exploitation phase has been approved. It is one of the largest mining projects approved in recent years by the Correa government, not without pressures and conflicts.
6. The La Colosa Project in the Tolima department is the second-largest gold deposit discovered in Colombia. It is a subject of important debate in Colombia because of its social, environmental and economic implications.

socioenvironmental conflicts have been reported. In Colombia and Ecuador, these are primarily related to processes of exploration. In Colombia, the La Colosa project in Tolima has encountered serious resistance from local communities. A similar situation occurred in Ecuador in the Fruta del Norte project, which has since been suspended. In Chile the mining project Doña Inés de Collahuasi, which is partly owned by Anglo American, has received complaints from surrounding communities about water problems, and its current expansion phase is controversial. The Alumbrera project in Argentina received the bulk of its complaints when overflows of its mineral pipeline and tailings dam contaminated the Vis-Vis River and the valley's agricultural communities.

The interdisciplinary strategy of this study relies on mixed methods. It is based on a literature review, analysis of primary and secondary sources of an institutional nature, and 65 semistructured interviews with members of so-called strategic actors in the mining sector: CEOs and high executives, senior government officials, political leaders, experts and leaders of NGOs, including community and environmental organizations. The discourse analysis (van Dijk, 2008) was based on semantic techniques. We used a structural discourse analysis, taking into account the overall logic of semantic speech articulation, narrative structures, semantic axes and paradigmatic axes, but focusing on the semiotic square (Greimas, 1966).

## Institutional views and actor views

The theoretical and institutional frameworks that have been developed in regard to the industrial consumption of water and energy in the mining sector come from various sources, primarily from international mining institutions and experts. These expert discourses and institutional discourses of companies, and of public and private institutions, show that the concept of efficiency – as applied to water and energy – is the most developed, extensive and referenced. This includes a set of good practices, procedures and technologies that point to an optimization of scarce resources in the diverse phases of the mining lifecycle. The concepts of ecoefficiency (WBCSD, 2013) and natural capitalism (Rábago, Lovins and Feiler, 2001) represent different perspectives on ecological interrelationships between resources. These two concepts have also been applied to the consumption of water and energy in mining, but they are almost inexistent in the discourses of the individual actors from the four case studies.

In regard to the efficient consumption of water and energy, and the incorporation of renewable sources of energy in mining, limited information is generated by corporate discourses. The production of knowledge about the consumption of water and energy in mining is relative to the degree of development of the mining sector in each country, being greater in Chile than in the other countries studied. Institutions such as the International Council on Mining and Metals (ICMM) – the most important corporate regulatory body – have developed a set of principles for sustainable mining development (MMSD, 2002; ICMM, 2003). However, only one of ICMM's 46 subprinciples refers to the responsible consumption of water and energy in mining. In the reference cases studied, the relevance of the consumption of water and mining is a theme of a "high level" and experts. It does not, however, seem to be picked up by other social actors. Similarly, references to water and energy consumption in the mining and environmental legislation of the countries studied are scarce (OCMAL, 2012). In the rules and regulations for environmental evaluation and monitoring, these issues are of secondary importance.

In short, the analysis of institutional discourses elucidates the importance of the principles and good practices driven by transnational companies. They emphasize the role of international financial agencies and institutions, such as the International Finance Corporation (IFC, 2012), and principles of environmental evaluation and report, such as the Global Reporting Initiative (GRI, 2011) and BellagioSTAMP (IISD and OECD, 2009).

This study underlines the existence of basic social representations that favour environmental considerations. The interviewed actors in the four countries were asked about the environment, climate change, development models and the relationship of man with nature. They responded in a few typical patterns that show their views on water and energy consumption. Some stress the role of policies of social and environmental responsibility of the mining companies and institutions, reflecting influential discourses at local and international levels. The alternative discourses, which oppose mining projects, resort to interpretational codes derived from a radical reconceptualization of the consumption of water and energy. They focus on their uses, meanings and valuations as associated with the notions of justice, and social and environmental rights.

The statement against which interviewees had to declare their preferences is taken from the mainstream discourses in public policy, saying that "'Sustainable development' in the context of my country's needs would be an economic growth model that mitigates negative environmental and social impacts."[8] The responses were primarily "strongly agree", which dominated among senior public officials and businessmen, and "disagree", which dominated among environmentalists and (college-educated) experts. We should take note of the emphasis on the idea of economic growth in this proposal, although it is certainly moderated by the idea of mitigating environmental and social impacts. Our results indicate that the concepts associated with growth that still dominate in public political and international institutional discourses are assumed by businessmen and experts, while even most politicians and some NGO leaders agree (77% of all interviewees were in agreement with the statement). Despite a common terminology, social representations of the environment and climate change, technology, the man–nature relationship and development models point to divergent positions. Yet there is a nuanced vision of the future. Asked about whether the future of the country would be clean or polluted, 54% declared that their country will be cleaner and 46% declared that it will be more polluted.

### Different views and discourse models

The interviews of strategic actors reveal important discourse structures, which can be classified into four models that express specific views on the consumption of water and energy in mining. However, this specific issue is linked to broader views related to mining and the national development model, which generate distinct perspectives on the environment and environmental policy (Dryzek, 2005). The aim of our analysis was to discover the elementary structures of the meaning in the discourses, followed by a linguistic and extralinguistic (social, political, cultural) interpretation of the discourse. The main elements of the four models are schematically presented in Table 6.2. These models are empirically reconstructed, built semantically through inductive and deductive steps.

#### Model 1: Indispensable but responsible mining with maximum efficiency

The first model assumes that the consumption of water and energy should be efficient within the context of responsible mining. Its point of departure is the unconditional affirmation of mining. In regard to water, it seeks to make its consumption efficient and to optimize its reuse:

> It seeks to reuse water, to utilize products that are biodegradable so that there is no pollution.[9]
> (Argentinean senior executive of a state-private mining company)

> The use of water in mining is so serious...that there is already technology to achieve it...(de-pollution).[10]
> (Ecuadorian senior executive of a transnational mining company)

171

Table 6.2 Overview of signifying content in the discourse models

| Themes | Discourse models | | | |
|---|---|---|---|---|
| | 1 | 2 | 3 | 4 |
| Water consumption | Efficiency and reuse | Efficiency, reuse and responsible consumption | Efficiency and recycling | Threatened water ecosystems |
| Energy consumption | Efficiency to reduce costs | Efficiency and responsible consumption | Efficiency and recycling | Evaluate carbon footprint |
| Renewable energy | Insufficient but complementary | Opening up | Indispensable | Change the overall energy grid |
| Mining | **Indispensable** | Necessary | Critical but necessary | Not sustainable, threatens people and ecosystems |
| Development | Growth | Sustainable growth | **Sustainable development** | **Other development, alternative development** |
| Technology | Fundamental | Optimal technology | Necessary, anti-technocracy | Green technologies |
| Management | Efficient | **Integrated management** | Monitoring, control | For change |
| Regulation | Market | Mixed forms and state regulation | Institutional control | Human security and life |
| The state | Should take step back | Subsidiary with clear policies | Should intervene, participatory citizenship | Should promote total change |
| Environmental responsibility | Responsible mining | Positive vs. irresponsible mining | Environmental control | Population and local communities |

The emphasis is on the fact that water as a resource has a low rate of consumption and is reused through technology. This discourse model seeks forms of efficient water and energy consumption in mining through its rational and balanced use.

> I return to the same issue, the consumption of water, the consumption of energy... The goal is to achieve that balance, but if you are a consumer the balance is about the question of how to mitigate consumption.[11]
> (Colombian senior executive of a transnational mining company)

The claim is made that the use of water in mining is considerably less than in other activities because of the funnel effect: large quantities of water are manipulated but little is consumed; recycling is very common. This also happens in regions where water resources are abundant (the tropical areas of Ecuador and Colombia, and even in some mountainous areas of Argentina). Water is accumulated in pools and recycled, thus a small amount of water is consumed and its quality is controlled.

> In other areas where water is consumed (agriculture), much of the water continues to evaporate.[12]
> (Chilean entrepreneur of the National Mining Society)

As for energy resources, this model considers them to be an absolute necessity for mining to function, but recognizes that they are a problem, and even a threat to competitiveness, given their cost. In particular, Chilean and Colombian interviewees problematize the issue of energy while the interviewees from Argentina and Ecuador tend to have a more optimistic perspective. The point of departure is that metal mining is recognized as intensive in terms of energy use, primarily derived from fossil fuels or hydroelectricity. However, this rhetoric downplays the volume of energy consumed.

> If mining consumes energy, then the price of energy should take into account the environmental impacts of generating that energy. Therefore, having paid your energy bill, you are fulfilling your role as a responsible consumer.[13]
> (Ecuadorian senior executive of a transnational mining company)

In this discourse model, the energy issue is commodified: it is necessary that the markets operate competitively.

> What is stronger in mining and more problematic is electric energy, this issue is very critical...[14]
> (Chilean senior official in the mining industry)
>
> Various projects... have been cancelled because of the high costs of energy...[15]
> (Chilean senior executive of a transnational mining company)

Furthermore, this model fits into a neoliberal conceptual framework that attaches greater relevance to the market than to the state.

> The market (should regulate), all of us want the market. I prefer the market...[16]
> (Argentinean senior executive of a transnational mining company)

Assuming that mining requires considerable energy for its processes, facilities and transportation, this discourse model recognizes that most energy comes from fossil fuels. Renewables, they claim, are not the best alternatives because they are expensive and are not processed continuously. Energy from fossil fuels (including electricity generated by gas and coal) is more convenient because of its low price. This discourse model proposes responsible mining that manages to establish a balance between the pursuit of profitability, the environment and social needs: in other words, a legitimate corporate mining activity. It privileges a market environmentalism that prioritizes private initiative but is aware that it should take care of certain environmental and social externalities. It therefore proposes the "rational use of resources", "responsible mining consumption" and "responsible growth".

## Model 2: Integrated management, regulation and responsible consumption

This second discourse model accepts mining as an important development tool. However, it also incorporates reservations about its negative environmental impacts, which can be repaired through proper regulation and institutional norms.

It is a multiplying activity... the local population had nothing to do in San Juan, but now there is mining that enhances other activities.[17]

(Argentinean expert and consultant on environmental issues)

Water and energy consumption are represented by a semantic axis of efficiency/inefficiency, where "efficient practices" oppose "inefficient consumption". Resources are scarce and often have high prices, especially energy, which is why efficiency must be promoted.

Being high-tech companies (big mines)... they should be as energy efficient as possible.[18]

(international expert)

The core idea is "efficiency". Unlike the previous model that emphasized technology as a transforming agent, here the emphasis is placed on integrated and efficient management. Its goal is the responsible consumption of water and energy. This "responsibility" should be assumed by private economic agents, but in case this does not happen the subsidiary state should determine its conditions.

It is the responsibility of the companies as much as of the authorities, how to develop, manage and implement the projects.[19]

(Argentinean expert and environmental consultant)

With respect to water... good mining is technically realized, economically profitable and it ensures the just participation of the Ecuadorian state, a socially responsible mining and mining environmentally managed with strict standards.[20]

(Ecuadorian director of state-owned mining company)

This discourse model favours regulation through "pricing mechanisms", among others, that stipulate mixed policies to enable the proper functioning of the market and forms of state regulation (environmental evaluation, laws and norms, effective fiscalization). The model proposes the establishment of clear energy policies that frame energy consumption in mining. This model seeks to "regulate" the energy grid with "rules" that are associated with "clear environmental policies". These "clear policies" must be given within the framework of a subsidiary state. The state should then intervene to adequately regulate and make the market function conveniently by establishing conditions for private investment in the form of laws, regulations and institutions.

> The incentives are well placed when the decision maker has internalized – to the greatest extent possible – all of the potential (environmental) costs that energy use represents.[21]
>
> (Chilean senior official and ex-minister of the state)

Regulation requires planning, evaluation and control of mining activity. From this perspective communities have to be prevented from deepening their opposition to mining projects and impeding the functioning of institutions and regulations. A clear policy that involves "integrated regulation" is fundamental for "legal certainty" to exist and to incentivize mining investment. The responsible consumption of water and energy in mining must point towards "sustainable growth".

In this discourse model there are some views and positions that are critical of the rationale of those who want to grow at any cost. It seeks to promote responsible growth at a responsible growth rate. For instance, an international expert recognizes that this is not necessarily a consistent practice of big mines, where emphasis is placed on the general discourse about CSR. This is not necessarily linked to a vision of integrated, efficient and responsible management of water and energy. In summary, this model generates a clear sense of the consumption of water and energy in mining with explicit central concepts, such as efficiency, recycling, integrated management and responsible consumption. Its second focus is on establishing the institutions and conditions that allow for better regulation and for the establishment of certain regulations that guarantee private investment and frame the responsible consumption of water and energy in mining.

### Model 3: Sustainable development and institutional control

This model makes strong statements about water and energy consumption in mining, focusing on the more political concept of sustainable development. It assumes that mining has negative environmental and health effects. This gives rise to various degrees of criticism of mining, but it agrees that – under certain conditions – mining is a necessary activity.

> I think that (mining) is worth the effort because the activity, if well developed, can be done with a relatively low level of environmental impact. I am talking about a mining at a scale... (that is) more human...[22]
>
> (Argentinean politician, advisor in Congress)

In this discourse, mining is problematic due to contamination from heavy metals. This is the origin of the need for efficiency and recycling in water and energy consumption, and – given the environmental crisis – the need for environmental control that guarantees sustainable mining.

> In the case of water... it must be addressed through strict control over available resources.[23]
> (Chilean politician, representative of the center-left)
>
> ... it should be, as I say, with the least environmental impact.[24]
> (Ecuadorian politician, progressive Congressman)

However, this control and monitoring supposes the existence of a state that clearly intervenes and regulates the market, and a democratic citizenship that participates, monitors and combats corruption. As for energy, the fundamental semantic axis resides in the contrast of "carbon energy" with "renewable energy".

> And in Chile... the energy grid is overly carbonized.[25]
> (Chilean politician, representative of the left)
>
> Our indigenous discourse has always been to defend the rights of nature... For that reason the president has decided to change the energy grid, for example from thermoelectric to hydroelectric energy...[26]
> (Ecuadorian indigenous leader, progressive representative)

In summary, this model is based on a political proposal of sustainable development, which criticizes the environmental impact of mining but includes mining as a factor of development. It subjects mining to controls, rules and regulations, and seeks to encourage the sustainable consumption of water and energy by promoting efficiency, recycling and a transition to renewable energy, including this transition within the mining sector itself. It proposes sustainable development with the clear intervention of the state in order to guarantee a market with clear and competitive rules. It intends to combat monopoly and corruption, and to stimulate citizen participation. In short, water and energy consumption is perceived as a political problem and not only one of technical management.

## Model 4: Alternative development for the protection of common goods

This discourse model departs from a critique of the environmental consequences of mining. It represents "immense risks", "environmental destruction", "water pollution", "wars" and even "death".

> Mining is non-viable or incompatible with the life of many human beings...[27]
>
> (Chilean environmental NGO leader)
>
> No mining is clean – it incites serious problems; the pollution is incredible.[28]
>
> (Ecuadorian senior official, ex-Minister of State)

The main semantic axis that stimulates the discourse is "life" versus "death"; mining has become "incompatible with life". Human life and nature would be in danger: peasants, indigenous and communities as well as ecosystems would be threatened.

In this discourse, the "rights of nature" are inextricably linked to the human rights of the affected populations, the communities and the indigenous people. Natural resources in this discursive view are meant for common use related to the rights of the community (residents, indigenous, etc.) and of society (the state). They are semantically disjointed from exchange values (the mining market), and should be neither commodified nor privatized. In general, the texts speak of the water-energy-mining complex as a whole, in sociotechnical and in sociopolitical terms. According to this view, as mineral reserves decrease, the intensive consumption of water and energy further increases. While the global mineral demand increases, the pressure for more intensive forms of production (in terms of capital employed) grows, along with policies to raise productive efficiency and efficacy in order to achieve maximum "competitiveness" and profitability in the global metal mining market.

In general, this discourse model goes beyond the references to specific issues such as water management and energy efficiency in terms of industrial mining consumption. Instead, concepts of greater abstraction, such as "ecosystems" and "carrying capacity", are used. The interviewees who fit into this model claim that both mining companies and the authorities have agreed to water consumption that is greater than nature's "carrying capacity", and that "overconsumption" of natural resources is fostered by the "extractivist model".

178  *Water-Energy-Mining Nexus*

> I insist, we should put the little resources that we have left (water and energy) towards alternatives for the future, not towards satisfying the needs of such a small percentage of the population...[29]
>
> (Argentinean leader of an environmentalist assembly, when referring to gold mining)

As for consumption, the discourse associates it with the "extractivist model" and opposes it to "another development" that is "non-consumerist". The latter is a mode of production that relies on mining "for the bare necessities" and that develops from values such as "solidarity" instead of "competitiveness".

> We do not call them "natural resources", but rather "common goods".[30]
>
> (Argentinean environmental leader)

With regard to energy consumption, this discourse model clearly favours the use of renewable energy, inclined towards non-conventional renewables but especially insisting on thinking about the global energy system in a different way.

> We are the country of the sun, the country of water, here we have potential and we have possibilities to generate a type of energy other than oil.[31]
>
> (Ecuadorian leader of an environmental NGO)

Compared with mining megaprojects, local projects with renewable energy at a "human scale" are favoured in the context of another (post-oil) energy system: hydroelectric energy and/or solar energy projects that can be developed along with communities and local governments. In that way, they could overcome the megaprojects' overconsumption of energy and water. There is talk of generating conditions so that the new mining projects would have a reduced "ecological footprint", "water footprint" and "carbon footprint". Perceptions of the intergenerational and long-term environmental impacts are present in this discourse model. It has strong Utopian connotations, an ideal that is inspired by values such as the "good life" and ecodevelopment from empowered local social actors.

In summary, this discourse model formulates social representations of water and energy consumption in the mining sector from a codification of meaning that proposes a systemic change with communities, especially those of the indigenous, as main points of reference. It is a critical look at the current development model and public policies, including those of the "progressive" governments. It advocates a change in capitalist modes of production – encouraging citizen participation, and decentralized and self-managed forms of production – with a clear preference for non-conventional renewable energy. In general, the alternatives to water and energy consumption in the mining sector are subordinate to issues of a greater magnitude. Mining should be rejected when it affects regions that are rich in biodiversity, water resources and ecosystems. This includes the risk of utilizing excessive amounts of water and energy.

## Conclusions: Governance of sustainable water and energy consumption in mining?

The interviews reveal that there is a consensus of "environmentalist" language with regard to common issues. Corporate environmental responsibility, protection of and care for the environment, concern about water and energy consumption, and an orientation towards sustainable development are mentioned as necessary by all interviewees.[32] But beyond the discursive rhetoric, deep code analysis reveals very different and even contradictory concepts about the following subjects: the environment, the responsibility of strategic actors for resources such as water and energy, the role of the government, and the water-energy-mining complex, which ultimately reflects different worldviews and *epistemes* about the relationship of man with nature.

In general, we observe that these different discourses are set forth and projected at different scales (transnational, national and local) and levels (business, government and politics, and civil society), and that there is little room for dialogue. They maintain positions in the social structure of elites: the first model is set forth mostly by CEO and high executives, and some senior government officials; the second model is affirmed by experts and also by senior officials; the third model is set forth by politicians and experts (but is slightly more important among politicians); and the fourth model pertains to environmental leaders and some politicians.

The consumption of water and energy in mining, seen in the light of the analysed discourses, is not an exclusively technical subject. The worldviews, linked to the social positions and interests of stakeholders, frame patterns of action and have an impact on the way in which the sustainable consumption of natural resources is represented. But they go beyond that because of the obvious practical consequences that they have and will have in social and political aspects. The first two discourse models have a technical bias (hard technologies and management technologies); the third and fourth discourse models have ideological-political biases, the last being traversed by ecological worldviews.

From the perspective of environmental governance, the positions behind the models point at disagreement. They will be a great source of conflict to the extent that some defend the thesis of economic growth, taking ecological factors into account only as secondary externalities (positions found in the first model). Others take an alternative stance, proposing an ecological perspective that focuses on avoiding economic growth and overconsumption in the neo-extractivist Third World (positions found in the fourth model).

The analysed discourses, with few exceptions, do not take long-term environmental risks into account. The central grid of this discursive logic is the capacity to control and intervene in water and energy consumption in mining, through technocratic (first model), normative and institutional (models two and three), or political-environmental (models three and four) means. The abstraction of the accumulative and latent effects of the long-term environmental impacts of the abovementioned consumption is proof that the autonomized effects of sociotechnical processes as a result of the increased extractive economy in the region are unknown.

Our study of discourses confirms that most of the stakeholders who are more likely to defend the expanded reproduction of the water-energy-mining complex – as the basis of the socioeconomic development of the region – do not take responsibility for the international and global implications of local environmental behaviour. The majority of these actors do not think in terms of a long-term global horizon. Consequently, the problems of climate change and the decisions that they implicate in terms of energy and water policy are considered without taking into account the reflexivity of local social processes in overall environmental risks. The structural positions of these stakeholders in developing countries, in the periphery of the world system, thus condition discourses with respect to these global implications.

In this chapter we first presented the problem of water and energy consumption in the mining sector, situated within the water-energy-mining complex. We sought to clarify linear, sectorial and reductionist perspectives and to approach a perspective that integrates synergies among discourses, rules, technologies, institutions and interpretations both diverse and contentious elements of environmental governance. The majority of the discourses ignore these interrelationships. Therefore the deeper significance of the overconsumption of resources, from which future scarcity of water and energy as well as consequences of climate change are expected, does not seem to be present in the majority of the analysed discourses.

In view of the transition towards more sustainable patterns, it is important to note that the first and second discourse models are associated explicitly with a confidence in technological innovation. The third and fourth model, on the other hand, introduce a more political and ecological logic in their vision of resource consumption in mining. The considerations about the intensity of use of water and energy resources in mining, as well as the technological structure with which they are associated, should be considered simultaneously as integrated systems that assume social, political and ecological connotations. The analysis of the processes of technological innovation linked to the shift towards sustainable consumption of water and energy in mining cannot neglect the associated social and political variables. In addition, this study of social representations of water and energy consumption of strategic South American actors demonstrates the recent increase in environmental consciousness.

In general, we observe that there is a struggle for legitimacy going on between conflicting discourses. The contradictory positions are opposite poles in a space of dialogue that should be promoted by a public policy that seeks environmental sustainability and resource governance. The recognition of the conflict of interests and views, and the discourse models with divergent positions – whose possibility for dialogue is still an open question – clearly demonstrate that there is a series of challenges ahead for environmental governance and for achieving sustainability in the extractive industries.

## Notes

1. See the Environmental Justice Atlas of the EJOLT Project at http://ejatlas.org/
2. See http://www.anglogoldashanti.com/en/Pages/default.aspx
3. In accordance with the Center for Copper and Mining Studies (Cesco), with headquarters in Santiago, Latin America will become the most important region in the world for attracting investments for mining development, with a record number of US$327 billion between 2011 and 2020. See the Metals Economic Group (2013).
4. We understand social representations in accordance with Höijer (2011) and Moscovici (1981).

5. For more information on sustainable consumption, see Parker et al. (2012).
6. This refers to the mining-energy complex (Baker, 2012; Sharife and Bond, 2012). We have expanded this to the concept of the water-energy-mining complex from the sociotechnical and sociopolitical perspective. In South America, as is also shown in the case of South Africa (Sharife and Bond, 2012), it is furthermore a structure of power through which the elites have historically appropriated those resources.
7. For information about the mining sector in Argentina, see Svampa and Antonelli (2009); Walter and Martinez-Alier (2010); Baigorrotegui, Parker and Estenssoro (2014); in Chile, see Newbold (2006); in Colombia, see Garay (2013); in Ecuador, see Bustamante and Rommel (2010); van Teijlingen (2012).
8. "El 'desarrollo sustentable' en el contexto de las necesidades de mi país sería un modelo de crecimiento económico con medidas de mitigación de los impactos ambientales y sociales negativos".
9. "se busca reutilizar el agua, se busca utilizar productos que sean biodegradables de manera tal que no exista contaminación".
10. "el uso del agua en la minería no es tan grave...ya existe la tecnología para lograrlo...(la descontaminación)".
11. "Vuelvo a lo mismo, el consumo de agua, el consumo de energía...El objetivo es lograr ese equilibrio, donde si usted consume el equilibrio es ¿cómo mitiga ese consumo?".
12. "en las otras áreas (agricultura) que consumen agua, se sigue evaporando mucha agua".
13. "Si la minería consume energía, pues en el precio de la energía debe estar considerado los impactos ambientales de generar esa energía. Por lo tanto, habiendo pagado su factura de energía, está cumpliendo con su rol de consumidor responsable."
14. "lo que es más fuerte en la minería y es más problemático, es la energía eléctrica, ese tema es bastante crítico...".
15. "varios proyectos...se han estado cancelando por los altos costos de la energía...".
16. "El mercado (debe regular), todos queremos el mercado. Prefiero al mercado...".
17. "Es una actividad multiplicadora...el sanjuanino no tenía qué hacer en San Juan, en cambio, hay desarrollo minero que potencia otras actividades."
18. "siendo que son empresas de alta tecnología (grandes mineras)...deberían estar siendo lo más eficiente energéticamente posible".
19. "es responsabilidad tanto de las empresas como de las autoridades, cómo desarrollar, cómo manejar, cómo hacer el implemento de los proyectos".
20. "en lo que concierne al agua...una buena minería que sea técnicamente realizada, económicamente rentable y que garantice una justa participación del Estado Ecuatoriano, una minería socialmente responsable y una minería ambientalmente manejada con rígidos estándares".
21. "Los incentivos están bien puestos cuando el que toma la decisión, tiene lo más internalizado posible todos los costos (ambientales) que representa el que use energía."

22. "Creo que vale la pena (la minería) porque la actividad, bien desarrollada, puede hacerse con un nivel de impacto ambiental relativamente bajo, o sea, hablo de una minería a escala...más humana...".
23. "En el caso del agua...tiene que ser abordada a través de un control estricto y eso con los recursos disponibles".
24. "debe ser como digo con el menor impacto ambiental...".
25. "Y en Chile...la matriz energética está demasiado carbonizada".
26. "Nuestro discurso indígena siempre ha sido defender los derechos de la naturaleza...Por eso también el presidente ha decidido cambiar la matriz energética, por ejemplo de energía termoeléctrica a energía hidroeléctrica...".
27. "está siendo inviable o incompatible la vida de mucha gente con la minería...".
28. "Ninguna minería es limpia...ocasiona gravísimos problemas, las contaminaciones son increíbles."
29. "Insisto, los pocos recursos que nos quedan (agua y energía) debemos emplearlos en alternativas para un futuro, no en satisfacer las necesidades de un porcentaje tan bajo de la población..."
30. "nosotros no les llamamos 'recursos naturales', sino 'bienes comunes' ".
31. "Somos el país del sol, el país del agua, aquí tenemos posibilidades y tenemos posibilidades de generar otro tipo de energías que en términos de petróleo."
32. This subject is nothing new. Beck observed – in his analysis of the subjectivity of politics since the 1980s, and then accelerated since the collapse of the Berlin Wall – the environmental concern for a threatened world that haunted Europe. Beck said that this concern, which "united conservatives with socialists...is only appearance, programmatic opportunism, perhaps occasionally an authentic reassessment" (Beck et al., 2001: 34–35).

# References

Baigorrotegui, G., Parker, C. and Estenssoro, F. (2014) "Visiones sobre los Tránsitos Socio-Técnicos hacia Patrones de Consumo Sustentable en Agua y Energía en la Minería Sudamericana. Dos Controversias Argentinas", *Sociologías* 16(37): 72–111.
Baker, L. (2012) "Power Shifts In South Africa's Minerals-Energy Complex: From Coal Crunch To Wind Rush?", Paper for Political Economy and the Outlook for Capitalism, Congress, 5–7 July 2012, Paris.
Bebbington, A.J. (ed.) (2012) *Social Conflict, Economic Development and Extractive Industry: Evidence from South America* (London: Routledge).
Beck, U., Giddens, A. and Lash, S. (2001) *Modernización Reflexiva Política, Tradición y Estética en el Orden Social Moderno* (Madrid: Alianza).
Bruzzone, E. (2010) *Las Guerras del Agua. América del Sur en la Mira de las Grandes Potencias* (Buenos Aires: Capital Intelectual).
Bustamante, T. and Rommel, L. (eds) (2010) *El Dorado o la Caja de Pandora. Matices para Pensar la Minería en Ecuador* (Quito: Flacso-Sede Ecuador).
Dijk, T. van (2008) *El Discurso como Estructura y Proceso* (Barcelona: Gedisa).
Dryzek, J.S. (2005) *The Politics of the Earth, Environmental Discourses* (New York: Oxford University Press).

Garay Salamanca, J.L. (2013) *Minería en Colombia. Fundamentos para Superar el Modelo Extractivista* (Bogotá: Contraloría General de la República).
Greimas, A.J. (1966) *Sémantique Structurale* (Paris: Larousse).
GRI (2011) "RG, Sustainability Reporting Guidelines 2000–2006", version 3.0, www.globalreporting.org, date accessed 2 February 2015.
Hajer, M. and Versteeg, W. (2005) "A Decade of Discourse Analysis of Environmental Politics: Achievements, Challenges, Perspectives", *Journal of Environmental Policy and Planning* 7(3): 175–184.
Höijer, B. (2011) "Social Representations Theory, a New Theory for Media Research", *Nordicom Review* 32(2): 3–16.
ICMM (2003) *Marco Conceptual sobre Desarrollo Sustentable del ICMM. Principios del ICMM* (Londres: Consejo Internacional de Minería y Metales).
IFC (2012) *Performance Standards on Environmental and Social Sustainability* (Washington, DC).
IISD and OECD (2009) *BellagioSTAMP: Sustainability Assessment and Measurement Principles* (Winnipeg: International Institute for Sustainable Development and Organisation for Economic Co-operation and Development).
Metals Economics Group (2013) Worldwide Explorations Trends 2013, http://www.metaleconomics.com/sites/default/files/uploads/PDFs/meg_wetbrochure2013.pdf, date accessed 6 January 2015.
MMSD (2002) *Breaking New Ground: Mining, Minerals and Sustainable Development* (Londres: IIED).
Moscovici, S. (1981) "On Social Representation", in J.P. Forgas (ed.), *Social Cognition. Perspectives in Everyday Life* (London: Academic Press).
Mudd, G.M. (2008) "Sustainability Reporting and Water Resources: A Preliminary Assessment of Embodied Water and Sustainable Mining", *Mine Water and the Environment* 27(3): 136–144.
Newbold, J. (2006) "Chile's Environmental Momentum: ISO 14001 and the Large-Scale Mining Industry – Case Studies from the State and Private Sector", *Journal of Cleaner Production* 14(3–4): 248–261.
Norgate, T. and Haque, N (2010) "Energy and Greenhouse Gas Impacts of Mining and Mineral Processing Operations", *Journal of Cleaner Production* 18(3): 266–274.
OCMAL (2012) *Legislación Minera en el Derecho Comparado: los Casos de Chile, Ecuador, Perú, Guatemala y el Salvador* (Observatorio Latinoamericano de Conflictos Mineros, FEDEPAZ).
Parker, C., Baigorrotegui, G., Estenssoro, F., Muñoz, J. and Bull, B. (2012) "Strategic Actors and Sustainable Consumption in Latin America and the Caribbean (LAC). Case Studies in the Mining Sector", ENGOV Project, Analytical Framework Report, www.engov.eu
Pizarro, N. (2012) "Agua Recurso Estratégico en Minería", C. CEO SCM Minera Lumina Copper Chile, Caserones, Santiago. Presentación PP. en EXPOMIN, Santiago, marzo 2012. www.expomin.cl/marketing/pdf/.../presentacion_nelson_pizarro_c.pdf, date accessed 15 December 2014.
PNUMA, SEMARNAT (2006) *El Cambio Climático en América Latina y el Caribe* (México, DF: Secretaría de Medio Ambiente y Recursos Naturales y PNUMA – Programa de las Naciones Unidas para el Medio Ambiente).
Rábago, K.R., Lovins, A.B. and Feiler, T.E. (2001) "Energy and Sustainable Development in the Mining and Minerals Industries", *Mining, Minerals and*

Sustainable Development (MMSD) N° 41 (London: Internacional Institute for Environment and Development (IIED), World Business Council for Sustainable Development, (WBCSD)), http://pubs.iied.org/G00540.html

Sharife, K. and Bond, P. (2012) "South Africa's Minerals Energy Complex", in H. Healy, J. Martinez-Alier, L. Temper, M. Walter and J. Gerber (eds), *Ecological Economics from the Ground Up* (London: Routledge), 163–188.

Sunkel, O. (2011) "América Latina entre el Cuidado y la Dependencia de sus Recursos Naturales", in A. Cubillos and F. Estenssoro (eds), *Energía y Medio Ambiente. Una Ecuación Difícil para América Latina* (Santiago: Idea, Instituto Igualdad).

Superneau, L. (2012) "Problemas en las Alturas: Conflictos por el Agua en las Industrias Mineras de Chile y Perú", *Business News Americas*, Mining Intelligence Series, http://member.bnamericas.com/webstore/es/intelligence-series/high-and-dry-water-issues-in-chile-and-perus-mining-industries, date accessed 3 December 2014.

Svampa, M. and Antonelli, M (eds) (2009) *Minería Transnacional. Narrativas del Desarrollo y Resistencias Sociales* (Buenos Aires: Editorial Biblos).

Teijlingen, Karolien van (2012) *Negotiating Values and Development at the Mining Frontier: Private, Public and Civil Society Interactions over El Mirador Mine in South East Ecuador*, Master Thesis (Amsterdam: University of Amsterdam).

Walter, M. and Martinez-Alier, J. (2010) "How To Be Heard When Nobody Wants to Listen. Community Action Against Mining in Argentina", *Canadian Journal of Development Studies* 30(1–2): 281–303.

WBCSD (2013) "Eco-Efficiency Learning Module", *World Business Council for Sustainable Development*, http://www.wbcsd.org/Pages/EDocument/EDocumentDetails.aspx?ID=13593&NoSearchContextKey=true, date accessed 29 March 2013.

Whitmore, A. (2006) "The Emperor's New Clothes: Sustainable Mining?", *Journal of Cleaner Production* 14(3–4): 309–314.

Wu, M., Mintz, M., Wang, M., Arora, S. and Peng, J.K. (2013) "Water Is Key for Sustainability for Energy Production", *Argonne Nacional Laboratory, US Department of Energy, Chicago*. http://www.transportation.anl.gov/pdfs/AF/620.PDF, date accessed 29 March 2014.

Zúñiga S. and Ana I. (2009) "Consumo de Energía y Emisiones de Gases de Efecto Invernadero de la Minería del Cobre de Chile", Departamento de Estudios y Políticas Públicas, Comisión Chilena del Cobre ponencia en Seminario Vulnerabilidad, Adaptación y Mitigación para el Cambio Climático en Chile, Santiago, 24 de septiembre de 2009.

 Except where otherwise noted, this work is licensed under a Creative Commons Attribution 3.0 Unported License. To view a copy of this license, visit http://creativecommons.org/licenses/by/3.0/

OPEN

# 7
# Overcoming Poverty Through Sustainable Development

Héctor Sejenovich

## Introduction

Latin America is home to alarming poverty rates and the greatest inequality gap in the world (ECLAC, 2010). The concentration of wealth has disadvantaged local populations and their needs, while simultaneously driving the degradation and destruction of natural resources. This process has rendered serious implications for climate change (IPCC, 2007; Sandberg and Sandberg, 2010). While economic constraints are perhaps the most important aspect of poverty, they are only one of many that impede the personal development of the population (Cimadamore and Cattani, 2008; Cimadamore and Sejenovich 2010).

Latin America accounts for only 8% of the world's population, but it is home to a significant portion of the planet's natural resources. This includes 46% of rainforests, 23% of forests and savannahs, 30% of freshwater (from a stable potable source), 30% of permanent crops, 23% of arable land, 17.7% of grassland and 16% of cattle-ranching land (Sejenovich and Panario, 1997). At the same time, as a geographical area it has shown significant industrial, infrastructural and financial development. This means that it has the potential to improve its productive activities in order to meet the needs of the population. However, there is a significant degree of social exclusion and poverty due to systematic disparities in income, possession of resources and power. While the rhythm of productive development has kept momentum, so has the destruction of natural resources and unsustainable use of biodiversity. As a result, the high concentration of monocrops has displaced populations, thereby accelerating the intensification of rural and urban poverty alike. Similarly, urban development has not followed environmental principles and has therefore contributed to pollution, habitat degradation and adverse effects on the health of the population (Alimonda, 2006). In order to reverse this situation, we must analyse the relationship between the concepts of poverty and sustainable development.

Poverty levels in Latin America depend not only on monetary income but also on the natural, infrastructural and social context in which the poor live and which does not allow them to reverse the situation. That would require much more than just increasing their income level; it would require sociocultural and health measures, new homes, participation in environmental governance and so forth. While the World Bank (2014) predicted less poverty in Latin American in the near future, the reports of the Economic Commission for Latin America (CEPAL, 2008; ECLAC, 2010, 2013) tell a contrasting story. Since the era of neoliberalism there has been a relative decline in poverty, but the region has maintained a high absolute level of poverty. In 2010, according to ECLAC, there were 177 million poor people, 70 million of whom were indigent (people whose income did not cover their subsistence).

We can indeed see that the poverty rate reached 48% of the total population in 1990 and 44% in 2002, and only in 2011 did it drop significantly to 27% (ECLAC, 2013). In the case of active, socially inclusive, redistributive states, asset levels decreased dramatically. Although there are positive aspects to this new situation, it should be noted that overall funding for these actions comes from the overexploitation of nature. Furthermore, they are contingency measures that are not based on stable yields and can therefore be reversed. According to the ECLAC report, the changes in poverty rate from 2002 to 2011 are as follows: Ecuador dropped from 49% to 32%; Argentina from 35% to 5.7% (this is debatable due to the evolution of prices within the country); and Venezuela from 40% to 32%. These changes were primarily due to subsidized employment that was reduced with the onset of the crisis. This situation is especially serious for children. According to CEPAL (2008), in 2000 it was estimated that approximately 36% of Latin American children under the age of two years were at high nutritional risk (i.e. their minimum subsistence needs were not being covered). Even in Argentina, which can produce food for a much larger population than it has, the Pan American Health Organization (PAHO, 2010) estimated that in greater Buenos Aires one in five children was malnourished. This situation has improved somewhat in subsequent years, according to ECLAC (2013: 14):

> These measurements are encouraging, with all countries reporting a decrease in the percentage of children under 18 who are deprived

of some basic rights (overall poverty). In the region as a whole (14 countries, comparable over time at national level), overall child poverty fell by over 14 percentage points over the period, from 55.3% of children in around 2000 to 41.2% around 2011.

For the abovementioned reasons it is difficult to resolve the structural poverty that plagues Latin America and which is the result of production patterns, which fail to absorb the quantum and dynamics of the economically active population and seem unable to reverse the concentration of production and income. Rather than identify and analysing these facts, though necessary, we should concentrate on analysing the costs of past damage and how to reverse this situation structurally and quickly.

## The conflict between poverty and sustainable development

In Latin America the relationship between poverty, environmental crisis and short-term accumulation in this age of globalization presents a particular complexity. The environmental issue is a fundamental part of the inequality and dependency issue of the development model (Martinez-Alier et al., 2010). In search of alternatives, theory can play an important role by showing that we have the resources and capabilities to change the situation. This requires better distribution and organization, which can give us a sustainable and socially just development (Salvia, 2011).

Development indicators such as gross domestic product (GDP) only highlight the productive face of development and ignore the degradation and waste that it causes. The social destination of production is geared towards those who can manifest themselves in the market, thereby satisfying needs while also generating poverty and misery for those who do not meet the minimum necessary income. The lifetime of products is reduced to avoid market saturation, leading to a significant generation of waste and pollution. Therefore, development indicators must be reworked.

The development of equity accounts indicates a fruitful path (Sejenovich and Gallo Mendoza, 1997). As a result of this conceptualization, an integrated and sustainable management of natural resources, habitat conservation, and energy and human capacity finally seems to be possible. It is essential, however, to consider all the negative externalities of state development projects. The production process neither begins with the traditional natural resource (because tasks must be performed in order to regenerate the resource in an integral manner) nor ends with the production of goods (the provision of its "return to nature", in terms of waste, must be analysed in order to prevent additional contamination). The eradication of poverty – and development of quality of life – implies a dynamic link between the individual, the community and the environment. The satisfaction of human needs is strongly associated with the continuous and creative participation of social partners and public policy in the transformation of the material, socioenvironmental and cultural conditions of production and of life. Social struggles energize and drive both individual and social development around situations that are changing and where there are projects for the future.

To achieve a reduction in poverty by ensuring equality requires a rethinking and modification of the current relationship between society and nature. This implies, among other things, a change in the technological pattern of production and consumption as well as a more equitable distribution of income (Anguelovski and Martinez Alier, 2014). Although the task seems difficult, there is really no alternative. The random occupation of space, the gigantic and uncontrolled scale of technology, and the destructive forms of short-term and unplanned use of natural assets and the habitat will exceed the limits of the biosphere. The effects of such activity are already manifested in climate change, food crises, structural poverty and social insecurity worldwide.

To the abovementioned income inequality we can add discrimination based on gender, age, language, identity, religion and different capabilities. This gives the dominant sector an excuse to pay lower wages to unskilled workers, thus yielding additional income. Therefore it is crucial that the state implements redistributive policies in order to improve employment rates and quality of life for the overall population. It is important to keep in mind that, depending on the year, 70–80% of the population possesses no more than 20–40% of the GDP in Latin America. It is also essential that the government implements a socioenvironmental system for land use, which should control the application of social as well as environmental legislation.

However, the state does not always apply these policies. As a result, the population suffers unmet needs and environmental degradation. The perception of this situation and the desire for change generate social and environmental movements that demand specific or more profound changes. The sciences provide tools for understanding these complex phenomena and for exploring potential alternatives, thus generating theoretical movements. In response to these social and scientific demands, the state typically begins with the implementation of changes and the definition of some policies. The relative strength of these actors determines the kind of change that is generated, as well as its future stability. In this way, grassroots environmental governance is created (Cimadamore and Cattani, 2008; García Linera, 2008).

In recent decades, changes in environmental governance in Latin America have stimulated the participation of different social actors who strive to implement environmental policies to improve quality of life and environment. Environmental governance can advance this development by ensuring the greatest participation of different social actors with conflicting interests. This is undoubtedly the axis from which different problems can be resolved (Kooiman 2005). It demands that the social sciences – in both theory and practice – deepen their concepts from multiple interactive perspectives in thematic, temporal and spatial respects. This line of action reinforces a more comprehensive view of the relationship between society and nature, and strengthens the intervention methodologies that allow for its implementation. In this way, social science research can collaborate with social movements and the state (and the actors involved in it) to more clearly visualize contradictions and challenges. Although success is not guaranteed, this is a vehicle that environmental movements should use intensively and that the state should permit and promote. It is an integral part of the democratization of the state.

This spectrum of environmental movement actions commits academic researchers to social sensitivity. It allows for their positive interference in conflicts and enriches natural and social sciences by incorporating research and action in the face of environmental challenges. Especially in Latin America, it is essential to rethink development issues in order to make the concept of sustainability more holistic. To do so, we must overcome the economic and social constraints to accessing products and services. The poor do not reach the minimally required threshold, and as a response they form social movements to demand more jobs and income. If the struggles are truly economic, they are integrated into a situation of greater social and cultural marginalization. At the same time, they attempt to address the overaccumulation of capital and power, taking advantage of a number of disparities among the population. This is the case for gender (where women are remunerated with lower wages than men and demand real equality); ethnicity (by claiming equal treatment); language (allowing for a multinational society); age (developing a policy of inclusion and protection for children and the elderly); difference in religion (where freedom of conscience is claimed); nationality (equal treatment); identity (where several concepts that address the history, their relationship with nature, and their relationship with a diversity of worldviews, society and nature are articulated); and different capabilities (respecting apparent limitations and enhancing capabilities).

The same fundamental categories that allow us to analyse the transformation of nature and its relationships will reveal the obstacles that inhibit the sustainable management of natural resources and the improvement of the quality of life of the population. This process demonstrates how the "organic whole" works – production, distribution, exchange and consumption. Instead of meeting the needs of the population, it only increases the income of the wealthiest. This generates negative externalities in both ecological and socioeconomic terms (Sejenovich, 2012).

Therefore, in order to increase the quality of life, we must implement different policies, actions and strategies that allow us to achieve our goal of sustainable development (Redclift, 1987). These objectives must overcome the myths about development that have been generated over several decades in Latin America – they must become countermyths or "fallacies", as demonstrated by Kliksberg (2014).

## The role of social rights

The definition of "poverty" – always normative by nature – is relative. It depends on the epistemic frame in which the minimum conditions and life needs required for their survival, development and reproduction are set. In contrast with the economist perspective of "welfare" – which is rooted in neoclassical (welfare economics) and developmentalist (favouring the gross output and income share) approaches – the concept of "quality of life" recognizes poverty not only as an unfair deprivation of basic human necessities but also as directly related to sustainable development.

Sustainable development is highly sensitive to the relationship among environmental dynamics, socioeconomic processes, sociocultural orientations and the sociopolitical actions of those who are subject to these conditions (Stahler-Sholk, Vanden and Kuecker, 2008; UNDP, 2014). In this regard it should be noted that improving the quality of life implies a dynamic link between the individuals and their environment. The satisfaction of human needs is strongly associated with the continuous and creative participation of social partners in transforming reality. This means a process in which the conflict energizes and drives development, both individual and social, around changing situations.

It is worth noting that – for individuals as well as for the collective – needs and satisfactions are perceived from within a frame of representations. Likewise, values are determined by the place occupied in the social structure, at a particular time and in a given society. We must also consider that the struggle for adequate quality of life refers to relationships with objects and with a potentially peremptory and changing nature. Considering that individuals are driven also by subjective perceptions, a range of meanings emerge as the subject is formed from the material as well as the imaginary aspects of the object (Salvia, 2011).

Therefore, rather than material gains (goods) that we obtain from a better quality of life, we should consider the struggle among the involved social sectors and the ways people can develop their capabilities. The latter could be a greater objective – to strive for the comprehensive development of the population. Therefore, the processes of each ecosystem are analysed through three different sets of satisfaction rights necessary to "sustain" the relationship between development, environment and quality of life.

*Right to livelihood:* This right establishes the need to ensure the items or natural, technological and social processes that allow people to construct a convivial society. This includes a conservative and productive management of one's habitat to maintain overall health.

*Right to protection:* This is the right to personal development by way of a productive, healthy, satisfying and creative job, striking a dynamic balance with the environment. This includes the right to be protected in a legal and material sense against acts of aggression, abuse or discrimination (economic, ethnic, social, cultural, religious or related to gender). It likewise addresses the full integration of women into society and the triumph over the exclusive assignment of reproductive responsibilities to women, thereby ensuring equal access to productive resources and benefits.

*Rights to levels of understanding and to participation:* In this case it is the ability to develop and pursue personal, familial and community projects in search of a sustainable better life within an active and growing system of environmental governance as an efficient instrument (Asotorga, Ame and Valpy, 2004). This law also takes into account autonomous political and community participation in matters of public order, without restriction or constraint. This entails overcoming the condition of a mere consumer, adopting the multiple physical and cultural functions of an individual and his or her interpersonal relationships.

By taking into account the interdisciplinary and multiscale methodology of environmental governance, we present short illustrative cases of four projects in Argentina and Uruguay which meet the criteria of ecosystem representation and progressive levels of social rights.

## Illustrative cases

To analyse how society transforms nature to improve quality of life, one must employ an interpretative framework that can be accessed through interdisciplinary exercises. In this section we briefly describe four case studies to define the needs/rights that can overcome levels of poverty based on different analytical levels and territorial characteristics (rural, urban and extractive).

The transformation of society – the systemic relationship among production, distribution and consumption – is always the result of the rationality imposed by a historical social formation. The latter imprints a particular modality on the process of transformation and then determines the social destination of production (for whom it is produced), the technological form (how it is produced), a certain level of production (where it occurs) and a demand for natural resources and a particular habitat (with what natural and social resources it is produced). It gives priority to cases that obtain short-term gain and generates concrete products that meet certain criteria, negative externalities that are generally not considered (Sejenovich, 2000). All nature is socially mediated and social relations operate in a natural structure with which they constantly interact, in such a way that all sectors form part of the manifestation of the society–nature relationship.

An example of the integrated and sustainable use of natural resources is the ecosystem of the basin of the Angostura river, where the village of Tafí del Valle is located. In the mountainous area of northwestern Argentina, in the province of Tucumán, this area is similar to the Peruvian highlands (Valdivia and Gilles, 2006; Gonzalez et al., 2010). The socioeconomic process comprises an integrated management of a protected territory to overcome the existing grazing area. Environmental production policies attempted to replace the introduced fauna with a native species, such as the camelid.

As for food and nutrition security, it is evident that subsistence rights are being regularly met. Stable employment, however, has not been guaranteed. However, the use of the landscape for activities of responsible tourism is also an important potential source of employment. Regarding the pressure (both tangible and intangible) on natural resources, yearly and seasonal population increase give rise to rural districts. In terms of rights of protection and participation, the guarantees maintained for the original population should be kept in mind. This includes access to means of communication – and participation in general community initiatives – for the original population and the native communities. Through the integrated and sustainable management of resources in a highly fragile area (48,000 Ha), 130 people have come to permanently occupy the land. This therefore guarantees the eradication of poverty, considering the cultural and communal aspects that already exist in the area.

A second example explores the strategy for sustainable development pursued by Gualeguaychu, and the impact of pulp mills in the community of Fray Bentos, Uruguay. The city of Gualeguaychu is developing a number of important industrial activities and implements agricultural and service activities – especially those involving tourism – in its ecosystems. The development of tourism and agriculture cemented the foundation for a more comprehensive and prolonged growth. This same growth has been threatened by the installation of two cellulose complexes on the Uruguayan shore, which have had negative impacts since 2003. In response, the population protested through legal and not-so-legal means, such as the occupation of highways and border bridges. Multiple studies have been conducted to demonstrate and quantify the environmental damage and lost profits that these projects would generate. They are not limited to ecosystems, infrastructure and urban areas; they also have direct effects on the population itself. The environmental costs are calculated according to the reduction of assets, which is measured on the basis of the harm to nature (Sejenovich et al., 2008).

The calculation of environmental damage and profit loss was not developed in hopes of retribution but as a strategy to put pressure on the international capital that supported the contaminating initiative. The population resorted to all legal means, including claims to international agencies and banks. They even went so far as to get the executive, legislative and judicial branches of the Government of Argentina to appeal before the International Court in The Hague. Although they were not entirely successful, they did prevent a company from being established and were responsible for the diffusion of the methodology of the Environmental Citizens' Organization Assemblies throughout the entire Southern Cone. They were sprouts of the environmental governance movement, where all sectors were expressed. This project is an initiative in the country with the highest incidence of identity crisis among the native population. In the struggle against the impacts of pulp mills in Fray Bentos, the population has essentially been fighting for the right to maintain a healthy environment and a stable landscape with little intervention since the time of their ancestors, who wanted to bequeath them the land.

Given the environmental damage, it was essential to develop activities related to the environment in order to value knowledge about local products. Environmental damage and lost profits would exceed the allowable amount of the investment. In each of the ecozones, the potential for integrated and sustainable management can be analysed against the potential loss of biodiversity, costs (of managing natural resources) and benefits (considering the integral and sustainable use of biodiversity), and the lost profits that are its result. If calculated as negative externalities of the project, the total of the land value damages (US$172,037,600), the value of homes (US$320,000,000) and the damage to health (US$68,726,000) should reduce the companies' profits to the extent that the project could be economically unviable. At the very least, it should offer incentives for a more sustainable implementation. Despite the pressure exerted at every level, the huge power of international capital managed to ignore the externalities (and not pay for them), and instead to install the pulp mill with a very high rate of return.

Another instance of the nature–society contradiction can be found in the soyabean industry in Argentina. Concentrated in the Pampas region, the nucleus of the most fertile land in the country, it is another example of an oligopolistic accumulation of natural resources. The soyabean monoculture brings with it high productivity and a series of direct and indirect negative impacts. These include degradation and waste of natural resources, habitat pollution, and impacts on the population in economic, social, cultural and especially health terms. In fact, an increased incidence of cancer has been found and is likely due to the effect of the agrochemical glyphosate (Carrasco, 2012; Dougnac Martínez, 2013; IARC 2015). This danger was recently echoed by the World Health Organisation (WHO).

The monoculture of soyabean (Slutzky, 2011) – currently the primary export crop of Argentina – has replaced cattle grazing and other crops, such as cotton, lentils, milk, meat and rice. As a result, there have been shortages and increases in the Argentinean food basket. This expansion is made possible by the hegemony of financial capital that rents fields and machinery for monoproduction, thus displacing small and medium farmers. This ultimately results in poverty and displacement to urban areas, and furthermore to the expansion of the agricultural frontier into land that is not meant for agricultural use (Bustamante and Maldonado, 2008). Given that soyabean production has displaced traditional foods, cultivation directly affects the Argentinean food structure and the right to subsistence. Much has been written about the alternatives to soyabean production, oriented towards comprehensive and sustainable resource management and poverty alleviation. For example, agroecology can be a highly productive process on a per-hectare basis. This maintains diversity, ensures the full use of land and provides an answer to rural poverty. This strategy will enable widespread environmental governance in rural ecosystems precisely because it involves the grouping together of occupations to be able to research, monitor and manage all of the plants. In turn, this generates significant revenue for the producer group. It also entails potential advantages in terms of the nutrition and diversity of food supply. However, a change of this nature would involve major changes in the line of interest within their respective elites.

The Matanza-Riachuelo Basin (Cuenca Matanza-Riachuelo (CMR)) project serves as our final example. The CMR spans part of the city of Buenos Aires and 15 surrounding municipalities, encompassing an area of 2,338 km$^2$ (the length of the main channel is 70 km). It is estimated that 5.3 million permanent residents and at least 3 million more commuters use CMR for transit. It is considered to be the centre of Argentine industrial development, but 23,523 companies which are active in the region have been registered as potential sources of pollution.

Ever since the colonial period, the contamination of the basin has generated significant actions, such as moving salt production to improve the water quality. It then suffered a second contamination from new industries, which affected the health of the population (ACUMAR, 2007). As a result, the state was sued by the direct victims in a case involving the Supreme Court of Argentina (2006). Known as the "Mendoza Cause", it was based on the implementation of court orders to restore the watershed and to improve the quality of life of the population. To meet this objective an interinstitutional body called the Matanza-Riachuelo Basin Authority (ACUMAR) was established. According to official data (2014), 459 industries have been converted; 289 have been closed; 1,364 have initiated a restructuring process; and 1,436 have presented plans to expand.

The right to livelihood is being met through decontamination to improve the health of the population. This includes the installation of sewage and clean-water pipelines, and the building of new homes and villas to eliminate slums and precarious housing. As of now, 17,771 people have benefited and 85% of the area's population will have clean water, better satisfying their needs and improving their quality of life.

A greater participation process has also been observed in the advisory body of the ACUMAR, which includes universities and NGOs. It is safe to say that environmental governance is becoming more effective by improving environmental conditions, poverty and subsistence issues, but they are changes that need to be accelerated (AySA, 2009).

The construction of major infrastructural projects allows us to visualize the fundamental aspects of watershed management. With the decontamination of the CMR and the solution of the problems of housing and services, there is no doubt that one of the main obstacles associated with poverty will be overcome.

## Key trends and the struggle for sustainable development

From the illustrative cases discussed here, the experiences and expertise of consultants, and other global studies, several general considerations about the relationship between environmental governance and poverty in Latin America have arisen.

Powerful economic groups continue to adopt highly concentrated exploitation and environmental degradation policies that violate not only socioeconomic resources of local livelihoods but also the sociopolitical capacity to design, plan and implement sustainable socioenvironmental development. Environmental policies are often not heeded. At different levels of government the state has failed to define the full potential and limitations that should regulate monocultures, especially that of soyabean cultivation. In fact, tax returns generated by this activity are privileged and no existing laws apply to regulate them. An example of successful environmental regulation is the case of the Forest Act and the Environment Act in Argentina, where environmental planning is legislated. Agricultural confederations such as the Rural Society, the Agrarian Federation, Coninagro, CRA, various trade associations and the Chamber of Commerce have succeeded in imposing their interests on developing soyabean activity. This often overshadows the claims and legal actions of other social actors, including those of the state.

The soyabean expansion case in Argentina can be expanded to the whole region as the ecosystems that have already been transformed (the humid pampas grasslands, subtropical jungle, scrubland, gallery forest) occupy a critical percentage of arable land in Latin America. The organization of the state apparatus is not neutral. The institutional legal structure in Latin America is essentially developmentalist and will therefore favour the amount and dynamics of production, regardless of its impact, if environmental policies are not enforced. Although the impacts of economies of scale have generated cost reductions, they are not translated into prices. Rather than improving the welfare of the general population, the oligopolistic market conditions have allowed large companies to increase profits. As a result, the "progressive spill" did not occur. In general, the rate of accumulation appears to impose oligopolistic structural rules of impoverishment, inequality and social exclusion. This is in addition to the processes of ecological impoverishment that result from ecosystem homogenization and environmental degradation.

These negative outcomes drive disputes in all illustrative cases. In Matanza-Riachuelo Basin, the creation of an intergovernmental body – ACUMAR, which has as its first priority to preserve and restore the Matanza-Riachuelo watershed with a range of public and non-public organisms – shows positive developments. In urban ecosystems the situation is not very different, but there are sociopolitical conditions that make the control, regulation and guidance towards socioenvironmentally sustainable projects more feasible.

However, overall urbanization in Latin America exceeds the guidelines of environmental planning, and this is reflected in almost all countries. The operation of watersheds and respect for their characteristics, under the onslaught of settlement infrastructure, remains a deficit that is frequently raised by environmental movements. One of the most serious problems is the political-economic action carried out by national governments in such projects – along with public action to develop megainfrastructural projects – to resolve problems that have been generated by the improper management of watersheds and ecosystems.

The developmental-productivist paradigm remains hegemonic when it comes to great solutions that affect most of the socioeconomic and sociopolitical regulatory institutions of social reproduction at local, regional and national levels. Many of the presidents' speeches at the Community of Latin American and Caribbean States (CELAC) Summit, in February of 2013 in Santiago de Chile, showed excessive optimism in regard to development actions without exploring certain limits that they really must consider. In any case, they should outline the progress that has been made in mobilizing public awareness and institutional improvements. This is the result of forces within and outside the governments, which fight for a solidary management of nature and among social sectors. To advance, it is important to overcome the temptation of criminalizing protest and for movements to deepen their creativity in action. If these aspects are satisfied, a better quality of life for disadvantaged sectors is possible.

## Conclusion

The study of different socioeconomic environmental scenarios, under a rights-focused approach, provides purposeful lessons for envisioning the relationship between environmental governance and poverty in Latin America. The organizational forms of the state and its operations should be reoriented to better achieve sustainable development (Kliksberg, 2014).

We observe fundamental contradictions between society and nature. The most general is between economic cycles (short term) and ecological cycles (longer term), presenting incompatibility between temporal horizons. Now is the time to respect the timeframes of regenerative mechanisms. Other contradictions arise from the heterogeneity of ecosystems versus the trend to homogenize exploitation for maximum profit through economies of scale. Following a short-term economic objective, only natural resources with competitive global (and sometimes national) advantage are being used; a comprehensive and appropriate use of resources could prevent the existing diversity from being wasted. This practice is widespread in Latin America, where the generation of short-term income generates ecological, social and cultural impoverishment in the long term.

Furthermore, the administrative structures of the state are predominantly defined by a sectorial vision: production and short-term efficiency are privileged, the importance of interaction is minimized, and there is generally little room granted to the participation and articulation of science, technology and the quality of life of the population. Integrated and sustainable management of nature in environmental governance eventually overcomes the apparent contradiction between protecting the environment and stimulating production.

It is clear that, taking comprehensive production into account, there is a vast increase in production, income, employment, tax base and financial jurisdiction. At the same time, the environment is taken into account in an active and integrated manner, thus preventing the loss of biodiversity.

The ability to generate these productive strategies requires, without a doubt, a training programme to understand the techniques of integrated resource management. Actually, all countries with complex ecosystems – and specially those whose forested areas are predominant – apply this principle. We must keep in mind that Latin America possesses nearly half of the world's tropical forests. It increasingly requires new planning processes that incorporate the population from the outset, along with the development of scientific interdisciplinary analysis. More and better participation, and a substantial improvement in training, are a priority for all governments. This aspect was particularly stressed at Rio+20.

Environmental governance of the land, patrimonial accounts, environmental assessment of investment projects, evaluation of environmental impacts, strategic environmental assessments and so forth are emerging as important alternatives. According to ECLAC (2010:140), "Territorial heterogeneity in Latin America calls for selective and targeted strategies. Local development, understood as a bottom-up process, mobilizes endogenous potential to build territories that are better able to create and drive their own capacities." The objectives of the National Environmental Governance Project in each country must reverse the process of poverty generation and, in turn, give more momentum to tasks already under way to directly improve the situation. Habitat improvement and policies to combat environmental degradation are systematically integrated with the possibility for a better life. In addition, the use of unusual environmental policies in Latin American countries – such as tax policies, credit, tariff or integration – all signify that there is a long way to go.

While these ideas are technologically plausible, and are also key for the sustainability of the planet, it is worth reiterating critical doubts that arise from both historical experience and theory. They question the ability of the current model of accumulation and the political regime of domination to advance socioeconomic and environmental sustainable development, without significant changes. The historical scenario seems to prolong an insurmountable contradiction between the interests to produce, distribute and consume, and the need to ensure social and environmental human life. Therefore, a greater organization and activity of environmental social movements emerges as a possible alternative.

An organizational form for sustainable development within environmental governance involves a holistic view, a direct relationship between research and action. It is a combination of the short, medium and long term, and of a generally high level of participation among civil society and social movements. It proposes implementing the necessary changes and taking actions that can lead to more successful forms of environmental governance and a better quality of life. Economic understanding must be open to all necessary actors, which requires reformulating the conditions for recovery and reproduction of capital with ecological, economic, social, technological and political implications. Only then do the desired reduction of poverty and reconciliation with nature truly begin.

# References

ACUMAR (2007) "Plan Integral Cuenca Matanza-Riachuelo (PISA)", in http://www.acumar.gov.ar/pdf/PLAN_INTEGRAL_DE_SANEAMIENTO_AMBIENTAL_DE_LA_CUENCA_MATANZA_RIACHUELO_MARZO_2010.pdf, date accessed 10 January 2015.
Alimonda, H. (ed.) (2006) *Los Tormentos de la Materia: Aportes para una Ecología Política Latinoamericana* (Buenos Aires: CLACSO).
Anguelovski, I. and Martinez Alier, J. (2014) "The Environmentalism of the Poor: Territory and Place in Disconnected Global Struggles", *Ecological Economics* 102: 167–1.
Asotorga, P., Ame, R. and Valpy, F. (2004) *The Standard of Living in Latin America during the Twentieth Century*, University of Oxford Discussion Papers in Economic and Social History 54.
AySA (2009) *Estudio Socioeconómico y Ambiental en la Cuenca Matanza Riachuelo. Vol. I, II, III y IV* (Buenos Aires: Publicaciones AySA).
Bustamante, M. and Maldonado, G. (2008) "Actores Sociales en el Agro Pampeano Argentino Hoy. Algunos Aportes para su Tipificación", *Cuadernos Geográficos* 44 (1): 171–191.
Carrasco, A. (2012) *Los Efectos Teratogénicos y Genotóxicos del Glifosato*. Jefe de Laboratorio de Embriología Molecular de la Facultad de Medicina de Buenos Aires (Argentina: CONICET).
CEPAL (2008) *La Transformación Productiva 20 Años Después. Viejos Problemas, Nuevas Oportunidades* (Santiago de Chile: CEPAL/Naciones Unidas).
Cimadamore, A. and Cattani, A.D. (eds) (2008) *Producción de Pobreza y Desigualdad en América Latina* (Bogotá: Siglo del Hombre Editores).
Cimadamore, A. and Sejenovich, H. (2010) "Cambio Climático y Pobreza", in *Voces en el Fénix* 1(2) (Buenos Aires: UBA/Plan Fénix).
Corte Suprema de Justicia Argentina (2006) "Mendoza, Beatriz Silvia y Otros c/ Estado Nacional y Otros s/ Daños y Perjuicios (Daños Derivados de la Contaminación Ambiental del Río Matanza – Riachuelo)", Buenos Aires. M 1569. XL. Originario.
Dougnac Martínez, G. (ed.) (2013) *De Especie Exótica a Monocultivo. Estudios sobre la Expansión de la Soja en Argentina* (Buenos Aires: Imago Mundi).
ECLAC (2010) *Time for Equality: Closing Gaps, Opening Trails* (Santiago de Chile: ECLAC/United Nations).
ECLAC (2013) *Social Panorama of Latin America* (Santiago de Chile: ECLAC/United Nations).
García Linera, Á. (2008) *La Potencia Plebeya. Acción Colectiva e Identidades Indígenas, Obreras y Populares en Bolivia* (Buenos Aires: CLACSO/Prometeo).
González, J., Sejenovich, H. et al. (2010) *El Uso Integral y Sustentable de los Recursos Naturales a partir de Estudios de Proyectos Productivos Aplicados a la Zona de Tafí del Valle, Provincia de Tucumán, Argentina* (Ministerio de Recursos Hídricos y Medio Ambiente de la Provincia de Tucumán/DINAPREI) in http://www.socioambiente.com.ar/index1.htm, date accessed 13 December 2014.
Intergovernmental Panel on Climate Change (2007) *Climate Change: Impacts, Adaptation and Vulnerability: Working Group II Contribution to the Fourth Assessment. Report of the IPCC* (Cambridge: Cambridge University Press).

International Agency for Research on Cancer (IARC) (2015) (Guyton, Kathryn Z. et al., "Carcinogenicity of tetrachlorvinphos, parathion, malathion, diazinon, and glyphosate", *The Lancet Oncology*). 16(5): 490–491

Kliksberg, B. (2014) *Otra Economía es Posible. Desde el Consenso de Washington a la Visión de una Nueva Economía* (Buenos Aires: La Página).

Kooiman, J. (2005) "Gobernar en Gobernanza", in A. Cerrillo i Martínez (ed.), *La Gobernanza Hoy: 10 Textos de Referencia* (Madrid: Instituto Nacional de Administración Pública).

Redclift, M. (1987) *Sustainable Development. Exploring the Contradictions.* (London: Methuen & Co. Ltd.).

Salvia, A. (2011) "La Medición del Progreso Humano en la Dimensión Social como una Medida de Cumplimiento de Derechos", in M. Rojas (ed.), *La Medición del Progreso y del Bienestar. Propuestas para América Latina* (México: Foro Consultivo Científico y Tecnológico, A.C.).

Sandberg, A. and Sandberg, T. (2010) "Introduction", in A. Sandberg and T. Sandberg (eds), *Climate Change. Who Is Carrying the Burden?* (Toronto: Canadian Centre for Policy Alternatives), 11–22.

Sejenovich, H. (2000) "Pobreza y Ambiente: Hacia una Nueva Relación Sociedad-Naturaleza", Ponencia en el *Seminario sobre Desarrollo, Equidad y Cambio Climático*, IPCC, La Habana, Cuba.

Sejenovich, H. (2012) "La Calidad de Vida, la Cuestión Ambiental y sus Interrelaciones", in H.I. Farah and L.Vasapollo (eds), *Vivir Bien: ¿Paradigma no Capitalista?* (Bolivia: CIDES/UMSA).

Sejenovich, H. and Gallo Mendoza, G. (1997) *Manual de Cuentas Patrimoniales* (México: PNUMA/Fundación Bariloche).

Sejenovich, H. and Panario, D. (1997) *Hacia Otro Desarrollo: Una Perspectiva Ambiental* (Buenos Aires: Nordan).

Slutzky, D. (2011) *Los Cambios Recientes en la Distribución y Tenencia de la Tierra en el País con Especial Referencia a la Región Pampeana: Nuevos y Viejos Actores Sociales* (Buenos Aires: Centro de Estudios Urbanos y Regionales, CEUR).

Stahler-Sholk, R., Vanden, H. and Kuecker, D. (eds) (2008) *Latin American Social Movements in the Twenty-First Century: Resistance, Power, and Democracy* (Lanham MD: Rowman and Lettlefield).

UNDP (2014) "Poverty Reduction Global Programme2014/2017", in http://www.undp.org/content/undp/en/home/librarypage/poverty-reduction/EmpoweringLivesBuildingResilience.html, date accessed 13 November 2014.

Valdivia, C. and Gilles, J. (2006) *Adapting to Change in the Andes: Practices and Strategies to Address Climate and Market Risks in Vulnerable Agro-Ecosystems* (SANREM CUSP Andes Project).

World Bank (2014) "Regional Indicators", in http://povertydata.worldbank.org/poverty/region/LAC, date accessed 10 December 2014.

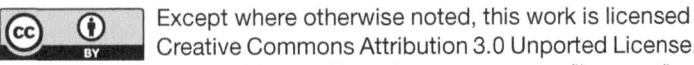 Except where otherwise noted, this work is licensed under a Creative Commons Attribution 3.0 Unported License. To view a copy of this license, visit http://creativecommons.org/licenses/by/3.0/

# Part III

# New Projects of Environmental Governance

OPEN

# 8
# Forest Governance in Latin America: Strategies for Implementing REDD

*Mariel Aguilar-Støen, Fabiano Toni and Cecilie Hirsch*

## Introduction

Global interest in and attention to forests have grown as concerns about global warming and climate change have taken a heightened position in international policy debates. Forests have been repositioned in international arenas as repositories of global value for their contribution to carbon sequestration and climate mitigation (Fairhead and Leach, 2003; Peet, Robbins and Watts, 2011). In this context, Latin American forests are seen as globally important in fighting climate change.

Carbon emissions in developing countries, particularly in Latin America, are related mostly to land-use and land-cover change. In Latin America, energy accounts for only 28% of regional emissions, whereas land use, land-use change and forestry (LULUCF) accounts for 67% (Barcena et al., 2010). Forests cover about 11.1 million km$^2$ and savannahs 3.3 million km$^2$, comprising several different types of vegetation. The region as a whole has the world's greatest forest loss (Pacheco et al., 2010). Most of the forest conversion in Latin America occurs in the Amazon basin. Some countries are already being pressed to reduce emissions related to land-cover change, particularly deforestation. Political pressure comes from the international arena in many forms and is exerted by several actors: sovereign states, international organizations, media, civil society networks and others.

Several Latin American governments have turned to climate policies as an opportunity to improve environmental governance. Current discussions focus on a set of policies known as REDD in developing countries plus carbon-sequestering forest activities. REDD was originally designed as a payment for environmental services – that is, a voluntary transaction where a well-defined service (or a land-use system likely to secure that service) is being "bought by a buyer from a provider, if and only if the provider secures the service provision" (Wunder, 2005). REDD is based on the idea that it is possible to reduce deforestation by offering economic compensation to forest users for not changing the use of forestlands. It is seen as a win–win approach that would potentially address the trade-offs between forest conservation and economic development. Some analysts claim that REDD projects have the potential to generate enough money to end deforestation in tropical countries (Nepstad et al., 2009).

Although originally presented as an "apolitical" technological fix (cf. Li, 2007), REDD has encountered much criticism, and early proposals faced fierce political resistance. The neoliberal idea of the commodification of nature seemed repellent to individuals and even to countries, which fear that developed countries would use their economic power to increase or leave unaddressed their carbon emissions at the expense of developing countries. There were also fears that REDD would benefit actors who have historically been responsible for deforestation, such as ranchers and large-scale farmers, while excluding the less privileged forest-dwellers, who cannot bear the transaction costs of carbon markets and do not even have the title to their lands (Boyd, Gutierrez and Chang, 2007; Blom, Sunderland and Murdiyarso, 2010).

REDD proved to be much more complex than a simple carbon-market arrangement. Since it is a project "in the making", it necessarily leaves room for bargaining and negotiations as to how forest and climate policies will take shape in specific contexts. As a result, REDD quickly moved from strictly carbon storage to having multiple objectives, including biodiversity conservation and the enhancement of local livelihoods (Angelsen and McNeill, 2012). This even more complex mechanism is not yet settled. There are important struggles at international, national and local levels to define how REDD should be implemented.

REDD can be seen as a multilevel project of environmental governance. By environmental governance we mean "a set of mechanisms, formal and informal institutions and practices by way of which social order is produced through controlling that which is related to the environment and natural resources" (Bull and Aguilar-Støen, 2015: 5). Some decisions regarding REDD are taken at the global level, other decisions are taken at the national level and finally actions, projects and initiatives are implemented at the local level. This complexity might result in the hybridization of REDD, and, as the idea is appropriated by different actors, such hybridization might also result in subtle or open power struggles among actors at the different levels.

REDD emerged as a global initiative from the climate negotiations, but it is going to be implemented in countries with very different approaches to combating deforestation, technical capacity, institutional and political settings, levels of decentralization of forest governance, budgets and so forth. Therefore it is possible to expect REDD to unfold in quite different ways across the region. To understand and analyse the diversity in which REDD is evolving in Latin America, in this chapter our analytical focus will move across different scales and will make use of some paradigmatic examples, with special emphasis on the countries representing such cases. Our analysis will show that despite their initial opposition, some groups of actors support REDD and are taking advantage of the new opportunities that the scheme offers. REDD initiatives, for example, have become an economic opportunity for both state and national governments as well as for international and regional environmental NGOs.

This chapter is organized as follows. After this introduction, we present our main analytical argument. The following section examines the phased approach to implement REDD in Latin America. In the third section, we present what we have identified as three general strategies to implement and shape REDD across the region. In the next section, we discuss some examples of how pilot projects are taking off in the region. Finally, we present our conclusions.

## Hybrid environmental governance and REDD

Forests in Latin America are territories where several conflictive interests meet. However, there is no consensus on the conceptualization of the causes and consequences of deforestation. Diverse conceptualizations of deforestation are closely related to claims over forest management and over resources (Fairhead and Leach, 2003). Forests are socially, culturally, ecologically, economically and symbolically valuable to different actors, including indigenous peoples, local users, governments, corporations, illegal cartels, NGOs, nations and the globe, albeit in different ways and for different reasons (Fairhead and Leach, 2003). All these actors have different potentials to exert power and access arenas to influence REDD-related policy-making.

The very notion of "environmental governance" implies that there is some sort of hybridity in terms of the actors, and in the mechanisms and practices it involves. This means that both public and private actors participate on various scales, in producing models and frames for governance. By focusing on REDD we pay attention to emergent governance arrangements that include state actors, subnational governments, multilateral institutions, scientists, NGOs and business (Karkkainen, 2004).

The conceptualization of REDD, its formulation, negotiation and implementation involve a range of actors because the necessary resources for such tasks are not controlled by a single entity. As our analysis will suggest, these resources function as sources of legitimacy for the participation of different actors in REDD. By legitimacy, we mean who is making "the rules of the game" in REDD preparations and negotiations. We see legitimacy as a source of power to create and support certain policies and practices, while simultaneously hindering others. Legitimacy rests, among other things, on the shared acceptance of rules by different groups of actors with shared interests on the issue to be governed (Bernstein, 2004).

REDD, however, is still a project "in the making". Because of that, this chapter only aims to examine two processes: (1) how different countries engage with REDD; and (2) how different actors within these countries get involved in a range of activities seen as necessary for the future implementation of REDD on the ground. In other words, our analysis will not focus on the outcomes of the REDD initiative because such outcomes are still uncertain.

Our proposition in this chapter is that REDD as a concept has been "black-boxed" (Latour, 1987; Forsyth, 2003; Goldman, Nadasdy and Turner, 2011). By that we mean that those engaged in REDD do not consider it necessary to further discuss or question what REDD means. This does not imply, however, that there are no other actors – who perhaps are not directly involved in REDD negotiations – who actually question and challenge the initiative. REDD policy-making reflects how different interests are negotiated between different actors on various geographical scales. In this chapter we will argue that a "distortion" of REDD – from a simple market mechanism to a complex multistakeholder, contested political processes – is one of the ways that the idea gets wide support from a range of actors and makes the hybridization we refer to above possible. REDD as a concept is broad and vague enough to permit different interpretations that would fit the goals of different actors (Angelsen and McNeill, 2012). This has allowed countries in Latin America to pursue different paths regarding the emphasis given to how to finance REDD (fund based or carbon markets) and what issues should be addressed before REDD actions are implemented.

To support our proposition we discuss three different strategies used by Latin American countries to engage or resist the REDD initiative. Also, the "distortion" works at more local levels by allowing different actors to get involved in planning activities. We will also discuss planning activities in the Amazon region to support our proposition and will show how there are some key resources that galvanize the participation of certain actors in REDD preparations. By key resources, we mean resources that can be "traded" to gain legitimacy to participate in REDD processes at local levels. As we will show below, access to networks and knowledge production are among such key resources.

## REDD in Latin America and the phased approach

In 2010, during the conference of the parties of the United Nations Framework Convention on Climate Change (UNFCC), governments agreed to adopt a phased approach for REDD. The idea of a phased approach came from a report (Angelsen et al. 2009) prepared by the Meridian Institute for the Government of Norway. The idea put forward by the report by Angelsen et al. (2009) was adopted by the UNFCC Cancun agreement[1] (Agrawal, Nepstad and Chhatre, 2011). The Cancun agreement stipulates that countries participating in REDD should implement activities by phases. These phases are (1) development of national REDD strategy plans and capacity-building; (2) implementation of national plan and demonstration activities; and (3) results-based actions with full measuring, reporting and verification. So far, most Latin American countries involved in REDD are in Phase 1. Guyana is in Phase 1 but has already received funding from Norway that would correspond to phases 2 and 3; Brazil is in Phase 2, entering Phase 3 (Figure 8.1).

There are many mechanisms for financing Phase 1, including public funds from the countries implementing REDD or from donors: the Forest Investment Programme supported by the Climate Investment (Multilateral Investment Banks), the UN-REDD programme, and the Forest Carbon Partnership Facility (FCPF) of the World Bank. The latter two are the main sources of funding, and some countries such as Bolivia,[2] Peru and Ecuador have applied to both. On the other hand, Brazil established its own Amazon Fund in 2008, through which reduced deforestation is going to be financed in the country. Guyana established the Guyana REDD investment fund (GRIF) in 2010 as part of a cooperation agreement with Norway in the framework of the Low Carbon Development Strategy (LCDS) of Guyana.[3] The LCDS of Guyana was prepared by the consultancy firm McKinsey,

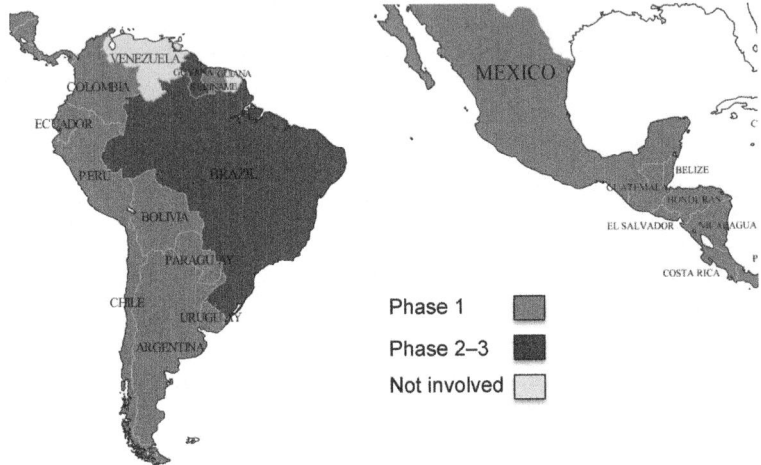

*Figure 8.1* Latin American countries in relation to their participation in REDD and the phased approach

and Guyana's president embarked upon an international campaign to attract funding for the initiative. Venezuela and French Guyana do not participate in any REDD initiatives under the United Nations or the World Bank.

In 2013, Norway was the single major financial contributor to the UN-REDD Programme, FCPF, the Brazilian Amazon Fund and the GRIF. Norway contributes 82% of the total budget of the UN-REDD Programme, 44% of the total budget of FCPF, 87% of the total budget of the Amazon Fund, and 100% of the GRIF.[4] The country is one of the major players in defining REDD at the global level and has some influence on the way in which REDD is advancing at national levels.

The incorporation of the phased approach launched by the Meridian report in the UNFCC's Cancun agreement contributes to stressing a particular way of prioritizing the activities necessary for the implementation of REDD. This particular approach is being reproduced in national contexts because its proponents believe in the technical superiority of the approach and because it promotes comparability and compatibility between countries, but not necessarily a solution to the problem of deforestation (Fairhead and Leach, 2003). As it might seem obvious to most, the driving forces behind deforestation vary enormously, as do the political and economic settings in each country, the interests and alliances among different actors, and the roles played by the state and non-state actors. The challenges associated with deforestation in the region are as political as technical, but the phased approach de-emphasizes other dimensions of the problem.

In the phased approach, institutional arrangements and technical capacity to measure deforestation are emphasized. REDD will rely on the specific target of measuring reduced emissions from deforestation. In Latin America, in addition to Brazil, only Mexico and Costa Rica have comparable technical capacity in place to measure forest-cover change. Consequently, a strong emphasis in readiness preparations in all other countries in Latin America is currently placed on strengthening technical infrastructure to monitor forest change.[5] A strong emphasis on measuring and monitoring forest cover has a depoliticizing effect on the understanding of deforestation's causes, consequences and risks to impose control mechanisms that might harm local livelihoods (Scott, 1998). If the causes and consequences of deforestation are not properly understood in each country, it might be that those who live closer to forested areas bear the blame for deforestation and the responsibility for avoiding it.

## The three REDD strategies in Latin America

Several Latin American countries (Brazil, Argentina, Paraguay, Peru and Venezuela) have been sceptical about offsets from carbon emissions trading, as declared by the countries at the BASIC Ministerial Meeting on Climate Change in Foz do Iguaçu, Brazil, in September 2013.[6] The ministers called for environmental integrity and stressed that "results-based payments shall not be used to offset mitigation commitments by Annex I countries [industrialized countries]". The ALBA[7] countries have held the same position.

Although the ideas that led to the intellectual elaboration of REDD in part emerged in Brazil (Santilli et al., 2005), the country opposed any attempts to include forests and deforestation under the scope of the Kyoto Clean Development Mechanisms. Without Brazil, any such mechanism would be doomed to fail, considering the magnitude of the country's tropical forests and its rate of deforestation. It is argued that, because of the long history of early initiatives to conserve forests in the region, Latin American countries are in the lead of early efforts to implement REDD (Hall, 2011).

Governments in Latin America have taken different approaches to implement and shape REDD efforts. We have identified three strategies. The first, which we will refer to as the "assertive strategy", is characterized by efforts made by the central government to frame REDD within an existing or emerging forest-climate policy framework. Brazil, Mexico and Guyana, for example, are employing this strategy. Countries following guidelines or directions decided at the global level and efforts to accommodate such guidelines in the national context characterize the second strategy, which we will call the "accommodating strategy". Costa Rica, Guatemala, Argentina, Chile, Honduras, Panama, Paraguay, Uruguay, Peru, Colombia, Ecuador and Suriname are pursuing this strategy. Open opposition to certain aspects of REDD or a lack of initiative to implement REDD characterize the third and last strategy, which we will call the "resisting strategy". The countries following this path are Nicaragua, Bolivia, Venezuela and French Guyana. In the paragraphs below we will use one or two countries to illustrate each of the strategies. First we present the assertive strategy because this represents one pole in the continuum between taking the lead and resisting a project. Next we present the accommodating strategy, which represents the situation of most Latin American countries and thus represents the middle ground of the continuum. We finish with the resisting strategy at the other end of the continuum.

### The assertive strategy: Brazil

While most other countries in Latin America were still working to put human capacity in place to deal with REDD within their ministries of the environment, Brazil launched the Amazon Fund in 2008. This, however, represents the way in which the position of Brazil evolved from resistance to leadership.

For many years the Brazilian Government was a fierce opponent of any attempts to include forest- and land-use change in the international negotiations to reduce carbon emissions. This position was justified on the grounds that developed and developing countries share common but differentiated responsibilities concerning global warming. Many opponents of such proposals were afraid that carbon credits would allow rich countries to keep pouring carbon into the atmosphere at the expense of developing countries. Furthermore, Brazil was concerned with any potential threats to its sovereignty and control of its forests resources, particularly in the Amazon. Any clause addressing deforestation could be interpreted as an obstacle to developing the region as the state saw fit.

Even though President Lula himself supported this realist view, as he made clear in 2007 during the opening of the UN General Assembly (Hall, 2008), change in the Brazilian position came from within the government. When President Lula took office in 2003, he appointed Marina Silva, a former senator and rubber tapper leader, as minister of the environment. She promoted some institutional changes that ultimately led to a turnaround in the Brazilian official position. The first change came by opening up new opportunities for participation of civil society organizations in policy-making. Knowledge networks formed by activists and scientists developed stronger ties with government officials and became more influential. A related second change was an administrative reform in the Ministry of the Environment. In 2007, Silva created the Secretariat of Climate Change and Environmental Quality, whose top officials were committed to the creation of carbon compensation mechanisms.

Activists and scientists had been discussing proposals to create compensation mechanisms to pay for avoided deforestation since the early 2000s (Santilli et al., 2005). By the time their peers ascended to the new secretariat, the government's efforts to control deforestation were already paying off. Therefore the idea of being compensated by reducing deforestation made much more sense to government officials.

Another crucial component of the policy network supporting compensation was Amazonian state governments. As proposals evolved towards compensating carbon stocks, governors saw an opportunity to channel resources into their states, particularly where there are vast areas under protection. Protected Areas (PAs) have traditionally been considered a burden for state and municipal governments. The benefits of conservation are global, but the perceived costs are local, particularly due to land-use restrictions. The economic losses imposed on states could therefore be, at least partially, offset by this new source of revenue. In 2009, a few months before the UNFCCC COP 15, the governors of all nine Amazonia states met and wrote a letter to the president, pointing out that Brazil was lagging behind other developing countries in the carbon market. They argued that if Brazil was to receive more funds from carbon credits and to reduce its own carbon emissions, REDD mechanisms had to be included in the international carbon market under the UNFCCC (Toni, 2011).

The Amazon Fund was launched as a means to obtain funding from donors to finance the Plan of Action for Protection and Control of Deforestation in the Legal Amazon. The Amazon Fund was created within the Brazilian National Bank of Social and Economic Development (BNDES). The mobilization of civil society, particularly international NGOs[8] and other environmentalists since the 1990s, and the engagement of politicians at the state and federal levels have been important for the advancement of REDD-like ideas based on assumptions of the efficiency of economic payments for environmental services to curb deforestation (Hall, 2011). These ideas are also supported by several governors in the

Brazilian Amazon and coincide with those of the president and the minister of the environment, contributing to create conditions necessary for the Brazilian involvement in REDD. For the Amazon Fund, the government of Brazil pledged to allocate US$500 million, but it is estimated that an additional investment of US$1 billion per year would be required to fully implement the plan (Meyer, 2010).

Brazil has the technical capacity to monitor changes in forest cover through remote-sensing technology and to ensure transparency to deal with the fund through institutional structures and mechanisms. By 2008, Brazil had already put in place some of the conditions to be enabled by Phase 1. This in part explains Norway's support of the Amazon Fund, which placed Brazil in phases 2 and 3. The Norwegian support of the Fund is contingent on demonstrating avoided deforestation against a historical baseline (results-based payments). Norway's involvement is also based on ideas of economic rationality, altruism and self-interest[9] as a humanitarian/environmental protection actor.

The establishment of the Brazilian Amazon Fund can be explained by the combined effect of the activities and initiatives of NGOs, state governors in the Amazon region, and politicians in key positions (the president and the minister of the environment). Norwegian support through Norway's International Forest and Climate Initiative (NIFCI) gave the scheme the final thrust to get the fund started. The Amazon Fund is important for advancing the Brazilian approach to REDD. This approach is well established in existing Brazilian institutions and is in accord with the country's views and priorities.

Brazil's REDD strategy has been characterized by a strong involvement of the central government, but NGOs and lower levels of the public administration have also played a role. The advanced technical capacity of Brazil in terms of remote-sensing and the establishment of a historical baseline of forest cover place the country in a privileged position in regard to the phased approach promoted at the international level. The alliance of Brazil and Norway for financing the Amazon Fund has given Brazil's strategy a very advantageous starting point.

Brazil's approach to financing REDD efforts has been based on the idea of a centralized fund that would allow the country to avoid the voluntary carbon market for financing reduced deforestation. However, the growing involvements of other networks, particularly those in which governors of the Amazon states are involved, have pushed the country towards additional mechanisms for financing avoided deforestation, particularly through their partnership with the governors of California and Illinois.[10]

In the following subsection we present the accommodating strategy, which is used by most countries in the Latin American region as mentioned above. To illustrate we use the cases of Colombia and Costa Rica.

### The accommodating strategy: Colombia and Costa Rica

REDD preparation activities in Costa Rica and Colombia have advanced quite differently from those in Brazil. Colombia has the most decentralized public administration in Latin America. Over 40% of total government spending is allocated by subnational governments against an average of 15% in the rest of Latin America (Alesina, Carrasquilla and Echavarria, 2005). The administration of forest and other natural resources is also decentralized (Alvarez, 2003). Costa Rica, on the other hand, represents a case of highly centralized forest governance. We will first describe Colombia and subsequently Costa Rica.

The lead for the REDD process in Colombia has been taken by the private sector, particularly business-friendly international NGOs (BINGOs), and not by the central government. Colombia has one of the most decentralized environmental administrations in Latin America. Local environmental authorities (Regional Autonomous Corporations (CARs)) are in charge of the management and administration of all natural resources and environmental issues in the area of their jurisdiction. Although CARs receive a portion of their budget from the central government, they also generate income through tax revenues that come from projects implemented in their jurisdiction. In this way CARs hold significant power to decide the direction of both environmental conservation and development projects.

The Colombian Government highlights the involvement of the private sector in the financing of environmental conservation efforts in various white papers (e.g. the National Strategic Plan for Green Markets produced by the Ministry of the Environment and the National Development Plan 2005–2010). A general perception from the Colombian Government is that private investments with little state regulation in remote forest regions are more economically efficient because they lower their intervention costs and could also offer better-adapted development options. A quote from an official of the Ministry of the Environment illustrates the position:

> The market in a way takes care of redistributing the resources at local levels. It is a lot simpler... it lower our costs... so, if the state does not receive the [REDD] money it does not need to invest in the regions

where they are receiving the money...well that is good...the government does not need to invest in those regions; in a way they take care of themselves.

All BINGOs operating in Colombia and some local NGOs expressed the same view during our interviews; they too want to increasingly involve private funds in current forestry and development mechanisms.

Within this context, REDD preparations have been largely led by NGOs. The BINGOs working in the country (WWF, Conservation International (CI), The Nature Conservancy (TNC)),[11] in collaboration with the United States Agency for International Development (USAID) and one local NGO/consulting firm (Corporación Ecoversa), created the Colombia REDD Table in 2008 (Mesa REDD-Colombia). Other private organizations (the Fund for Environmental Action and Children (FAAN), the Natural Patrimony Fund and the Nature Foundation) as well as the Ministry of the Environment and the Institute for Environmental and Meteorological Studies (IDEAM) joined the Colombia REDD table a year after its creation. Participation in the REDD table was not open to all those who were interested. Instead, the control of certain resources (i.e. knowledge, networks and technologies) legitimate and facilitate their participation. Civil society organizations, universities and others who are not considered "REDD experts" by the terms established by the REDD table are excluded.

The REDD table in Colombia has positioned itself as a legitimate network to be consulted or to provide inputs on various REDD-related issues. For instance, the funds provided by the FCPF for REDD preparation activities are administered on behalf of the government by an NGO (FAAN). The REDD table is the most active and important network that disseminates information concerning REDD in Colombia and that reports to the World Bank.[12]

The Colombia REDD table strongly supports the inclusion of carbon markets in the mechanisms to finance REDD. This has also been the position of Colombia in the international climate negotiations, in which it has insisted on countries' freedom to choose between different financial sources, markets and/or an international fund. The voluntary carbon market is a salient project among members of the Colombia REDD table, partially due to the engagement of international and some local NGOs with actors interested in, connected to or involved with the carbon business. These actors include the local public environmental authorities (CARs), national and international business partners (i.e. mining and energy-producing companies, plantation companies, forest companies, carbon-marketing companies), international research organizations, development cooperation agencies, and indigenous and Afro-Colombian leaders. These engagements would allow the channelling of funds from a range of private businesses directly into carbon-market projects that could eventually become part of REDD.

The REDD programme in Costa Rica is seen as a means to strengthen and broaden the Payment for Environmental Services (PES) programme. PES emerged in Costa Rica in the 1990s as a response to the perceived problem of deforestation and forest loss. Between 1986 and 1991, the country lost 4.2% of forest cover per year (Sanchez-Azofeifa, Harriss and Skole, 2001), suggesting that Costa Rica had one of the highest deforestation rates in the world. The launching of REDD occurred ten years after Costa Rica became the first country in the world to establish a system of PES in 1997. The financial structure of the Costa Rican PES programme is a hybrid of market-like mechanisms, subsidies and state regulations. This is evident in the way that the programme is funded: while it receives 3.5% of the revenues from a tax on fossil fuels, it also depends on loans from the World Bank, from a series of grants from the Global Environmental Facility (GEF), from NGOs, from contracts with national companies (Pagiola, 2008) and from international governments. The German Government, through the German Reconstruction Credit Institution (Kreditanstalt für Wiederaufbau (KfW)), provided US$12 million for a five-year contract in 2007, and in 1996, Norway bought 200,000 tonnes of carbon-emission reduction credits for US$10 per tonne (Russo and Candela, 2006). The REDD national strategy is being discussed within the framework of the national PES programme. Because the current PES programme is unable to cover the demand for payments for environmental services, which is very high, REDD is seen as an avenue to increase the coverage of the national PES.

Costa Rica applied to the FCPF in 2008 to fund the REDD readiness preparations.[13] A grant was approved in 2010. In Costa Rica, public institutions are leading the REDD readiness preparations. The PES experience and Fondo Nacional de Finaniciamiento Forestal (FONAFIFO) largely shape the REDD process. FONAFIFO's board of directors is the REDD coordinating entity in Costa Rica. The board will include one representative from indigenous people's organizations and one representative from civil society.

FONAFIFO carried out a series of dissemination and outreach activities to engage with different stakeholder groups. As for indigenous peoples, it has invited the Indigenous Integral Development Associations (Asociación de Desarrollo Integral Indígenas (ADIIs)) to participate in information meetings and activities. Indigenous leaders contest the legitimacy of the ADIIs in representing indigenous peoples. In 1982, in an effort to make the indigenous territories legible to the state (cf. Scott, 1998), the Government of Costa Rica established the ADIIs as the legal representative bodies of indigenous peoples.

To carry out PES in indigenous territories, the government designated the ADIIs as the collective representative institutions of indigenous peoples vis-à-vis FONAFIFO. The ADIIs became responsible for distributing the benefits from PES in indigenous territories and for helping FONAFIFO to implement PES in the indigenous *resguardos*. Currently, indigenous leaders challenge this decision, arguing that the ADIIs are official government bodies that "represent" and govern each indigenous territory by law, but do not necessarily represent or respect traditional ways of organization and are not accountable to indigenous peoples. FONAFIFO carried out a series of early information dissemination workshops and it has engaged in an initial dialogue about the REDD process with a range of stakeholder groups, and with indigenous peoples in the Atlantic and Pacific areas through the structure of the ADIIs.

Costa Rica recognizes carbon, insofar as it is considered an environmental service, as property of the landowner, by law. The country has chosen a national approach to reduced emissions accounting and the development of a national baseline for avoided deforestation. At the international level, Costa Rica, similar to Colombia, advocates for a mix of funding for REDD. The approach in Costa Rica is towards a centralized REDD programme. In Colombia, on the other hand, the approach is towards a decentralized system. These two different approaches reflect the way in which forest governance is understood in the two countries. In the following subsection we will analyse the third and last strategy, using Bolivia as the example.

### The resisting strategy: Bolivia

Bolivia has resisted REDD as part of carbon markets and offsets, based on the idea of environmental justice and the non-commodification of nature. The current Bolivian position on REDD was first communicated in a letter to the General Assembly of the United Nations in 2008, emphasizing "direct compensation from developed to developing countries, through a sovereign implementation that ensures broad participation of local communities...". In its second communication to the UNFCCC in 2009, Bolivia stated that the country did not support carbon markets "or the possibility of developing new flexibility in this area", and called for domestic action for emissions reduction, under the argument that the "carbon market allows developed countries to continue to pollute at home while developing countries face unfair restrictions".

The position was not a complete rejection of REDD but rather an attempt to reshape it and to broaden the international perspective on both forests and carbon. Different actors were involved in the planning of a national joint programme in Bolivia, beginning in 2008, and Bolivia was one of the first pilot countries in the UN-REDD programme from 2009 onwards. A REDD team was set up in the Ministry of the Environment (MAYA) as part of a larger national strategy for curbing deforestation (Estrategia Nacional de Bosque y Cambio Climatico, MAYA, 2009). The setting up of a national REDD programme was supported by German (Deutsche Gesellschaft für Internationale Zusammenarbeit (GIZ)) and Danish cooperation at the time, and a parallel process was started with the FCPF of the World Bank. The UN-REDD programme was presented for civil society actors in 2010, and four indigenous and peasant organizations approved a capacity-building plan.

Beginning in 2010, different currents both inside and outside the government caused confusion about the Bolivian position. At the People's Conference for Climate Change and the Rights of Mother Earth in Cochabamba in April 2010, where many Bolivian officials also participated, a declaration rejecting all forms of REDD/REDD+/REDD++ was presented.[14] Following the conference, the negotiation team from the Ministry of Foreign Affairs (with representatives from the Unidad de Madre Tierra) brought the Cochabamba position to the climate negotiations in Cancun as promised, while the Ministry of the Environment signed off on the UN-REDD programme on the condition that UN-REDD would respect the Bolivian position against carbon markets.[15] The collaboration with the World Bank was halted, and Bolivia never handed in a signed version of the formal document Readiness Plan Idea Notes (R-PIN).

The confusion and lack of advancement of the UN-REDD programme in the 2008–2011 period also opened up the arena for private actors and NGOs to get involved in REDD-like activities. Local communities have reported that private actors (represented by NGOs, a Santa Cruz-based company and local businessmen) contacted communities, asking them to sign "REDD contracts" that involved the lease of land for 90–100 years, in exchange for untouched conservation areas and the "selling of oxygen". The government later stopped the attempts.

In 2008 the national NGO Friends of Nature Foundation (FAN), with support from the Gordon and Betty Moore Foundation, set up an indigenous REDD project in the Amazon (Beni Department). The government, originally a partner in the project, withdrew in 2010. Several regional and local indigenous organizations also withdrew, making the argument that the NGO would have too much power over the project and the resources involved. Furthermore, the local communities participating in the project rejected the component regarding quantifying emissions reductions, and the project was left only with select components that addressed sustainable forest management, the enforcement of Brazil nut collection and enhanced control of the area against illegal logging. The project was in operation until 2012.

Later in 2011, a conflict between the central government and the lowland indigenous organization Confederación de Pueblos Indígenas de Bolivia (CIDOB) over a road-building project through the national park TIPNIS led to a rupture in contact among the ministries, public agencies and the indigenous organization, hampering the possibilities for further dialogue about the UN-REDD project. The plan for initiating the participatory planning process for the UN-REDD programme was set on hold. Meanwhile, CIDOB called for direct REDD funding to indigenous areas and for the self-management of funds.

A parallel process was started in 2011 to develop a mechanism for the sustainable management of forests, and joint climate-change mitigation and adaptation efforts. The process involved a number of national NGOs, academics and public entities, such as the Authority for Forest and Land (ABT), the National Institute for Agricultural Innovation (Iniaf) and the Forest Directorate in MAYA. Bolivia hoped that the mechanism could be supported through an alternative REDD scheme outside the carbon market. The mechanism was included in the Law of Mother Earth in 2012, with an emphasis on holistic management of the forests. A team was set up to facilitate the exchange of information and meeting arenas. As public entities had poor official records of deforestation in Bolivia, the participation of the NGOs (e.g. FAN) with such expertise was crucial for the team. Former officials, the Noel Kempff Museum of Natural History and representatives from research institutions and social organizations contributed with important experience and information, forming a final project document that was presented to the UN-REDD in 2012.

In 2011, Bolivia informed the policy board of the UN-REDD programme about its desire to modify its original National Programme document. Two contradictory communications, which were sent from Bolivian officials to the policy board in December 2011 and March 2012, led the board to freeze the funds and send a high-level mission to Bolivia in June 2012. The mission concluded that there were several challenges concerning the mechanism (e.g. the lack of an incentive system based on verified reductions of emissions, the targeting of drivers, and the lack of full participation from the indigenous organization CIDOB in the making of the mechanism) and that the project was not eligible for full financing by the UN-REDD programme. Later, contrasting declarations about the participation of indigenous organizations in the making of the mechanism were also communicated to the UN-REDD policy board. The mission finally recommended that the National Joint Programme be implemented in its original form, and that it neither be redrafted nor replaced with the new Bolivian mechanism. Bolivia agreed to continue with the programme, and a small part of the UN-REDD financing was channelled to the mechanism (such as the register of all forest initiatives, forest inventory and the mapping of land-use change).[16]

The proposal for an alternative mechanism was marginalized by powerful REDD donor countries in the international negotiations, claiming it would lead to the fragmentation of the REDD project. Finally, in 2013, Denmark, Switzerland and the EU granted support of over US$43 million to the Bolivian mechanism. At the international level, Bolivia has worked insistently with the inclusion of non-market-based approaches, such as joint mitigation and adaptation – methodological issues related to non-carbon benefits – and it continues with its strong opposition to carbon-market mechanisms.

Due to opposing currents both within and outside the Bolivian Government, different actors in Bolivia have pursued slightly different strategies to influence and shape REDD, from complete rejection to the reshaping of the initiatives, locally, nationally and internationally. However, the rejection of carbon markets has been a common position across the majority of actors involved, as well as the integration of indigenous rights and the recognition of different functions of the forests. The role of indigenous organizations and indigenous autonomy is still to be defined in the Bolivian mechanism, along with clear strategies to work with the drivers of deforestation.

In the following section, we shift our focus to analyse ongoing efforts at local and national levels. We will focus on demonstration and readiness activities, and the actors involved in them.

## REDD projects in Latin America

An important component of the planning phase of REDD is demonstration and readiness activities. These are projects implemented at the local level to test the options available for countries and communities. REDD projects can be seen as a means to understand how REDD will unfold on the ground; REDD demonstration activities are seen as means to learn lessons for future REDD implementation. These early implementation projects influence debates about REDD, the ways in which so-called co-benefits are being addressed, and who is involved and who benefits from REDD.

In principle, REDD country strategies to be defined in Phase 1 are the first step in the implementation of REDD national policies. National REDD strategies would define the current situation in each country and the direction in which the country is going to move in terms of reduced carbon emissions from deforestation, addressing so-called co-benefits and defining who would benefit from economic payments. In practice, however, numerous REDD projects are taking place before the design of a country's REDD strategy is finished or in parallel with its development. Early implementation projects are informing the policy-making process in each country and at the global level. Proponents of REDD projects stand in a better position than other actors, who do not have any experience with such projects, to influence REDD debates because not having knowledge about REDD is a barrier for being included in the official debates.

We have identified three approaches employed by actors involved in early REDD planning, implementation and readiness projects, and the consequences of such approaches. The first one is knowledge production and dissemination. Second is the creation of technologies or standards to legitimize or validate projects. The third approach is enrolment in new, emerging or alternative networks. In what follows we analyse these three approaches by highlighting who is involved, the resources mobilized to employ each approach, and the outcome. It is worth saying that these approaches are not mutually exclusive, and different actors within each country put distinct emphasis on each of these approaches.

**Creation of knowledge and dissemination of information**

Our findings indicate that, to a great degree, networks involving NGOs and international research institutions with support from development cooperation agencies and private actors are creating and disseminating knowledge about REDD in the region. These networks systematize information about REDD in Latin America and at the global level. They are having a great influence in defining what a REDD project is, who the legitimate implementers are, who will benefit from it and how. The Center for International Forestry Research (CIFOR), the NGO Global

Canopy Programme,[17] and the voluntary REDD database[18] created at the Oslo Climate and Forest Conference in 2010 produce compilations and databases that include all types of REDD-like projects.

The majority of REDD projects are being initiated or planned by private actors in private lands, including national and international private companies, and local and international NGOs (WWF, CI, WCS, TNC, IUCN and Rainforest Alliance). In some cases, pilot projects are executed with the participation of state governments in coalition with BINGOs. Fair-trade cooperatives, carbon certifiers and research institutions are also involved in pilot projects. Pilot project proponents act as de facto researchers, testing REDD implementation modalities, and producing information and knowledge about the projects.

As for funding sources for the projects, development cooperation aid money, particularly from Norway and Germany, as well as private funds, is the most important source. But here it is necessary to explain in more detail what types of private fund are involved. The range is wide and includes (1) direct investments in particular projects from investors from the USA, Europe, China and India; (2) direct investments from companies (e.g. the largest Brazilian mining company, Vale); (3) investments that private companies make in BINGOs; and, similarly, (4) partnerships between local NGOs and private companies as part of their CSR portfolio; (5) a plethora of alliances among domestic NGOs and local-level environmental authorities (CARs), national and international business partners (mining and energy-producing companies, plantation companies, forest companies and carbon-marketing companies), international research organizations, development cooperation agencies and indigenous leaders.[19] These alliances influence the emphasis given to particular components in the projects.

The outcome of this approach is that private actors and research institutions, which are often international organizations, are creating knowledge and disseminating information about REDD in Latin America. The consequence of this is that these actors position themselves better than public institutions or national research centres and have better resources to influence the international debate. Even Bolivia, with a government strongly sceptical about NGOs, saw the need to include these actors as they have better forest data (e.g. maps) than the government. The way in which they gain this privileged position is by accessing funding from private sources or international development cooperation agencies, coupled with the privileged position in neoliberal environmental governance that they have maintained since the 1990s. To overcome complex issues such as those related to ownership of the land, most projects are initiated or planned on private lands. In the following subsection, we focus on measurements to validate REDD projects.

**Measures to validate projects**

NGOs, corporations and research institutions are involved in creating standards to certify carbon offsets that can be traded in the voluntary carbon market or in a future REDD carbon market. Organizations involved in pilot projects are also creating standards to demonstrate how they involve local populations in REDD projects.

An illustrative example of this is the Rainforest Standard™ (RST). This was developed by Columbia University in New York in collaboration with private environmental funds from Bolivia, Peru, Brazil, Ecuador and Colombia. According to its proponents, "this standard integrates carbon-accounting, socio-cultural/socio-economic impacts and biodiversity outcomes into one single REDD standard[20]". Projects certified with Royal Forest Society (RFS) can be registered in the Climate Community and Biodiversity Alliance (CCBA)[21] and in the Verified Carbon Standards (VCS),[22] to be traded in the voluntary carbon market.

The alliances and associations built among NGOs, the private sector and research institutions contribute to the creation of facts, standards, knowledge and concepts seen as accepted "truths" (cf. Goldman and Turner, 2012). These accepted truths are shaping the direction of REDD in the Amazon basin before governments have managed to put a plan of action into place. For example, in Colombia, where the readiness process is still incipient, BINGOs and local NGOs managed to include the RST as a standard to certify REDD projects by the government in the national REDD strategy. Projects that do not comply with the RST will not be included in the national REDD register of Colombia, and their proponents will not be invited to participate in the debate.

In the following subsection, we focus on alternative channels that different actors are using to engage in REDD. These are particularly relevant in creating a counterbalance to mainstream views and values.

**Alternative channels**

REDD networks as described above, in which BINGOs and local NGOs, development cooperation agencies, private actors, government agencies and research institutions participate, are channels where REDD knowledge is being produced and circulated. Such networks have a form of agency in the creation of environmental knowledge that is validated and re-enforced at different levels. Access to REDD networks is not open to all of those who could be interested or affected by REDD policies and projects. Participation in REDD networks is conditioned by overriding narratives on deforestation and by the role of monetary incentives in tackling deforestation (see Forsyth, 2003). Activists seeking to influence existing networks may have to decide between working within such dominant rules and establishing alternative and competing networks (Forsyth, 2003; Taylor, 2012). In this way, networks become important resources to advance alternative views and values.

Initially, indigenous peoples were sceptical about REDD and rejected carbon markets because they did not consider them to be offering real solutions to climate change (see the Anchorage declaration adopted by the participants at the indigenous people's global summit on climate change in 2009).[23] Indigenous organizations in the global South criticize carbon markets and carbon-sequestration projects for their oversimplified portrayal of ecosystems and forests, and for ignoring the socioeconomic, political and institutional implications of carbon sequestration for indigenous peoples.

Indigenous people's organizations in Latin America, and particularly in the Amazon basin countries, have since engaged in existing networks that support REDD, or in alternative networks that are sceptical about REDD and carbon markets. The different paths taken by different indigenous people's organizations are in part explained by previous engagements with other organizations and by their own experiences with REDD. Indigenous people's organizations' choice of position is also influenced by their experiences of negotiating with their governments, and the organization's own visions and priorities.

During the 12th session of the UN Permanent Forum on Indigenous Peoples in 2013, indigenous people's organizations presented two opposing views on REDD, later communicated at COP19 in Warsaw. Some organizations oppose REDD on the grounds that it weakens existing national legal frameworks to protect indigenous people's rights, particularly in regard to territorial and collective land rights, consultation and autonomy, and their opposition to carbon markets and the commodification and fragmentation of nature. Other organizations look at REDD as an opportunity to strengthen the land rights of indigenous peoples and their local management, and to control their territories with the help of direct funding.

The experience of some indigenous people's organizations with so-called "carbon cowboys", particularly in Brazil, Peru, Bolivia and Colombia, has made them extremely aware of some of the risks that REDD projects might entail. Peruvian, Brazilian, Bolivian and Colombian indigenous organizations denounced the fact that indigenous leaders signed disadvantageous contracts with private companies. On the other hand, some groups are already developing long-term land-use plans that involve REDD mechanisms defined in their own terms. That is the case of the Suruí in Brazil (Toni, 2011).

The Suruí live in a 247,000 Ha reserve in the state of Rondonia, and 93% of their land is still preserved (Suruí, 2009). The Suruí population was 5,000 people when they first made contact with non-indigenous Brazilians, but currently only about 1,000 individuals live inside their lands or in the nearby cities. During the 1980s an intense migration of non-indigenous people to the Western Amazonia took place. By the end of that decade, the population had decreased to roughly 250 members.

Despite this drastic reduction of their population, the Suruí started to organize themselves in the 1980s. They created the Metareilá Suruí Association in 1989 to defend and preserve the Suruí's cultural and territorial patrimony.

In 2000, Metareilá started a participatory diagnosis to assess the potential of the Suruís and their territory. Based on this diagnosis, it designed a plan for the use of the territory for coffee cultivation (one of the crops introduced to their land by the invaders), for the management of Brazil nuts, and for the restoration of areas degraded by illegal logging.

With the support of other NGOs (Associação de Defesa Etnoambiental Kanindé, Amazon Conservation Team, Forest Trends, Idesam), the Suruís decided to set aside 13,575.3 Ha of forests for 30 years, which will avoid emissions that average 7,423,806.2 tonnes of $CO_2$. The project was validated in conformance with the Climate, Community and Biodiversity Standards in 2012 (RA-VAL-CCB) and with the Verified Carbon Standard in 2013. Despite the broad alliance that prepared the project, Metareilá has full rights over carbon credits and will be the sole recipient of the financial benefits.

The design of the Suruí Carbon Project included an extensive consultation process, training for community members, development of a baseline for carbon accounting, and analysis of the legal framework regarding indigenous peoples and forest carbon. The Suruís initiated this process in accordance with their own demands; they saw the sale of carbon credits as an opportunity to complement a long-term plan for the development of their community.

## Conclusion

In this chapter we have looked at different strategies employed by Latin American countries and actors in their meeting with the global forest and climate initiative, REDD, from resistance to accommodating to assertive strategies. Brazil has been one of the major actors in the initiative after it changed its strategy from resistance to a more offensive approach and managed to align REDD with its own domestic interests. A strong actor such as Brazil has the resources, knowledge and power to shape REDD in its interests, and with the focus on results-based payments, the country is in a privileged position. It has also succeeded in sovereignty issues in international negotiations, such as those related to monitoring, reporting and verification/national forest monitoring systems.

The experiences of the countries that have followed the accommodating strategy show how the history of environmental governance in each country affects the implementation of the REDD initiative. Colombia has, to a large extent, left the initiative in the hands of private actors and local authorities, while Costa Rica has applied a model of "hybrid" governance and a centralized REDD programme. Bolivia has stood out in Latin America as one of the fiercest opponents of carbon markets, something that has affected its possibilities and willingness to take part in the initiative. Bolivia's commitment to the inclusion of civil society demands in environmental governance and the anti-commodification rhetoric has formed its responses to the global initiative. However, there are divergent opinions, especially among the indigenous organizations, about the right path to follow. Indigenous organizations with recognized titles to their land believe that REDD can bring new opportunities. However, although Bolivia's position has been similar to that of Brazil to a large extent, with national sovereignty and opposition to offsets as focal points, Bolivia has instead been seen as the "activist state" that is trying to fragment REDD. It was not until 2013 that Bolivia won support for its alternative mechanism to forest and climate efforts.

These three strategies illustrate how the "black-boxing" of REDD has allowed for the emergence of quite different hybrid models of negotiating environmental governance at the international level.

Our research reveals that there is a constellation of actors shaping the direction of REDD+ in Latin America. That constellation varies from country to country and includes among others, donors, BINGOs and national NGOs, research institutions, and in some cases different levels of government. Through their engagements in networks that promote and advance a narrative in which markets and monetary compensations offer the solution to deforestation, these actors are in a privileged position to participate in the co-production of knowledge and policy, and to advance their agendas.

For some governments, engaging in REDD – at least at the discursive level – does not conflict with their priorities in other sectors, such as oil exploitation, soy expansion, the expansion of large-scale cattle-ranching, and mining and infrastructure development, which all represent threats to the forests and further deforestation. REDD is seen as an alternative that will allow for the ending of trade-offs between forest conservation, poverty alleviation and economic development. A good example of how this change is unfolding can be found in the partnership between Norway and Brazil. Thanks to REDD, Brazil became the largest receiver of Norwegian development cooperation aid, which is an enormous paradox given that Brazil is one of the fastest-growing economies in the world. At the same time, but not necessarily as a consequence of such collaboration, Brazil has drastically decreased deforestation in the Amazon.

NGOs have the technical and rhetorical expertise to participate in negotiations in national and international arenas. They also have connections with farmers, indigenous and traditional populations, government officials and bureaucrats. That makes them a privileged set of boundary organizations (Guston, 2001) that can help to break resistance against REDD and to open channels for the implementation of pilot projects. They have been particularly strengthened by REDD due to this role. They are becoming knowledge-providers to governments, donors and local organizations, which has opened the doors for them to policy-making forums. Environmental NGOs are now in a better position to offer business alternatives to corporations and other private actors. Aside from their role as boundary organizations, they are also brokers in REDD implementation and have a direct stake in the negotiations.

The black-boxing of REDD has allowed for the construction of a large and diverse network that supports the initiative. The widespread questioning of the market premises of REDD has led to a broadening of the concept to accommodate disparate interests, ideologies and representations of what forests are and why they should be conserved. That is why countries that have been vocal against REDD, such as Brazil until the mid-2000s, are engaging in REDD preparedness. Accordingly, some groups that initially opposed the mechanism, such as indigenous populations, have pilot projects in their lands as REDD might offer an alternative to strengthen their land rights. However, many indigenous organizations remain critical of carbon markets.

The way in which REDD is going to be financed is still an open question. Although it was born as a market mechanism to trade carbon, political mobilization from different actors has resulted in discussions that challenge the market orientation of REDD, and many actors in the Latin American region advocate for a global public fund to finance the initiative. The political opposition of several actors in Latin America has also resulted in a broadening of the focus of REDD to multiple aspects of forests and their related environmental services. In some countries, at the domestic level, it is increasingly assuming the format of a public policy, whereas in the global arena it resembles what Angelsen (2013) has called a "performance-based aid" mechanism. This means that development cooperation funds are used to finance REDD on the condition that countries demonstrate that they achieve certain levels of performance in terms of reduced deforestation.

## Notes

1. http://unfccc.int/resource/docs/2010/cop16/eng/07a01.pdf. See also Angelsen et al. (2009: 3).
2. The final Readiness Plan Idea Note (R-PIN) was never signed by the Bolivian authorities.
3. http://www.lcds.gov.gy.
4. Other donors contributing to UN-REDD are, in order of the size of their contribution, the EU, Denmark, Spain, Japan and Luxembourg. Germany provides 34% of the total budget of the FCPF. Other donors include Australia, the UK, the USA, Canada, the European Commission, the Nature Conservancy and two private companies: BP Technology Ventures, an alternative energy company with venture investments in projects specific to biofuels, wind and solar energy; and CDC Climat, a company that includes emissions trading and energy investments in its portfolio. The other contributors to the Amazon Fund are Germany and the Brazilian oil company, Petrobras. Sources: http://mptf.undp.org/factsheet/fund/CCF00; http://www.forestcarbonpartnership.org/sites/fcp/files/2013/FCPF%20Carbon%20Fund%20Contributions%20as%20of%20Dec%2031_2012.pdf; http://www.amazonfund.gov.br/FundoAmazonia/fam/site_en/Esquerdo/doacoes/; http://www.guyanareddfund.org/index.php?option=com_content&view=article&id=101&Itemid=116.
5. See Readiness Preparation Plans of Colombia, Peru, Ecuador, Guyana and Suriname.
6. In addition to the four BASIC countries (Brazil, South Africa, India and China), representatives from Argentina, Fiji (as chair of the G77 and China), Paraguay, Peru and Venezuela were at the BASIC meeting. http://www.twnside.org.sg/title2/climate/info.service/2013/climate130904.html
7. The Bolivarian Alliance for the Peoples of Our America is a regional organization launched in 2004 and is made up of eight countries: Antigua and Barbuda, Bolivia, Cuba, Dominica, Ecuador, Nicaragua, Saint Vincent and the Grenadines, and Venezuela.
8. Brazilian environmentalists and NGOs (Instituto Socio Ambiental (ISA), Greenpeace, Instituto Centro de Vida (ICV), Instituto de Pesquisa Ambiental

da Amazonia (IPAM), TNC, CI, Amigos da Terra Amazonia Brasileira (AdT), Instituto do Homen e Medio Ambiente (IMAZON) and WWF-Brazil) launched the Zero Deforestation Campaign. This was based on ideas of strengthening the participation of state governments in forest governance, payments for environmental services, strengthening of protected areas and support for indigenous peoples.

9. According to the former Norwegian oil and energy minister Terje Riis-Johansen, the allocation of Norwegian money to the Amazon Fund contributes to opening doors for the Norwegian oil industry in Brazil. Paradoxically, thanks to the commitment to the Amazon Fund, Brazil – one of the largest and fastest-growing economies in the world – has since 2009 become the largest recipient of Norwegian foreign development aid. http://www.dn.no/energi/article1975276.ece «rainforest millions open oil doors».

10. The Governors Climate and Forest Task Force (GCFT) brings together subnational-level authorities from Brazil, Mexico, Peru, Indonesia, countries in Africa, and the governors' offices of California and Illinois. In this project, California and Illinois will potentially be able to purchase carbon offsets from projects in developing countries, as part of the cap-and-trade programme of these states, which will use a market-based mechanism to reduce greenhouse gases. The GCFT receives funding from the Gordon and Betty Moore Foundation, ClimateWorks, the Climate and Land Use Alliance, the Norwegian Agency for Development Cooperation (Norad), and the David and Lucile Packard Foundation. Collaborating partners include NGOs from Brazil (Institute for the Conservation and Sustainable Development of Amazonas -DESAM and Amazon Environmental Research Institute – IPAM), Indonesia (Kemitraan), Mexico (ProNatura), a transnational private company (ClimateFocus), and the US-based private research organizations the Carnegie Institution for Science and the Woods Hole Research Center.

11. WWF, CI, TNC.

12. See the report of the due diligence mission of the World Bank to Colombia, 15–27 January and 22–23 March 2012. http://documents.worldbank.org/curated/en/2012/04/16508452/colombia-fcpf-redd-readiness-project-aide-memoire-april-18th-25th-2012

13. In addition to the FCPF, other sources of funding include GIZ through the REDD-CCAD-GIZ programme, which has financed different activities in Costa Rica with special emphasis on forest reference level; the Norwegian development agency (Norad); and USAID.

14. Later it turned out that the Bolivian officials were against the total rejection of REDD.

15. The UN-REDD team respected the Bolivian position at the time and said they would not intervene in the funding for the Bolivian programme.

16. In total, US$1.4 million. *Source:* Diego Pacheco.

17. The REDD desk is funded by the Gordon and Betty Moore Foundation, the Climate and Land Use Alliance, the Department of Climate Change and Energy Efficiency of the Australian Government, GIZ and USAID.

18. http://reddplusdatabase.org.

19. Interview FAN; interviews Colombia.

20. http://cees.columbia.edu/the-rainforest-standard and interview FAN.

21. The CCBA is a partnership between research institutions (CATIE, CIFOR, and ICRAF), corporations (the Blue Moon Fund, The Kraft Fund, BP, Hyundai, Intel, SC Johnson, Sustainable Forestry Management, and Weyerhaeuser) and NGOs (CARE, CI, TNC, the Rainforest Alliance and WCS).
22. The VCS was established in 2005 by the Climate Group, the International Trading Association and the World Business Council for Sustainable Development. It is one of the world's most widely used carbon-accounting standards. Projects across the world have issued more than 100 million carbon credits using VCS standards. VCS headquarters are in Washington, DC, with offices in China and South America.
23. http://www.unutki.org/downloads/File/Events/2009-04_Climate_Change_Summit/Anchorage_Declaration.pdf

# References

Agrawal, A., Nepstad, D. and Chhatre, A. (2011) "Reducing Emissions from Deforestation and Forest Degradation", *Annual Review of Environment and Resources* 36: 373–396.

Alesina, A., Carrasquilla, A. and Echavarria, J.J. (2005) "Decentralization in Colombia" in A. Alesina (ed.), *Institutional Reforms. The Case of Colombia* (Cambridge, MA: MIT Press).

Alvarez, M.D. (2003) "Forests in the Time of Violence", *Journal of Sustainable Forestry* 16(3–4): 47–68.

Angelsen, A. (2013) *REDD+ as Performance-Based Aid: General Lessons and Bilateral Agreements of Norway*. Working Paper (s.d.:WIDER).

Angelsen, A., Brown, S., Loisel, C., Peskett, L., Streck, C. and Zarin, D. (2009) *Reducing Emissions from Deforestation and Forest Degradation (REDD): An Options Assessment Report. Prepared for the Government of Norway* (Washington: Meridian Institute).

Angelsen, A. and McNeill, D. (2012) "The Evolution of REDD+", in A. Angelsen, M. Brockhaus, W.D. Sunderlin and L.V. Verchot (eds), *Analysing REDD+ Challenges and Choices* (Bogor, Indonesia: CIFOR).

Barcena, A., Prado, A., Samaniego, J. and Malchik, S. (2010) "Climate Change: A Regional Perspective", *Unity Summit of Latin America and the Caribbean Riviera Maya*. ECLAC; IDB. Riviera Maya, Mexico.

Bernstein, S. (2004) "Legitimacy in Global Environmental Governance", *Journal of International Law and International Relations* 1(1–2): 139–166.

Blom, B., Sunderland, T. and Murdiyarso, D. (2010) "Getting REDD to Work Locally: Lessons Learned from Integrated Conservation and Development Projects", *Environmental Science and Policy* 13(2): 164–172.

Boyd, E., Gutierrez, M. and Chang, M. (2007) "Small-Scale Forest Carbon Projects: Adapting CDM to Low-Income Communities", *Global Environmental Change* 17(2): 250–259.

Bull, B. and Aguilar-Støen, M.C. (eds) (2015) *Environmental Politics in Latin America: Elite Dynamics, the Left Tide and Sustainable Development* (London: Routledge Studies in Sustainable Development).

Fairhead, J. and Leach, M. (2003) *Science Society and Power: Environmental Knowledge and Policy in West Africa and the Caribbean* (Cambridge: Cambridge University Press).

Forsyth, T. (2003) *Critical Political Ecology. The Politics of Environmental Science* (New York: Routledge).

Goldman, M., Nadasdy, P. and Turner, M.D. (2011) *Knowing Nature: Conversations at the Intersection of Political Ecology and Science Studies* (Chicago: University of Chicago Press).

Guston, D.H. (2001) "Boundary Organizations in Environmental Policy and Science: An Introduction", *Science, Technology, and Human Values* 26(4): 399–408.

Hall, A. (2008) "Paying for Environmental Services: The Case of Brazilian Amazonia", *Journal of International Development* 20(7): 965–981.

Hall, A. (2011) "GETTING REDD-Y Conservation and Climate Change in Latin America", *Latin American Research Review* 46: 184–210.

Karkkainen, B.C. (2004) "Post-Sovereign Environmental Governance", *Global Environmental Politics* 4(1): 72–96.

Latour, B. (1987) *Science in Action: How to Follow Scientists and Engineers Through Society* (Cambridge: Harvard University Press).

Li, T.M. (2007) *The Will to Improve. Governmentality, Development and the Practice of Politics* (Durham: Duke University Press).

Meyer, P.J. (2010) *Brazil-US Relations* (Washington, DC: Library of Congress Congressional Research Service).

Nepstad, D., Soares-Filho, B.S., Merry, F., Lima, A., Moutinho, P., Carter, J., Bowman, M., Cattaneo, A., Rodrigues, H., Schwartzman, S., McGrath, D.G., Stickler, C.M., Lubowski, R., Piris-Cabezas, P., Rivero, S., Alencar, A., Almeida, O. and Stella, O. (2009) The end of deforestation in the Brazilian Amazon. *Science* 326:1350–1351.

Pacheco, P., Aguilar-Støen, M., Börner, J., Etter, A., Putzel, L. and Diaz, M.d.C.V. (2010) "Landscape Transformation in Tropical Latin America: Assessing Trends and Policy Implications for REDD+", *Forests* 2(1): 1–29.

Pagiola, S. (2008) "Payments for Environmental Services in Costa Rica", *Ecological Economics* 65(4): 712–724.

Peet, R., Robbins, P. and Watts, M. (2011) "Global Nature", in R. Peet; P. Robbins and M. Watts (eds), *Global Political Ecology* (Abingdon UK: Routledge), 1–48.

Russo, R.O. and Candela, G. (2006) "Payment for Environmental Services in Costa Rica: Evaluating Impact and Possibilities", *Tierra Tropical* 2(1): 1–13.

Sanchez-Azofeifa, G., Harriss, R.C. and Skole, D.L. (2001) "Deforestation in Costa Rica: A Quantitative Analysis Using Remote Sensing Imagery", *Biotropica* 33(3): 378–384.

Santilli, M., Moutinho, P., Schwartzman, S., Nepstad, D., Curran, L. and Nobre, C. (2005) "Tropical Deforestation and the Kyoto Protocol", *Climatic Change* 71(3): 267–276.

Scott, J. (1998) *Seeing Like a State: How Certain Schemes to Improve the Human Condition have Failed* (London/New Haven: Yale University Press).

Suruí, A.N. (2009) "Projeto Carbono Suruí", Presented at the *Curso para Lideranças Comunitárias, Pagamentos por Serviços Ambientais*. http://www.katoombagroup.org/documents/events/event33/Apresentacoes/EstudodeCaso_Almir_Surui_1.pdf, date accessed 13 September 2014.

Taylor, P.J. (2012) "Agency, Structuredness, and the Production of Knowledge within Intersecting Processes", in D. Goldman; P. Nadasdy and M. Turner (eds), *Knowing Nature. Conversations at the Intersection of Political Ecology and Science Studies* (Chicago: The University of Chicago Press), 81–98.

Toni, F. (2011) "Decentralization and REDD+ in Brazil", *Forests* 2(1): 66–85.

Wunder, S. (2005) *Payments for Environmental Services: Some Nuts and Bolts*, Occasional Paper No. 42 (Jakarta, Indonesia: CIFOR).

 Except where otherwise noted, this work is licensed under a Creative Commons Attribution 3.0 Unported License. To view a copy of this license, visit http://creativecommons.org/licenses/by/3.0/

OPEN

# 9
# Rights, Pressures and Conservation in Forest Regions of Mexico

*Leticia Merino*

## Introduction

The drivers of environmental degradation and the strategies to counter them are the subjects of heated debate. Several conceptual and policy approaches consider the key factors of this degradation to be the weakness and instability of property rights over natural resources. The *commons* perspective, on the other hand, emphasizes the viability and potential of the self-governance of shared resources such as forests. This perspective calls for a better understanding of the roles of local users and their institutions – understood as "rules in use" – with regard to natural resources (Ostrom, 1991; McKean, 2000; Berkes, 2006; McCay, 2007). In this literature, collective action is understood as cooperation and coordination to solve collective dilemmas related to the management of the commons (Cárdenas, 2008; Meinzen-Dick, 2010). The influence of the commons perspective goes beyond academia, gaining recognition among some international funders, environmental agencies and practitioners. It follows the repeated failures of previous efforts of international aid to halt deforestation through the support of governmental agencies.

This approach has led to two important policy proposals: (1) the decentralization of control over common resources to lower levels of government, including local user groups and stakeholders (Ribot, Agrawal and Larsson, 2006; Agrawal and Ashwini, 2009); and (2) the devolution of property rights to local users in order to create incentives and a commitment to sustainability (Whyte and Martin, 2003; Molnar and Alcorn, 2006; Barry, 2008). Although the "commons school" has had limited academic influence in Latin America, the region has been marked by proposals to decentralize forest governance and devolve rights to local communities. Such initiatives have been coherent with the struggles of local communities over land and natural resources all over the region. The "devolution" of forestlands to local populations has been an intense learning process, with a range of outcomes that need to be better documented and understood. This extends from the forest concessions to "comunitarios" in Petén Guatemala, the official recognition of traditional rights over the lands of the indigenous "mizquito" in Nicaragua, and the indigenous reserves in Panamá, Brazil and Bolivia; to the forest property of Afro-American communities in the Colombian Pacific and rubber tappers in Brazil. Experiences of collective action, local governance, rural development and conservation coexist with cases of conflict, elite capture and forest deterioration. "Community forestry" has been a positive option for conservation and local livelihoods in different regions.[1] However, community-based governance is neither a panacea nor a reality to be taken for granted. The outcomes of these experiences derive from a variety of historical as well as recent factors, on which public policy often has major impacts.

Mexico stands as a singular case of community-based forest governance in Latin America. Mexican communities gained legal rights over lands and forests long before anywhere else in the contemporary world. At the same time, Mexico's deforestation rates were some of the highest in the world for decades (1970–1990).[2] Forests cover more than 60% of the country, providing important ecosystemic services that benefit a range of actors.[3] During most of the twentieth century, the forest industrial sector searched for access to low-cost raw materials. Backed by government agencies that promoted economic growth, their position weakened with the implementation of NAFTA. Since 1994 the Secretaria de Medio Ambiente Recursos Naturales y Pesca (SEMARNAP) has promoted conservationist measures that sought to minimize forest use in order to protect the megabiodiversity within, often drastically limiting human presence in forested and wild areas.[4] Conservationist policies have gained influence in public opinion, backed by national and international environmental agencies. They tend to regard deforestation as a generalized process in the country, mainly driven by collective property regimes and rural poverty. Local communities are the legal owners of the majority of Mexico's forests. Although they were favoured by agrarian reform, they have rarely received coherent policy support to develop forest-based livelihoods and to become forest stewards. Most forest communities have weak political voices and exercise little influence on other social sectors. Community residents value and benefit directly from many of the forest's ecosystemic "services": goods for domestic consumption and the flow of goods harvested and processed for commercial purposes. For them, the forest has patrimonial value and represents a legacy to be passed down from their elders to their children.

The relationships between relevant stakeholders – including federal and state governments – are permeated by poor coordination, pronounced economic and political asymmetries, and misconceptions. Conservationists tend to dominate in the context of global concern over climate change and biodiversity loss. For international agencies and for the federal government, forests and climate-change policies are the fields of experts and the central government. They favour the recentralization of control over rural landscapes. Forest communities tend to be seen as obstacles to conservation and the mitigation of carbon emissions. Commoners' perceptions of environmental change,[5] their increasing "climatic" vulnerability, their livelihoods and governance are mostly disregarded. The accomplishment of general mitigation targets is often prioritized over local adaptation needs.

Based on empirical research carried out in 103 forest communities, this chapter will discuss some of the main demographic and socioeconomic conditions of Mexico's forest communities, land-tenure features, forest use and local perceptions of pressures on forest areas. In addition, the relationship between local institutions, forest economies and social capital is analysed. Although the analysis focuses on Mexico, the experiences of community forest tenure and community forestry may provide useful insights into the general interaction of local communities with forests. This could be applicable to forest regions and forest policies in other Latin American countries.

## Forests in Mexico

### Forest tenure and property rights

Mexico has ecological, social and historical features that are similar to those of many Latin American countries. Much of the lands are forested and mountainous, and most forest areas are inhabited spaces.[6] Forest regions are home to nearly 12 million people in Mexico, many of whom are indigenous and are living under conditions of extreme poverty (INEGI, 2010). Community property remained in place during the three centuries of the colonial rule (16th and 19th centuries) and continued to exist in areas where colonial control remained incomplete due to difficult access. During the nineteenth century, many communities lost their lands as privatization policies were imposed by the central government. Indigenous presence and collective property were regarded as backwards and saw privatization as imperative for economic modernity (see Chapter 1).

Despite these similarities, Mexico is a unique case in Latin America because powerful social movements brought about extensive land-tenure reform when the state tried to resolve popular claims. The government's recognition of collective tenure as the basis for agrarian reform was guaranteed in the 1917 federal constitution. Today, 70% of forestland is under collective tenure, while more than 50% of communities have forest cover (Warman, 2000; Bray and Merino, 2004; Merino, 2004; Bray, Merino and Barry, 2005).

Across the entire world, public property of forests (often under concessions to third parties) is the prevailing institutional arrangement. Only from the late 1980s to the early 2000s did communities and local groups obtain rights over forests in other Latin American countries, such as Nicaragua, Bolivia, Brazil, Guatemala, Peru and Colombia.

There are two legal types of collective holdings in Mexico: *ejidos* and *comunidades agrarias*.[7] *Ejidos* were created when the government granted land to groups who demanded it, including former hacienda workers. *Comunidades agrarias* resulted from the official recognition of the historical rights of indigenous communities. CONAFOR estimates that today 30,305 communities collectively own 105 million Ha of forest. Legally, *comunidades agrarias* are able to incorporate young members at their will, while *ejido* members can only pass on their rights to a single successor. Community forests have to be commonly managed, and their division or sale is legally prohibited.[8]

In spite of the legal status of communities, their rights are clearly limited: the Mexican state maintains the right to regulate forest use.[9] Second, as in most of Latin America, water and underground resources are legally public property, giving governments the right to directly use these resources or to grant them in concession to third parties. Finally, according to Mexican legislation, mining holds a national priority status over conservation and mitigation of global greenhouse gas emissions.

### Forest policies

Since the late 1940s, industrial development based on an import substitution model became a national priority in Mexico. As in other large Latin American countries such as Brazil and Argentina, strong centralized governments assumed the role of directly promoting this model.

Small rural producers were given the role of providing staple foods at low cost, thereby enabling low industrial salaries.[10] The state did not consider members of forest communities to be capable of managing forest operations. Instead, long-term concessions of community forests were imposed in different regions in order to give industries access to raw materials. Similarly, communities lost their rights in almost half of the forestland where logging bans were imposed to protect river basins. From the perspective of some commoners, these policies made forests an obstacle to real ownership of the land. Confronted by continuous local resistance, forest industries were organized for short-term profits and they kept their rate of reinvestment as low as possible. In the 1970s, forest industries were nationalized. The Mexican state used the profits of forest exploitation for investments in other economic sectors. Forest regions under logging bans suffered strong deterioration as private forest industries continued their operation in those areas. Logging activities were carried out without any restrictions or provisional measures (Bray and Merino, 2004; Merino, 2004; Boyer, 2005; Merino and Segura-Warnholtz, 2005; Bautista, 2007).[11]

In the 1970s this economic and political model started to show signs of exhaustion. With regard to forest policies, neither concessions nor logging bans accomplished their economic or environmental goals. Forest deterioration increased while most industrial logging plants operated below their installed capacity.[12] Concessions favoured "rent extraction" over sustained exploitation, reinvestment in forest protection and long-term management systems. This led to a "disinvestment" and consequently resulted in the loss of forest resources, value and productivity.

By the early 1980s, when the concessions were close to expiring, many communities claimed the right to regain the use and control of their forests. Social mobilization, support of civil society groups and the closing of many state-owned industries enabled communities in Mexico to win this struggle. After having worked for concessionaries for many years, community members realized that timber extraction could be profitable and sustainable. Some communities engaged in community forest production. Their initiatives were supported by a progressive group within the federal administration: the Dirección de Desarrollo Forestal (DDF), which held the view that communities could be both efficient producers and forest stewards. The DDF promoted the organization of community unions to create economies of scale that would enable communities to hire technical advisors who were previously provided by the federal government. Through these unions, communities gained a stronger presence both in politics and in the market (Alatorre, 2000; Bray and Merino, 2004; Chapela, 2005). In 1986 a new forest law banned concessions and granted communities the right to be consulted on the implementation of any policy that restrained their property rights.

Communities with the most valuable forest assets and good organization showed remarkable achievements. They reinvested most of their profits from forest businesses in improved forest-management systems, building and providing for the maintenance of forest roads. They also acquired industrial equipment, and organized their own technical and administrative teams. Not only did forestry provide employment and income to local residents but its profits were invested in local public goods: schools, clinics, community celebrations, roads and transport. Some communities adopted environmental agendas to promote sustainable harvests, minimize environmental impacts and diversify forest use. Since the 1990s a group of communities were granted forest certification under the Forest Stewardship Council scheme. There are currently 39 certified community forests in Mexico, amounting to 655,206 Ha.[13]

These successful forms of community forest management created local incentives for conservation, improved quality of life in marginalized regions, and favoured democratic governance of forest commons. Some certified communities have even gained international recognition.[14] The experience of the *ejidos* of southern Quintana Roo was replicated in the neighbouring tropical forests of Petén-Guatemala, where Mexicans trained local user groups and thus supported the establishment of community forestry operations.

Sustainable forest management and production – one of the strategies proposed to halt deforestation – require coherent and continuous long-term support to local users. However, government support for community forestry faded during the late 1980s and early 1990s. With the implementation of NAFTA, the national market was abruptly opened and community producers were unable to compete with US and Canadian forest producers. At the same time, the over-regulation of forest activities by the Mexican state led to high transaction costs. Finally, subsidies for tropical agricultural, cattle holding[15] and the extension of illegal logging led to the disruption of some community forestry initiatives during that period.[16] NAFTA therefore put pressure on the productive initiatives of forest communities. While some export-oriented subsectors in agriculture, manufacturing and services benefited, many small and medium-sized (urban and rural) businesses failed in the absence of policies to protect and promote their productivity and competitiveness.

In 1994 the federal administration created the Ministry of the Environment, SEMARNAP, with forest management as part of its jurisdiction. SEMARNAP's main natural resources policy was the expansion of the protected areas system and a simultaneous increase in regulations for activities such as forestry. Nevertheless, SEMARNAP also renewed support for community forestry, creating two small programmes: PRODEFOR (Programa para el Desarrollo Forestal) and PROCYMAF (Programa de Conservación y Manejo Forestal), a joint initiative of SEMARNAP and the World Bank. PROCYMAF was a pilot project based on the recognition of a variety of socioeconomic and ecological conditions of forest communities, and on the need to continue to build finetuned strategies to address the diversity of local contexts.[17] PROCYMAF was influenced by the international advocacy in favour of participatory, decentralized, pro-poor forest policies that emerged during the 1990s and 2000s. Its main goals were to strengthen communities' social capital, and their productive and institutional capacities. After a few years, PROCYMAF presented some important achievements, such as a growing system of forest area under certification, the creation of numerous community forest enterprises, the adoption of participatory land-use planning, the definition of community rules for local forest governance, and the establishment of regional committees of forest communities.

After the Partido Acción Nacional (PAN) took over the national government in 2000, Mexico's economy and governability were increasingly characterized by corruption and authoritarian practices. Aiming to increase its legitimacy and to show political strength, the PAN government launched an extensive "anti-drug war". In the context of economic and institutional failures, widespread corruption, persistent poverty and inequality, this led to the spread of violence to many regions. Criminality against the population became common, involving criminal gangs but also the police and the army (Cendejas, 2015).

At the same time the PAN administrations (2000–2012) responded to environmental concerns that tried to give forest policy a high profile. Between 2000 and 2008, public investment in forests increased by 7000% (Figure 9.1). The distribution of these public funds expressed a conventional conservationist vision. About 70% was invested in the establishment of forest plantations and massive reforestation programmes. In general, these investments yielded poor results.[18] Some 12% of the funds were used for the Programme of "Payment for Environmental Services" (PES), and were given to forest owners who gave up forest use.[19] Only 10% of the federal forest budget was used to support community forestry (Merino and Ortiz, 2013).

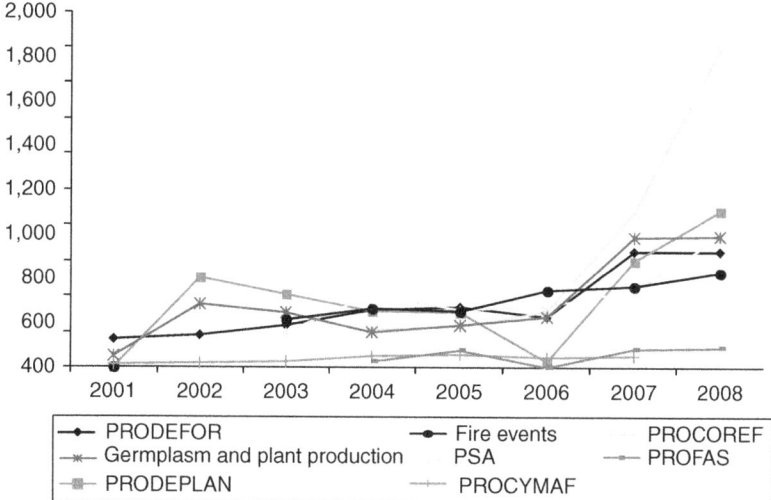

*Figure 9.1* National annual budget of CONAFOR according to different forest-related projects in Mexico (in million pesos), 2001–2008
Source: Merino and Ortiz (2013).

## Forest communities in Mexico

Forest communities in Mexico exist in a range of socioenvironmental contexts. This study is based on an analysis of a sample of 102 forest communities, which basically tested two main hypotheses (following Cárdenas, 2006):

- forest conditions and sustainable forest use depend largely on the robustness of local institutions;[20]
- institutional robustness relies on interlinked characteristics of forest users, namely, social capital and dependence on forest resources.[21]

The sampled communities are distributed across five states: (1) Oaxaca in the south, where 19 indigenous groups (mainly *Zapotecos* and *Mixtecos*) constitute the majority of the population; (2) Guerrero, also in the south, with an important presence of *Nahuas* and *Mixtecos*; (3) Michoacán in central Mexico, home of the *Purépechas*; (4) Jalisco in the west, whose mountains are home to the *Huichol* people; and (5) Durango in the north-centre, with five indigenous groups, mainly *Tepehuanes*.[22] Together these states add up to more than half of the forestland of

Mexico, where approximately 70% of the nationally produced timber comes from. As a whole, these forest regions have a lower population density and lower deforestation rates than the rest of the temperate forest areas.[23] The remainder of this chapter will provide a detailed analysis of forest conditions, use and governance in the studied areas.

## Forest types and uses

The sampled communities are located in the mountain range along the Pacific coast and the central neovolcanic axis, at high altitudes. These areas frequently have important altitudinal ranges of non-forested lands and different types of forest vegetation, including temperate forests (pine, pine-oak, oak, fir and cloud forests), as well as dry and humid tropical forests (below 1,500 to 500 m above sea level. Distinct types of forest are perceived, used and managed in different ways.

Forest uses vary according to forest type (Table 9.1). Firewood is the only type of wood collected in almost all type of forest. Commercial logging – the most important income-generating forest activity – takes place in about half of the pine and pine-oak community forest areas. Agriculture and grazing – sometimes based on the removal of the forest cover – take place in the dry and tropical forests. Interestingly, community conservation initiatives, sometimes supported by government programmes, are not present in tropical humid and tropical dry forests within the sample. The latter type of forest has the greatest biodiversity and number of endemic species, and it represents the most endangered forest type in Mexico.[24]

In summary, sustainable use options are limited or absent in most of the community forests analysed in this study. The lack of these

*Table 9.1* Different uses of forest by community residents in Mexico

| Type of forest | Firewood collection % | Grazing % | Agriculture % | Conservation/ PES % | Logging % |
|---|---|---|---|---|---|
| Pine | 65[25] | 60 | | 62 | 58 |
| Pine-oak | 81 | 60 | | 18 | 48 |
| Oak | 92 | | | 80 | |
| Fir | 45 | | | 70 | |
| Cloud[26] | 41 | | 30 | 80 | |
| Tropical dry | 61 | 75 | | | |
| Tropical humid | | 75 | | | |

*Source:* Survey on Forest Communities with Temperate Forests in Mexico, IIS-UNAM.

opportunities endows forests with low social value. Under such conditions, forest areas are prone to highly impactful activities, such as mining or commercial plantations.[27]

### Forest size and tenure issues

Forest resources are valuable assets for the majority of the communities, particularly those that own forest resources of commercial value.[28] In most cases, however, forest areas are relatively small. Only 10% of the communities have more than 10,000 Ha, while half of them are smaller than 2,000 Ha and 20% of communities possess forests of between 500 and 300 Ha.

The governance of communal forests has the potential to generate a range of social benefits, including more participation in forest protection (Merino, 2005; Agrawall and Ashwini, 2009). However, collective tenure does not necessarily lead to equal access to forest resources for all community members, or to equal incentives to protect them. The two types of communal forest in Mexico – *ejidos* and *comunidades agrarias* – present important differences. In *ejidos*, which are predominant in Durango, Jalisco and Michoacán, many families do not have property rights, while *comunidades agrarias* own 95% of the forests in Oaxaca. *Ejidos* also face serious ageing problems – 88% of the rights holders are at least 40 years old and 28% are over 60 years old. In contrast, 64% of the residents in *comunidades agrarias* are younger than 40. The different age structure results from the legal rights of *comunidades agrarias*, which facilitate the inclusion of new members.

Collective tenure in Mexico remains strong in spite of the many pressures that it faced before and after the 1991 legal reform that enabled the privatization of *ejido* lands (Wayne, 1998; Warman, 2000). Sales of *ejido* lands occurred in 30% of the sampled communities.[29] In 82% of the communities, local authorities declared that the majority of community members favour the maintenance of collective property.

Conflicts over land tenure are not rare in the case studies. In particular, intercommunity conflicts over borders occur in 34% of the cases. Intracommunity conflicts over the limits of individual plots were reported in 21% of the studied communities. Conflicts over borders are more frequent in *comunidades agrarias*, where these problems have remained unresolved for generations. Such conflicts usually have negative impacts on forest conditions due to unclear ownership. According to local authorities, they favour deforestation and illegal logging, and therefore create challenges for implementing protective measures.

### Forest Communities performance

The performance of forest communities was measured by five indices: (1) pressure on forest areas; (2) protection and conservation activities; (3) social capital and organization; (4) local institutional strength; and (5) community forest economy[30]. Table 9.2 summarizes the main results, divided into five rank categories.

*Table 9.2* Indices of forest communities' performance

| Index | Very low % | Low % | Moderate % | High % | Very high % |
|---|---|---|---|---|---|
| Pressure on forest areas | 10.7 | 26.2 | 26.2 | 12.6 | 24.3 |
| Protection/ conservation activities | 35.9 | 27.2 | 22.3 | 9.7 | 4.9 |
| Social capital and organization | 3.9 | 23.3 | 53.4 | 16.5 | 1 |
| Local institutional strength | 27 | 27 | 36 | 10 | 0 |
| Community forest economy | 69 | 13.6 | 7.8 | 5.4 | 3.9 |

N=103.
*Source*: Survey on Forest Communities in Mexico, IIS-UNAM.

The *pressures on forest areas index* combines (1) occurrence of illegal logging; (2) forest fires and pests; (3) grazing in forest areas; and (4) land-use change. The results show that pressures on forest areas are remarkable in nearly 37% of the sampled communities, while a very similar proportion of the community forests face low levels of pressure. It is worth mentioning that questions have arisen during the past decade about the perception of change (increase or decrease) in forest pressures. Most of these pressures have a socioecological basis. In particular, effects of global change add a level of uncertainty regarding the occurrence of fires and pest outbreaks, as well as in rainfall and drought patterns. A significant share of these communities (16.5%) reported recent forest losses.

*The conservation and protection activities* index combines variables related to (1) monitoring of forest areas in order to address forest fires, pests and illegal logging; (2) local organizational and technical capacities to face these pressures; (3) initiatives of reforestation; and (4) the existence of community conservation areas. This index was built to capture practices that favour conservation rather than actual conservation or degradation of forest areas.[31] Protection and conservation practices are low in the majority of the cases (63%). However, in 27% of the communities where conservation and protection activities were ranked as "low", communities perform basic protection activities such as fire-fighting. It is interesting to note that the proportion of communities with very poor conservation practices is similar to the percentage of communities where forest pressures are perceived as "high" and "very high". In communities where conservation and protection measures are moderate (22.3%), residents are engaged in the monitoring of forest areas.

Only in 14.6% of these communities were conservation and protection practices classified as "high" and "very high". A relevant finding is the presence of community conservation areas, particularly in the *comunidades agrarias* with indigenous background (61% of the sampled communities in Oaxaca, and 58% in Guerrero). Community conservation is also significant in 44% of the *ejidos* and *comunidades agrarias* in Michoacán, and among 38% of the *ejidos* in Durango. Many of these conservation areas are located in areas identified as water-capture sites and have been established as part of community projects to protect water sources.

The *social capital and organization* index includes (1) frequency of community meetings; (2) strength of local governance systems; (3) participation in community meetings; and (4) non-paid community work. This index is particularly important as social capital and organization are considered by the "commons school" to be preconditions for forest governance and sustainability (Ostrom, 2009). Social organization in *ejidos and comunidades agrarias* faces a variety of challenges and has important downsides: the exclusion of women and young people; conflicts created by the "elite capture" of benefits of common resources; and the "costs" of traditional practices of governance and reciprocity. Increasing out-migration puts social organization under additional stress, as it drains crucial human resources needed for local governance and generational replacement. This adds to the challenge of maintaining social capital across different generations. These pressures are particularly strong in about one-quarter of the sampled communities (27.2%), where the value of this index is "low" and "very low".

Governance based on local participation takes place through regular community assemblies to discuss collective issues, make decisions and formulate rules about the following issues: use and management of the forest commons, use of the profits of the communal productive initiatives, and relationships with government programmes. Community members take part in different positions of the local governance system, mostly on a voluntary basis. In addition, non-paid community work – which takes place in many cases – serves as a base on which to build and maintain collective infrastructure, public services, forest protection and forest-restoration activities.

Despite the organizational foundation observed in many forest communities, the low percentage of communities with a higher level of organization and social capital (17.5%) reflects the high costs of community and common forest governance. Within this sample, lower values of social capital and organization are often linked with the exclusion of *avecindados* (family heads living in communities), lack of property rights and little or no rights to take part in meetings or use common resources.

*The local institutional strength* index is based on (1) the existence of community rules for local governance; (2) rules related to the use and provision for local commons (e.g. public spaces, forests, infrastructure, community profits from forestry or other collectively held activities); (3) community participation in the definition of the agreement; (4) awareness and knowledge; (5) monitoring and sanctioning of compliance with the rules; and (6) community members' trust in rule compliance. Local institutions are considered fundamental for sustainability and governance by the "commons/collective action perspective" as they are the result of collective agreements for commons governance and use. Nevertheless, the definition and enforcement of local institutions are demanding tasks. Community participation and knowledge are required to legitimate local rules and better match the local context. Communities of users and/or owners of common resources coordinate to create collective institutions when they perceive the need and have the conditions that enable them to do so.

In most of the communities under study, local rules refer to local governance and, sometimes, to the extraction and use of firewood. The values of this index express a relative weakness of local institutions: a lower level in half of the communities' local institutions and high in only 10% of these communities (Table 9.2).

This pattern partially reflects the centralized forest governance in Mexico, in which local communities are completely excluded from the definition of use and management rules. As a result, national rules are often inadequate for particular forests or communities in a large and highly diverse country. In addition, frequent changes in laws and rules increase uncertainty and the ability of forest users to comply with government regulations.

Conflicts between local and national monitoring systems add to the challenges for local institutions. In Mexico a federal government agency (the Procuraduría de Protección Ambiental) is officially responsible for monitoring compliance with federal forest rules. Limited coordination between the Mexican environmental enforcement agency (PROFEPA) and the monitoring initiatives of local communities leads to conflict between the two institutional arrangements. Imposed rules, external – and often inefficient – monitoring and sanctioning, "crowd out" risk, and eroded local institutions have resulted in a favouring of local "open access" conditions (Cárdenas, 2008).

Finally, the *community forest economy* index combines (1) a level of vertical integration of forest production and the capacity to add value to forest products; (2) diversification of forest uses, taking commercial and domestic purposes into account; (3) productive forest assets owned by communities; and (4) ownership of financial assets. This index corresponds to "forest dependence", an important condition for the social value of common forests, and the incentives to commit to their governance and conservation (Ostrom, 2009).

The level of development of the communal forest economy was considered "very low" in 69% of the sampled communities. In half of the communities, forest only provides firewood for domestic use. In the other half, residents harvest and sell non-timber forest products (NTFP) such as mushrooms, resin, medicinal plants and firewood.[32] Individuals or family groups who take part in these activities are often the poorest members of the community. These products deliver very low profits due to market control by intermediaries.

Logging remains the most important (legal) income-generating activity in forest regions. It takes place in one-third of these communities, of which 13.6% sell timber as "stump". In these cases, outsiders perform forest management and extractions with little community control. These operations, which often have a high impact on forests, deliver scarce local benefits, and create mistrust and opposition to commercial forestry.

Forest management and timber-harvesting operations are carried out in only 17% of the communities. About half of this last subgroup produced only raw material (logs) due to limited productive capacities and financial resources to cover production costs. Nearly 10% of these communities have achieved productive vertical integration, including their own forest mills and sale of primarily tables. However, only 4% of the sampled communities have achieved vertical and horizontal integration as forest producers. They have diversified commercial forest uses, combining timber products with NTFP and/or providing ecotourism services. This low performance reveals the challenges faced by the community forestry industry – namely, the organization of production, how to reach national and international markets, financial and fiscal tasks, efficiency, accountability to communities' assemblies, and the operation of entrepreneurial administrations in the context of local governance systems. Nevertheless, these communities have created local sources of employment and income. They have financed local infrastructure and public services with the profits of their own business. They have also contributed to developing and strengthening human resources, social capital and local governance (Bray, 2007). The following section shows how the community forestry economy is related to forests and communities.

### Forest communities and community forestry

Community forestry touches upon the socioeconomic conditions of the community as well as ecological conditions and pressures of the forest. The indexes analysed in the previous section reveal that social organization, forest conditions and forestry are closely related.[33]

Organization around local governance, and commons management and use, is present in many forest communities. However, communities with weak local institutions tend to report a higher level of pressures in their forests. Conflict over community borders is particularly related to increasing pressures on forest areas, which are almost four times as greater as for communities that do not face this problem. In contrast, communities that control forest management and forest production tend to be more involved in protection and conservation activities. Furthermore, communities with internal rules regarding the protection and management of forests tend to be more successful in addressing pressures on forests. Their members engage in the reduction of the risks of forest pests and fires. They also monitor forest areas in order to observe early signs of potential threats. Nevertheless, the level of pressure varies considerably according to the individual forest dynamics. Fires and pests, for example, are multifactorial, in which climatic events such as strong and/or longer dry seasons may play an important role. Therefore, as pressures on forests increase, local rules must be fine-tuned as well.

Not surprisingly, we found a strong relationship among social capital and organization and local institutional strength. Basic organizational practices, such as collective rules in use and trust, are important for promoting social capital. In general terms, communities with the strongest organization are also those where protection and conservation activities are more frequent and diverse. This pattern reveals the relevance high levels of social organization required to support local coordination and collective action addressing forest protection and conservation activities. A small number of communities, however, perform forest-protection activities at high intensity. This pattern is a possible outcome of government subsidies for reforestation. In contrast, a reduced number of highly organized communities showed very few protection and conservation practices. These cases reveal that communities may be organized for different purposes that do not necessarily coincide with forest conservation. In summary, social organization is an important requirement for performing conservation activities but it is not sufficient for creating incentives to engage in forest protection.

Protection and conservation activities are closely related to the development of community forestry. The forest economy tends to be low where local institutions are weak. As a result, limited protection and conservation activities lead to increased pressure on forests. Interestingly, pressure on forests drops considerably according to the increasing importance of commercial forest activity in communities. In cases where recent deforestation took place, forest economy in the communities is weak. In general, communities with lower levels of pressure on forests are those with the most consolidated economic forest activities. This information suggests that as the incentives, knowledge and technical skills increase – as a result of a more diversified community forest economy – community members are more able and willing to identify and address pressures on forests before their impacts grow out of control.

Communities with the highest level of local forest economy[34] tend to have lower institutional strength in comparison to communities with only vertical integration of forest production. The former needs stronger and more diversified institutions in order to manage industrial and commercial operations, diversify forest production and carry out multiple activities, such as timber extraction and processing. However, as these data reveal, new economic forest activities may lack the institutional support to scale up their commercial activities. If not properly addressed, these "institutional gaps" can undermine common natural resources used in production processes, collective initiatives and community governance itself.

## Conclusion: Community forestry beyond autonomy

In highly unequal societies such as those in Mexico and most of Latin American countries, governments and urban societies need to overcome the anti-rural, anti-community, anti-poor biases that are frequent in legal frameworks, and in environmental and economic policies. For decades, Mexican forest communities have faced adverse policies that constrain local initiatives. These have encouraged the abandonment of many forest and rural regions where local livelihoods have become difficult to sustain. Research on local forest use in Mexico and other developing countries (IFRI; Ribot, Agrawal and Larsson, 2009) shows a permanent tension between trends of decentralization and centralization of decision-making rights over natural and strategic resources.

Most Latin American forests are owned by central governments while logging concessions are given to international corporations (Whyte and Martin, 2001). Concessionaires tend to maximize short-term profits of forest operations to reinvest outside the country. As a result, forests become sources of revenue for national governments with limited local control over the impact of extractions.

Not surprisingly, this model of "mining forestry" leads to the marginalization of local people and high environmental impacts. The last two federal administrations in Mexico (2000–2012) responded to global environmental concerns, thereby attempting to give forest policy a high profile. Despite the increased public investment in forests between 2000 and 2008, this budget largely overlooked the needs to promote local productive and governance capacities, and the creation of stable incentives for conservation.

Successful experiences of community forestry have revealed important lessons that can change this trend. They reveal positive synergies not only among common forest management, local livelihoods and conservation but also with maintenance and the development of "commonality" based on local institutions and social capital. The results of this research show that social capital and institutional strength are key factors for the protection of forest commons, and for local capacities to face traditional and emergent pressures on forest ecosystems. Human resources and collective action are critical for resilience. The presence of communities with forest conservation, governance and local development in Mexico shows the viability of these initiatives, even if they still constitute a minority.

Forests are commons whose sustained management and use require high levels of cooperation among relevant actors. Collective action in Mexican communities is even more necessary due to the collective tenure of the vast majority of Mexican forests. Communal property can be an important possibility for favouring sustainability and the governance of complex ecosystems, such as forests. However, despite the relevance of legal recognition of property rights over lands and forests to local communities, it is hardly sufficient for forest communities to achieve their economic, environmental and social potential. The empowerment of local communities by acquiring technical and governance capacities is equally important in contemporary contexts. The results of this survey show the existence of many communities that suffer forest deterioration and limited social capital. In these communities, the contribution of forest activities to local livelihoods is often very limited. The development of a forestry economy is fundamental not only for supporting the social and institutional development of these communities but also for delivering protective measures for sustainable forest use.

The experience of forest communities in Mexico shows that the synergy between forest economy and conservation does not happen naturally; it requires favourable public policies as well as access to adequate training and technical advice.

The state has undermined community rights and livelihoods, favouring communities' dispossession. This entails a recentralization of land control and resource management, over-regulation of resource use, imposition of high transaction costs on legal forest use, and criminalization of many local uses of natural resources. But if local governance and environmental citizenship are regarded as assets for conservation and governability, the state can play a key role by recognizing communities' rights over natural resources. This would provide favourable legal frameworks for community forest use and governance – by coordinating with local actors to control illegal land use – and would favour markets able to internalize sustainable management costs.

Lessons from Mexico's community forestry experience are relevant for other Latin American countries, such as Guatemala, Nicaragua, Bolivia and Brazil – where governments recognize local collective rights. In countries where most of the forestlands are owned by governments and are used by private companies, local governance, incentives and recognition of communities' rights can be avenues for reversing environmental injustice and deterioration. In summary, community forestry is not a panacea or a fixed model that can simply be replicated inside or outside Mexico. Nevertheless, it represents an important alternative to combine goals of local empowerment, forest sustainability and rural development. While some communities in Mexico seek to distance themselves from the state and the traditional market (see

## Notes

1. I consider "community forestry" to be those cases in which local communities have and practise use and control rights over the forested areas (Schlager and Ostrom, 1992) and where they preserve the forest cover and have instituted use and management rules, regardless of the ways in which they use forest resources.
2. About 3% yearly.
3. The Millennium Ecosystem Assessment defines these "services" as provisional, regulatory, cultural and support services.
4. Mexico has the fifth greatest biodiversity in the world; the top ten megadiverse countries host 70% of the Earth's biological diversity.
5. The Stockholm Resilience Center defines interrelated dimensions of global environmental change as loss of biodiversity, ocean acidification, changes in the cycles of phosphorus and nitrogen, land-use change, depletion of the atmospheric ozone layer, pollution of soils and water, and aerosol atmospheric load (Rockstrom et al., 2009).
6. Some 73% of the land has forest cover, accounting for nearly 142 million Ha.
7. I use the term "community" when referring to both *ejidos* and *comunidades agrarias*.
8. De facto forest division is happening in many communities.
9. In terms of the "bundle of property rights" scheme proposed by Schlager and Ostrom (1992), community members have access, use, exclusion and some management rights over forests. The federal government maintains key control rights over them.
10. The Mexican diet was based on corn and beans, the prices of which were controlled by the federal government for decades.
11. From 1950 to 1970 the national demand for forest products grew continuously and the country's economy grew by 7% annually. From 1950 to 1989, the population growth rates were close to 3% per year. These were also years of strong expansion of the market economy in traditional rural communities.
12. Communities with forests under concessions were not legally able to use them, nor were they free to choose timber buyers or negotiate timber prices, which were fixed by the Ministry of Agrarian Affairs. Nevertheless, they kept the right to allow or refuse logging in their lands.
13. During the last decade, certified forest areas have decreased as certification poses strong demands without giving clear access to better marketing conditions.
14. San Juan Nuevo in Michoacán; Ixtlán, UZACHI, Ixtlán, Textitlán, Mancomunados and San Pedro el Alto in Oaxaca; Santiago Papasquiaro in Durango; el Balcón in Guerrero; el Largo in Chihuahua; Nohbec in Quintana Roo.

15. Some subsidies for small agriculture were created after the implementation of NAFTA, mainly with political purposes. This is the case for PROCAMPO, which provided resources per hectare planted with corn, regardless of its productivity. Subsidies to acquire cattle were maintained until recently, given mainly by state governments.
16. It has been estimated that illegal logging is at least as great as legal production (Consejo Civil Mexicano para la Silvicultura Sostenible – CCMSS, PROFEPA).
17. PROCYMAF worked in the states of Oaxaca, Guerrero, Michoacán, Jalisco, Quintana Roo and Durango.
18. By 2009, Greenpeace reported that official reforestation had a survival rate close to 10%.
19. Payments were established based on the average price of corn at an estimated national agricultural productivity average in areas with no irrigation.
20. Institutions are defined as "rules in use" (Ostrom, 2005).
21. This study was carried out as part of the international programme Forest Resources and Institutions (IFRI). A global database of forests and forest users around the world has been developed since the 1990s (Wollenberg et al., 2007). By focusing on two hypotheses of the IFRI programme (Cárdenas, 2006), a team from the Instituto de Investigaciones Sociales of the National University of Mexico (IIS-UNAM) applied a survey inspired by the IFRI conceptual approach.
22. The universe of the sample includes all the communities in these five states with at least 300 Ha of temperate forest. It is stratified based on the proportion of communities with this characteristic in each state, as compared to the total number of communities with 300 Ha (or more) of temperate forest in these five states.
23. The results of the survey are representative of half of Mexican temperate forests that face less pressure. We could not include the state of Chihuahua (the second largest timber producer in Mexico, which has the largest forest extension and where the conditions of forest regions are similar to those of the state of Durango).
24. Mexico's dry forests are rich in "neo-endemism" (new species that originated in a particular region and are only found there). This is currently the fastest-disappearing forest type. Mexican cloud forests are rich in "paleo-endemism".
25. This is the percentage of forests of each of type within the communities of the sample.
26. During 1970–1980, subsidies for sun coffee based on the removal of the forest were the main driver of the rapid disappearance of cloud forests; since 1990, many communities have grown shade coffee, preserving forests. Some of them are certified as organic sustainable coffee producers.
27. From 2006 to 2012 the areas subject to mining concessions in mountain forests increased by 30%. Much of the medium- and small-scale mining is now controlled by drug cartels, as in the southern sierra of Michoacán, which is rich in iron ore deposits. Other activities with a high environmental impact that are practised in dry forests include the establishment of plantations (e.g. *Agave cupreata*, used for the fabrication of tequila), illegal cropping and extensive cattle ranching.

28. Mainly pine.
29. These are primarily sales of plots among *ejido* dwellers. In most cases, they are not associated with privatization of the *ejidos*. They do not include sales of forestland.
30. See methodology used for the construction of these indices at the site http://www.ccmss.org.mx/documentacion/830-a-vuelo-de-pajaro-las-condiciones-de-las-comunidades-con-bosques-templados-en-mexico-borrador/
31. The assessment of forest conditions requires other types of research methodology and techniques.
32. These are wood products classified as NTFP.
33. For a more detailed description of these data, see Merino, Leticia and Martínez Ana Eugenia, "A vuelo de pájaro. Las condiciones de las comunidades con bosques templados en México", www.conabio.gob.mx.
34. Industrial capacities, diversification of forest production: NTFP, environmental and touristic services and so on.

## References

Agrawal, A. and Ashwini, C. (2009) 'Trade-Offs and Synergies Between Carbon Storage and Livelihood Benefits from Forest Commons', *Proceedings of the National Academy of Science* 106(42), 17667–17670.

Alatorre, G. (2000) *La Construcción de una Cultura Gerencial Democrática en las Empresas Forestales Comunitarias* (México, D.F: Editorial Juan Pablos, Procuraduría Agraria).

Bautista, L. (2000) *Las Vedas Forestales en México*. Tesis de Maestría en Estudios Regionales (México: Instituto Dr. José María Luis Mora).

Bevir, M. (2013) *Governance: A Very Short Introduction* (Oxford, UK: Oxford University Press).

Boyer, C. (2005) 'Contested Terrain: Forestry Regimes and Community Responses in Northeastern Michoacán, 1940–2000', in D. Bray, L. Merino and D. Barry (eds), *The Community Forests of Mexico: Managing for Sustainable Landscapes* (Austin: University of Texas Press), 27–48.

Bray, D. and Merino, L. (2004) *La Experiencia de las Comunidades Forestales Mexicanas* (México: Instituto Nacional de Ecología).

Bray, D., Merino, L. and Barry, D. (eds) (2005) *The Community Forests of Mexico: Managing for Sustainable Landscapes* (Austin: University of Texas Press).

Cárdenas Campo, J.C. (2009) *Dilemas de lo Colectivo: Instituciones, Pobreza y Cooperación en el Manejo Local de los Recursos de Uso Común* (Bogotá: Universidad de los Andes, Facultad de Economía, CEDE, Ediciones Uniandes).

Chapela, F. (2005) 'Indigenous Community Forest Management in Sierra de Juárez', in D. Bray; L. Merino and D. Barry (eds), *The Community Forests of Mexico: Managing for Sustainable Landscapes* (Austin: University of Texas Press), 91–110.

Durán, E., Mas, J. and Velásquez, A. (2005) 'Changes in Land Vegetation Cover and Land Use Change in Communities with Forest Management and Protected Areas', in D. Bray, L. Merino and D. Barry (eds), *The Community Forests of Mexico: Managing for Sustainable Landscapes* (Austin: University of Texas Press).

De Janvry, A., Salouet, E. and Gordillo, G. (1999) *La Segunda Reforma Agraria de México: Respuestas de Familias y Comunidades 1990–1994* (México: El Colegio de México).

Demzets, H. (2002) 'Towards a Theory of Property Rights II: The Competition Between Private and Collective Property', *The Journal of Legal STudies* 31(2):653–672.

Hardin, G. (1968) 'The Tragedy of the Commons', *Science* 162(3859): 1243–1248.

INEGI (2010) *Censo General de Población y Vivienda* (Instituto Nacional de Geografía y Estadística).

Larson, A., Cronkleton, P., Barry, D. and Pacheco P. (2008) *Tenure Rights and Beyond: Community Access to Forest Resources in Latin America*. Occasional Paper no. 50 (Bogor, Indonesia: CIFOR).

Manson, R. and Jardel, E. (2012) 'Perturbaciones y Desastres Naturales: Impactos sobre las Eco-Regiones, la Biodiversidad y el Bienestar Socioeconómico', in *Capital Natural de México, vol. II: Estado de Conservación y Tendencias de Cambio* (México: CONABIO), 131–184.

McCay, B.J. (2007) 'Introduction to Human Ecology. On Marine Resources', *Human Ecology* 35(5): 513–514.

McKean, M. (2000) *People and Forests: Communities, Institutions, and Governance* (Cambridge: MIT Press).

Meinzen-Dick, R. (2007) *Proceedings of the National Academy of Sciences* 104(39): 15200–15205.

Merino, L. (2004) *Conservación o Deterioro. El Impacto de las Políticas Públicas en las Comunidades y en los Bosques de México* (México: Instituto Nacional de Ecología).

Merino, L. (2010) Interview with Elinor Ostrom. The Commons Digest (also in youtube), www.iasc-commons.org, date accessed 12 January 2015.

Merino, L. and Martínez, A.E. A Vuelo de Pájaro. Las Condiciones de las Comunidades con Bosques Templados en México. www.conabio.org.mx, date accessed 22 January 2015.

Merino, L. and Ortiz, G. (2013) *Encuentros y Desencuentros: Las Comunidades Forestales y las Políticas Públicas en Tiempos de Transición* (Mexico: Miguel Angel Porrua/ Instituto de Investigaciones Sociales de la UNAM).

Merino, L. and Segura-Warnholtz (2005) 'Forest and Conservation Policies. Impacts on Forests Communities in Mexico', in D. Bray; L. Merino and D. Barry (eds), *The Community Forests of Mexico: Managing for Sustainable Landscapes* (Austin: University of Texas Press), 49–70.

Molnar, A., Scherr, S. and Khare, A. (2004) *Who Conserves the World's Forests. A New Assessment of Conservation and Investment Trends* (Washington, DC: Forest Trends).

Millennium Ecosystem Assessment (2003) *Ecosystems and Human Wellbeing* (San Francisco: Island Press).

Ostrom, E. (1990) *Governing the Commons. The Evolution of Institutions for Collective Action* (Cambridge University Press).

Ostrom, E. (2005) *Understanding Institutional Diversity* (Princeton University Press).

Ostrom, E. (2007) 'A Diagnostic Approach for Going Beyond Panaceas', *Proceedings of the National Academy of Sciences*, 104(39):15181–15187.

Ostrom, E. (2009) *A Polycentric Approach for Coping with Climate Change*. Policy Working Paper 1 (Washington, DC: The World Bank).

Ostrom, E., Dietz, T., Dolšak, N., Stern, P.C., Stonich, S. and Weber, E.U. (eds), (2001) *Drama of the Commons* (Washington, DC: National Academy of Science).
Pierce-Colfer, C. (eds) (1997) *Who Counts Most for Sustainable Forest Management?* (CIFOR).
Ribot, J., Agrawal, A. and Larson, A. (2006) 'Recentralizing While Decentralizing: How National Governments Re-appropriate Forest Resources', *World Development* 34(11): 1864–1886.
Rockström, J., Will, S. and Noone, K. et al. (2009) 'Feature: A Safe Operating Space for Humanity', *Nature* 461: 472–475.
Schlager, E. and Ostrom, E. (1992) 'Property-Rights Regimes and Natural Resources: A Conceptual Analysis', *Land Economics*, 68(3): 249–262.
Warman, A. (2000) *El Campo en México en el Siglo XX. Siglo de Luces y Sombras* (México: Fondo de Cultura Económica). Cornleius, W. *Ejido Reform: Stimulus or Alternative to Migration. The Transformation of Rural Mexico. Transforming the Ejido Sector* (San Diego: Center for US-Mexico Studies).
Wollenberg, E., Merino, L., Agrawal, A. and Ostrom, E. (2007) 'Fourteen Years of Monitoring Community-Managed Forests: Learning from IFRI's Experience', *International Forestry Review* 9(2): 670–684.

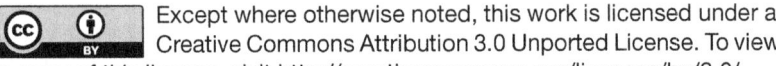 Except where otherwise noted, this work is licensed under a Creative Commons Attribution 3.0 Unported License. To view a copy of this license, visit http://creativecommons.org/licenses/by/3.0/

OPEN

# 10
# Local Solutions for Environmental Justice

*David Barkin and Blanca Lemus*

In the context of the prevailing abundance of diversity (biological, ethnic), the profound social inequalities, and the trends and attitudes of hegemonic forces in Latin America, a coherent process of environmental governance is proving difficult and environmental injustice is aggravated. In virtually every country in the region, increasing subordination to the global market has led to dramatic transformations in productive structures and processes along with the often violent opening of new territories to domestic and foreign investment in renewable energy projects, primary production for international markets, and natural resources exploitation. These changes are provoking direct confrontations between, on the one hand, domestic policy-makers, well-financed investors positioned to operate in international markets, purveyors of technologies, investors with concessions in regions and sectors recently opened to foreign investment, and, on the other hand, organized groups from many parts of society who see these penetrations as a menace to their productive systems, to their livelihoods and their health, while also being destructive of their communities, their cultures and the

---

We are deeply indebted to the members of the Local Solutions teams participating in the Environmental Governance in Latin America project for their contributions to this chapter; this formulation would not have been possible without the continuing exchanges in the communities over the course of the past three years. The contributions of Gustavo Esteva, Mario Fuente and Victor Toledo have also been important. Special thanks are due to the critical contributions of the participants in the seminar in heterodox economics in the doctoral program in economic sciences at the Universidad Autónoma Metropolitana, and the active participation of the specialists in ecological economics in the participating communities. Of course, responsibility for this text is exclusively that of the authors.

ecosystems on which they and we all depend. Regardless of where one turns in the region, there is an increase in the number and intensity of conflicts between groups committed to promoting economic development (i.e. growth), and those claiming to speak for the planet and/or the welfare of the large majority of the population or particular minorities, who feel excluded from these processes and are bearing the brunt of the negative impacts of these activities.

This chapter addresses some of the underlying causes of these conflicts by giving voice to some of the actors who are actually involved in developing their own alternatives to the development proposals of the hegemonic forces driving the transformations in their societies. These alternatives emerge from groups whose organizations are shaped by different cosmologies, products of their multiple ethnic origins, and by the profound philosophical and epistemological debates of the past half-century that emerged from numerous social movements proposing different strategies for achieving progress, improving wellbeing and conserving ecosystems. While many past confrontations among social groups have produced compromises modifying individual development projects, few have created some space for the emergence of alternative social and productive structures that respond to the demands for local control of the governance process to assure local wellbeing and responsible environmental management.

The analysis draws on an important emerging literature that proposes a different epistemology and methodology, reflecting the direct participation of a diversity of communities around the world in research about themselves and their possibilities for implementing different approaches to improving their wellbeing. In spite of the widely separated regions and traditions from which they come, there are striking commonalities in their reflections on how research should be conducted and how they might collaborate with "outsiders" in their search for ways to advance in their pursuit of an improved style of life and their ability to govern themselves. A notable early contribution from this intellectual and academic current was published by a Maori sociologist (Smith, 2012), reacting to the tendency of scholars from the principal academic institutions in New Zealand to make assumptions about local social structures, production possibilities, and the possibilities of and competence for innovations of their "aborigines". Since this early text, a burgeoning literature has emerged, not only emphasizing the methodological limitations of much Western scholarship in the Third World but also extending the critique to epistemological, ethical and cosmological planes. The contributors to this process argue that since social categories are deeply embedded in institutions, profound difficulties arise when trying to understand the discourse and proposals of peoples of other cultures, especially those distanced from societies rooted in the Judeo-Christian tradition; the obstacles can be traced back to the very essence of the differences in value systems and the relationship of society itself to the world which we inhabit (e.g. Apffel-Marglin and Marglin, 1996; Apffel-Marglin, Kumar and Misra, 2010; Venkateswar and Hughes, 2011; Stephen and Hale, 2013). The area of intercultural dialogue has proved particularly fruitful, going beyond both universalism and cultural relativism, to engage in cultural relativity and cultural pluralism for a democratic, just and peaceful harmonization of conflicting interests (Panikkar, 1979, 1995a, 1995b; Vachon, 1995; Dietrich et al., 2011). The increasing interest in the commons, as a world emerging beyond the market and the state, expresses the new protagonism in the social and political scene of old and new communities (Ostrom, 1985, 1986, 1990; Linebaugh, 2008; Walljasper, 2010; Bollier and Helfrich, 2012; Barkin and Lemus, 2014; McDermott, 2014).

This approach clarifies the difference between dominant concepts of environmental governance and our understanding of the problem, along with its applicability to the work of the communities with which we are collaborating. As generally understood in Western social science literature, and excellently set forth in the introductory materials in this book, environmental governance is an extension of the process of public deliberation and policy formulation, to integrate into the sociopolitical parameters additional considerations of the impact of society on ecosystems, locally and globally. This relatively new field of political and social action has become poignantly crucial in recent years, as the depths of the environmental crises that we are living have made their impact increasingly evident. In our work we have clearly identified the problem of governance with the challenge of assuring that we examine the origins of the problems and the proposed strategies to address the intimately related matter of social justice.

In this chapter, however, we focus on the contrasting conceptions of the functioning of the political process and the possibilities for change. The dominant conception derives from a vision in which the world economy is central, a behemoth comprising a variety of national and regional units forming a single interconnected network of markets that feed a process of capital accumulation. This network of markets is controlled by a small group of powerful economic interests, backed by their national governments within an international institutional framework that reinforces their control over national and international economies.

The prevailing model of international politics and environmental governance is firmly grounded in the dynamics of the global marketplace, the private ownership of property and the means of production, creating an increasingly unequal distribution of income, wealth and power within societies and on a global scale, as well as producing a devastating impact on the environment.

In contrast, our research identifies myriad local and regional groups trying to overcome centuries of repeatedly being relegated to ever more inhospitable regions while also being targets of oppression, as a result of an unequal form of integration, transforming them from independent peoples into victims of colonialism and (inter)national capitalist "development". By emphasizing their rejection of the market-driven forces that control and distribute resources, they are seeking to design and implement different approaches to decision-making, based on a set of values that generally privilege collective solutions and wellbeing over individual gains and assume a cosmocentric view. These approaches emerge from a different and more complex set of objectives, rooted in historical experience, cultural traditions, and intergenerational relationships and responsibilities that situate their choices in a longer time horizon than that typically considered by the dominant methodologies that guide environmental governance at present. Because they attempt to bring to the centre of social life politics and ethics, displacing from it the economy, they explicitly reject the primacy of an economic calculus in making fundamental decisions about society, economy or ecosystem management. As a consequence, their decisions often result in proposals that are at odds with the policy prescriptions offered by the institutions with which they must interact, whether it is for the management of specific natural resources or for addressing problems of political, social and/or economic dynamics. As a result, these communities are actively building alliances among themselves, regardless of whether they are located in contiguous regions or associated through sectoral or cultural organizations that offer platforms for strengthening their ability to negotiate with local and national authorities, or resist the imposition of policies or projects to which they are opposed. In the process, they are seeking to isolate themselves from the hegemony of these international forces and epistemologies, forging their own institutions to create spaces of greater autonomy, in political, social and productive spheres, defending their ways of life and their territory from assimilation into the international economy or its outright seizure/appropriation by international capital.

These communities, as examined in later sections of this chapter, are searching for new ways to strengthen their societies and improve their ability to govern themselves. In many cases, this involves a redefinition of their identities, combining knowledge of their cultural heritage with present-day understandings of the significance of their cultural roots and the history of their struggles against many of the numerous forms of injustice to which they continue to be subjected. These struggles have "never been a blind, spontaneous reflex to objective economic conditions. [Rather, they have] been a conscious struggle of ideas and values all the way" (Thompson, 1959: 110). As such, the communities have been able "to hold fast to the vision of collective good".[1]

It is striking that a common feature of solidarity in many of these communities is a growing realization of the importance of this heritage and history, its contribution to their own definitions as peoples, as communities, whose collective identities and belief systems have generated unique forms of organization and social dynamics. These organizations are discovering new ways of integrating their belief systems, their cultures and their relationships to their environments into cosmologies that lead to creating contrasting models of society, models that directly address the demands for social justice and sustainability while protecting the whole panoply of traits that define a people.[2] While the current uncertainties have encouraged the emergence of different forms of localism, isolationism and often violent fundamentalism, most communities are not trying to go back in history but to discover in their traditions inspiration, and wise and sensible alternatives for their current predicaments.

While forging these new models of society, the communities are actively engaged in a complex process of defining (or redefining) their identities. It no longer suffices to declare that they are of one or another ethnic origin, or that they are peasants of one or another tradition. This search for identity is complex, involving the combination of numerous concentric and competing contexts, coming from national and local or regional cultures, ethnic origins and environmental features that impact on social structures. Coming, as it does, from a different point of origin, the demand for social justice, for example, cannot consent to the idea that profound inequalities are part of the human condition; or that changes in the legal system can legitimate the plunder of community resources or planetary equilibriums. This discussion necessarily leads to a profound distinction between the nature of the social contract on which each society is constructed, posing the question of whether the individual has the right, in the ultimate instance, to assert his or her individual interest at the expense of the community's, a right which is generally questioned within the communities with which we are collaborating. For many of them, they are not individuals but singular persons, knots in nets of relations, for whom the community is the first layer of their personal being.

Of course, these discourses also define trajectories for social progress. The dominant market-based approach identifies an increase in material production as the leading indicator. Economic growth, as valued in the marketplace and measured by monetary units aggregated into indices of gross national product (GNP), clearly devalues changes in the status of women, the wellbeing, or the impact of production on natural resources and the ecosystems. In contrast, the version emerging from Latin American community initiatives generally incites broader discussions about lifestyles and community organization; approaches simplified as *Buen Vivir* ("good living"), *mandar obedeciendo* (govern through obedience, command by obeying) or *comunalidad* (communality) are concepts that imply moderation as part of complex strategies for constructing alternative organizations. Our consultations with the communities to which we refer in this chapter identified five basic principles for this process: autonomy, solidarity, self-sufficiency, productive diversification and regional sustainable management.[3]

In what follows, we summarize our direct collaboration with communities and alliances of local groups involved in the process of trying to consolidate their own governance structures capable of responding to their visions of an appropriate society consistent with assuring wellbeing and sustainability. It takes as its point of departure their struggles to consolidate alternative programmes to produce the basic goods needed to assure their livelihoods and to strengthen their ability for self-governance, while attempting to respect the possibilities and limits of their environments. What is striking about these collaborations is the extent to which the participants are well informed of the burgeoning discussions of epistemologies that explicitly question the logical structures of dominant governance and development models;[4] many of these seemingly academic debates have become an integral part of the discussions and design of strategic proposals by these local groups to understand and implement programmes for local and regional advance. If presented in clear and simple terms, complex theoretical debates produce in the communities an "Aha! effect": they have already been discussing the issues.

While most of the detailed fieldwork that we are documenting is based on intensive interactions with communities in the Mexican state of Oaxaca, the materials for this chapter draw on additional contributions produced by people actually involved in local and regional processes in other parts of the region, and with others who are emerging from resistance movements to implement their own proposals for consolidating a material and institutional basis for improving material wellbeing and assuring their capability for promoting ecosystem balance.

## An alternative understanding: A different point of departure

Forging their own solutions is an ambitious endeavour for peoples proposing to overcome discrimination, marginalization and systematic efforts by colonial powers of yore or by today's power elites to relegate them into ever more isolated corners of their territories. What is remarkable about the histories we are discovering and the collaborators we are fortunate enough to meet is the wealth of proposals with which they are experimenting and the tenacity with which they continue to resist efforts to integrate them into national and international economies as underprivileged individuals in increasingly polarized societies. Our efforts to invite various communities to collaborate, helping us to understand their approaches to governance and their aspirations, also added another dimension to our understanding of current day social dynamics, one that is not lost on the analysts shaping the process of globalization, but perhaps is underestimated or even misunderstood by academia. In its assessment of the likely global trends regarding national security in 2015, the director of Central Intelligence, as head of the United States Intelligence Community, was informed by a group of outside experts in 2000 that indigenous resistance movements in Latin America will be one of the principal challenges for national governments in the next 15 years:

> Indigenous protest movements...will increase, facilitated by transnational networks of indigenous rights activists and supported by well-funded international human rights and environmental groups. Tensions will intensify in the area from Mexico through the Amazon region... [It goes on to report:] Internal conflicts stemming from state repression, religious and ethnic grievances, increasing migration pressures, and/or indigenous protest movements will occur most frequently... in Central America and the Andean region.
> (Tenet, 2000: 46, 49)

Although we concentrated our efforts on collaborating with groups in a limited number of regions in Mexico with high concentrations of ethnic populations, it quickly became clear that resistance movements are proliferating throughout the hemisphere, partly in reaction to state policies to promote local integration into national and international development projects, by permitting outsiders privileged access to natural resources and to construction of infrastructure, in territories traditionally controlled by these peoples.[5] What we found, however, was that there are also positive developments motivating communities throughout the Americas to strengthen their abilities to govern their territories, by better understanding the relationships between themselves and their surroundings while also engaging in deliberate efforts to build alliances among themselves and transnational organizations capable of defending their claims in international arenas.

The need for this process of organization has become increasingly evident as conditions within each country, and, internationally, changed dramatically. A concerted effort to accelerate the region's internal integration and connectivity with the global economy, as well as to facilitate the access of international enterprises to domestic resources as part of a drive to promote domestic growth, is changing the map of Latin America (Bessi and Navarro, 2014), impacting first and foremost indigenous communities in the hemisphere. These analysts summarized the problem:

> The reordering of territory has blurred borders in both economic and political terms with projects such as the Mesoamerican Project (previously Plan Puebla-Panama) and the Initiative for Regional Infrastructure Integration of South America, which both entered into force after 2000.[6] Their primary objectives include the construction of transportation and telecommunication networks, as well as energy-generation projects such as hydroelectric dams and wind farms. They also plan to designate national parks, protected areas, Heritage for Humanity sites, cross border conservation areas, transnational parks (also called Parks for Peace), ecological and biological corridors and networks of protected areas... The design of these projects is indeed strategic, and 'progressive' governments are presenting them as a development opportunity.
>
> (in Navarro and Bessi, 2014)

Ana Ester Ceceña, a Mexican economist, added (in Bessi and Navarro, 2014):

What will happen with IIRSA is that local governments will be forced to be more disciplined because they will be brought in line with global markets. There are 500 transnational companies that produce half of global gross domestic product; when one looks at IIRSA's design and these companies' projects, they complement one another: The groundwork is being laid for the circulation of communication, merchandise, raw materials and energy... Capital needs a reordering of territory – considering this as a type of historical-social construction – in order to continue reproducing itself, as much in terms of materials as in power relations, of accumulation of capital and profits. The ordering enables access on a large scale to certain types of material from the earth.

In characterizing this latest form of neoliberal development, Gustavo Esteva (in Bessi and Navarro, 2014) observed: "Indigenous people are on the front lines of a battle, fighting a war that is on behalf of all of us, because it is there that the capitalist system looks to relaunch a new form of accumulation."

Indigenous peoples are increasingly insistent on demanding the recognition and integrity of their territories, many of which are threatened by the grandiose proposals of global capital; their actions are confronting directly these schemes, and changing the maps of the Americas in the process. They have strengthened their resolve to prosecute their historical claims as they become increasingly skilled in achieving the enforcement of the agreement ratified by the ILO to guarantee prior consent of native peoples with territorial claims for outsiders to undertake activities or exploit natural resources in their regions.[7] Accompanying the changing map is a new consciousness of the significant differences in understandings of even the most elemental concepts in their exchanges with their interlocutors in the states of which they are a part: although a significant discrepancy occurs throughout the Americas, as different social groups and peoples question governmental procedures to charge a single agency with implementing unified policies for the myriad ethnic groups in their countries,[8] an even more serious source of conflict involves the very notion of property and the apparent freedom with which outsiders (government agents) can discuss the possibility of alienating people's claims to land or natural resources. This problem arises because of the profound differences between the historical significance attached to the different concepts of property and territory; for many groups, territory is an all-encompassing term with complex implications that are not easily incorporated into prevailing market-based understandings of the significance of land or property. This is so essential that even the Organization of American States finds itself obliged to take note of its consequence in the context of the demand to draft an American Declaration of the Rights of Indigenous Peoples. This discussion is central to our understanding of the underlying basis of the prosecution of demands for autonomy by native peoples (OAS-CJPA, 2003: 1–2):[9]

> Territorial rights are a central claim for Indigenous Peoples in the world. Those rights are the physical substratum for their ability to survive as peoples, to reproduce their cultures, to maintain and develop their organizations and productive systems... Indigenous Peoples have strengthened their organizations and developed a more organized struggle to reclaim their rights. Central among those demands are the issues related to land, territories and natural resources... these rights are not merely a real estate issue... Rather indigenous land rights encompass a wider and different concept, that relates to the collective right to survival as an organized people, with control of their habitat as a necessary condition for the reproduction of their culture, and for their own development, or as Indigenous experts prefer, for carrying ahead "their plans for life" ("planes de vida") and their political and social institution.

Indigenous areas, then, are a complex amalgamation on which the very existence of these peoples depends. This is clearly defined in the Brazilian Constitution, which gives renewed strength to the ancestral possession as a basis for the territorial rights characterized by four significant traits: (1) permanent ancestral possession; (2) areas necessary for their productive activities, including the reproduction of flora and fauna; (3) areas necessary for their cultural reproduction, and for their survival as a collective; and (4) habitat with the physical capacity and shape to allow the full functioning of the mechanisms of authority and self-government of the indigenous people. These territories are the habitat necessary for their collective life, activities, self-government, and cultural and social reproduction.[10]

Problems arise when the state seeks to exercise its sovereignty or eminent domain, to build infrastructure, to exploit or license the exploitation of natural resources, or any other action or project that might affect indigenous lands and the use of their territory. International law now restricts this possibility, obliging the previous fair and serious consultation with the affected indigenous peoples (Convention

169, ILO, endnote vii). Since indigenous peoples are consolidating their constitutional and legislative demands to codify symbolic and political elements of autonomy and self-government, as elements of internal self-determination, governments are finding themselves treading on new "ground" as they attempt to reconcile global visions of "development" with local efforts to achieve wellbeing.

Throughout the Americas, governments continue to assume that prices of both landed property and natural resources can be fixed according to market processes, and in the best of circumstances negotiators of goodwill can arrive at mutually beneficial agreements for their exploitation, thus assuring their "unlocking" to promote national development by trading them in the global marketplace. In these circumstances it seems almost incomprehensible to the dominant powers that local groups might object to the terms of these negotiations, refusing to even discuss the possibility of placing a forest enterprise, a mine or a power-generating facility in their regions as it would upset a delicate historical and spiritual balance that they consider threatening to their social structure or cultural integrity, defined in terms of one or more many non-monetary dimensions for which financial compensation is inconceivable.

> The nature and scope of this struggle is very old. At the end of the colonial period, for example, in the XVIII century, the areas claimed by the indigenous peoples in Mexico were called "Indian Republics", meaning they did not represent only a piece of land but a whole way of life and government, in spite of being subordinated to the Spanish Crown. This struggle also has very old precedents: known as the *Magna Carta* and the *Charter of the Forests*, the King and the nobility in England agreed, at the end of "the long twelfth century", to establish limits on their power to assure the subsistence of the commoners (Linebaugh, 2008: Ch. 2). The traditional struggle for land provoked the first social revolution of the XX century, in Mexico, and played itself out with diverse intensity in all Latin American countries during the last hundred years. The upheaval of the last 20 years represents a political mutation from such tradition to a struggle of territorial defense, as expressed in the *Declaración de Quito* (2009) by the International Commission for Integral Agrarian Reform of Vía Campesina: "For the agrarian reform and the defense of land and territory". This implies a profound conceptual shift: "A specific form of relation to the land is claimed which is markedly different to the one imposed by public and private developers in the last 50 years.

It expresses a sovereign practice of the collective will, which does not contain separatist elements but openly challenges governmental institutions. The political form of this claim is usually presented as autonomy".

(Esteva, 2010: 65)

Territorial defence is also a new central theme in the cities. The old tradition of illegal settlement, which shaped most Latin American cities during the twentieth century, is today complemented by active movements to redefine urban life. The most spectacular case was Argentina (2001–2002), but from Oaxaca (2006) to Brazil (2014), vibrant movements express the vitality of new social subjects and new forms of social protagonism (Colectivo Situaciones, 2002; Mariotti et al., 2007; Zibechi, 2008; Giarraca and Teubal, 2009).

## Building the commons: Local solutions are collective endeavours

This complex process of differentiating territory from property and clarifying the significance and importance of social ownership and membership as distinct from individual activities encompasses yet another important dimension: the communities generally think of themselves as part of a regional, and even a global, commons. But unlike the formal discussions of the concept in much of the academic literature, their understanding of the commons cannot simply be reduced to a collection of "common pool resources", such as air, water and other natural resources shared by all that were the focus of the debate set off by Garret Hardin's "tragedy" (1968);[11] rather their activities are much more akin to what one of the leading historians of the process describes as the "active movements of human commoning and the worldwide demands to share wealth and safeguard common resources on every continent" (Linebaugh, 2008: 280). The organizations that are so engaged are not involved in shaping "an alternative economy, but rather an alternative to the economy" (Esteva, 2014: i149). The commons are extended to encompass the social and cultural components of collective life; they are not simply a set of things or resources. Rather, like many other aspects of the societies we are discussing, the organizations they are creating bestow great importance on social relations within the community, as well as a firm commitment to ensure the conservation and even the enlargement of the commons. This relationship reflects a collective and enduring transformation of the way in which society conceives and manages itself while also developing the basis for collective and communal management.

Protecting, defending and governing the commons are complex and risky processes. Complex, because they encompass all aspects of social and biological existence. Risky, because they involve challenging the de facto powers and questioning the legitimacy of their "rule of law" – that is, the legal system that is creating and perpetuating a profoundly unjust society, exacerbating social disparities and accelerating environmental destruction. This dispute about the nature of the state stems from a rejection of the philosophical underpinnings of the hegemonic order, based on the idea of a single "social contract" that presupposes the possibility of applying universal norms, such as "social justice", "equality" or even "democracy", impartially to attend to the needs of all social groups.[12] For this reason, it also involves a prima facie repudiation of the legitimacy of national "authorities", which assume their right to transfer community resources – the commons – to others, for whatever reason, without regard for the wellbeing of the people, local decisions, or historical and environmental considerations, as is common practice in mining, forestry and water management, although it now extends to complex issues of bio- and nanotechnology in many nations today.[13] Thus the efforts to promote solidarity among diverse social groups call for a political approach that requires each to extricate itself from the dominant social and political institutions that are incapable of attending their particular needs.

But consolidating the foundations of this society entails much more than undertaking specific activities or establishing appropriate institutions for governance or management. The solidarity society requires personal commitments from each member to assume responsibility for the wellbeing of others and for limiting individual claims for access to collective resources (Robles and Cardoso, 2008; Martinez Luna, 2010). To strengthen these foundations it is essential to begin with a common vision of society as a whole, whose point of departure is reversing the historical tendency for the personal enrichment of a few at the expense of the many; as such, they incorporate collective decisions to assure transparency and direct participation in decision-making, and universal responsibility for administration or implementation of this dynamic. This challenges the presumption of the freedom of the individual within the group, obliging each member to carefully measure their impact on others, and the whole, and be guided by reference to their impact on the collectivity in their decisions and actions. In historical terms, and specifically in the light of practice in today's globalized society, it calls for a redefinition of peoples' relationship with their society, rejecting the notion that one person has the unfettered right to withdraw from, or even oppose, the commonwealth after having participated in the process of arriving at a decision.

This point of departure has important implications for the way in which priorities are determined and activities are organized. Perhaps one of the most striking and demanding of these is the need to reverse the hierarchical organization of the workplace: of course, people should be paid for their work, but they should not have to submit to demeaning and authoritarian social relations to satisfy their basic needs. The existing proletarian organization of society is part of an underlying condition of the helplessness of the workers, unable even to survive without entering the labour force; the alternative under construction here starts from the presumption that all members of society enjoy the legitimate right to a socially determined way of life, independent of their contributions to production or output. Their participation in collective activities becomes rooted in a sense of duty and belonging to the community, but also an obligation that is explicitly enforced by communal authorities. Such an approach eliminates the double alienation of modern labour: from the fruits of work and from the logic of creative activity.

## Creating the foundations for communal governance: Generating and managing surplus

The decision to create autonomous forms of self-government within the framework of the nation state represents an audacious challenge to the prevailing model of governance, and of social and economic justice based on representative democracy and its marriage with the free market. Rooted in the commitment to define and defend their territories, the process involves creating new institutions and processes for the social appropriation of both the natural environment and the productive systems that they have created to assure their ability to maintain and strengthen their community, to provide for their basic needs, and to facilitate exchanges with partners (barter) and in the marketplace. The mechanisms established by the communities for management often involve complex dynamics for mutual consultation among different groups within the communities, as well as forms for delegating responsibilities to members on the basis of expertise and social commitment, or for assuring broad political participation and accountability. Thus it is not only the choice of activities themselves but also the implementation processes that are crucial to the design of the social mechanisms that contribute to the desired outcomes related to equity and sustainability.

In the following discussion of individual projects with which we have come into contact (see the next section), an interesting facet of the analysis is not only the choice of technique but also, and often just as important, the nature of the activities themselves; they speak to a concern for addressing the socially defined basic needs of people in the communities while also creating a balance between the use of natural resources and the restoration, regulation of land use, and conservation of the ecosystems from which they are drawn.

What makes these activities unique is that they are being organized by groups that come together on a voluntary basis to ensure their viability and continuity. In many cases they are trying to regenerate the social fabric eroded by both external and internal forces. While we focus on the collective nature of decision-making, it is just as significant to understand the mechanisms that make possible the consolidation of the community and its ability to advance. During our interactions with the communities in their search for solutions that provide the wherewithal for moving forward, we identified a central feature that contributed to this success – one that also explains their ability to consolidate the capacity to implement the collective governance model that is fundamental to society's continuity and its possibility to assure improvements in the lives of its members: the explicit organization of social and productive resources to generate surpluses for "reinvestment" and "redistribution" (Baran, 1957).

The centrality of surplus in community management is an often invisible and misunderstood facet of the administrative process. Much of the literature describes rural communities in general and indigenous groups in particular as living at the margins of subsistence, as the poverty in material means limits their ability to advance and reduces the scope for broadening the range of activities they can undertake. In contrast, our dealings with communities throughout the Americas reveal the ability and commitment of many to produce this surplus and manage it collectively, using it to reward members who have made important contributions in producing it and channelling the rest for collective purposes.

By focusing attention on the processes of producing and managing surplus within the limits for satisfying socially defined needs and the possibilities of their ecosystems, this collective management structure of the diverse local projects has proved effective in constructing a framework for environmental justice that is proving so elusive in the larger societies of which they are a part. Unlike those other parts of society closely tied into the global market economy, these communities have created possibilities for organizing themselves to ensure that their members need not suffer from extreme poverty and unemployment. As a result, they are generating a productive potential far greater than might be appreciated by a simple accounting of the financial resources that they have at their command. Some of this potential is well documented in the literature, as is the case of the "voluntary" labour that is expected from all members for collective tasks involving building and maintaining infrastructure or conserving ecosystems (e.g. *tequio, minga*). The social mechanism for assigning and rotating administrative and political positions so important for governance is another way in which resources that are often invisible in the market economy or formal accounting calculus are generated in these communal organizations. But, just as important, the commitment to universal inclusion or participation also creates a corresponding responsibility from the members to contribute to collective tasks – assuring that most individuals will be involved in a multiplicity of activities for their own benefit and that of the community.

Surplus has existed in human organization from time immemorial. Even when there were no formal institutions for exchange and accumulation, the construction of large and small projects to channel water or create monuments is testimony to the ability of societies to advance beyond their immediate needs, building projects to increase productive capabilities or the grandeur of their "leaders". What distinguishes the myriad communities guided by cosmologies removed from those based on material gain and individual benefit at the expense of the whole is their ability to promote a broad participation for advancing the general welfare. Most recently, these societies have improved their possibilities for implementing new projects, taking advantage of advances in science and technology while also critically incorporating knowledge and contributions from the past, generating opportunities for increased or more efficient production as well as more effective means for improving their wellbeing and ability to protect their ecosystems. By examining the availability and mobilization of surplus, the communities are better equipped to consider how best to implement their long-term visions. What is striking about the individual experiences with which we have been associated is the clear understanding by many of the participants and the leadership of the ways in which particular activities may contribute to overall goals.

## Communal approaches to environmental justice

Communities across the Americas are involved in designing and implementing local solutions that contribute to their broad struggle for environmental justice under circumstances of harassment and overt violence exercised by state powers in the societies of which they are a part. While a great deal of energy must be devoted to protecting themselves from encroachment by forces attempting to control their natural resources and subject them to the various disciplines of markets and political systems, it is remarkable that they continue to mobilize locally and nationally while associating internationally with other communities and NGOs to consolidate new lines and technologies of production, and experiment with ways to improve existing activities.

These actions are the product of the complex interaction of dynamic forces within the communities and reactions to outside pressures. They are part of a search for a unique identity that has become increasingly important as these peoples assert their legally binding rights to self-determination as defined by their varied histories and their understanding of the privileges accorded them by the ILO Convention 169 and similar agreements promulgated by other international bodies, and the ongoing efforts in the Organization of American States (2003) to draft a similar commitment (endnote ix). In Mexico, as elsewhere, this process has a long history, which was codified in its constitution of 1917, as indigenous communities were recognized and granted collective rights by the agrarian reform.[14]

During the last half of the twentieth century, Mexican communities waged an unrelenting and difficult battle to assert their rights to control the lands over which they were able to retain or regain control after the revolution. They were particularly effective in wresting exploitation contracts for their communal forests from private firms that had been given concessions to manage them (Bray and Merino-Pérez, 2004). Today there are a variety of management plans in effect, testimony to skills that the communities have acquired as they attempt to reconcile pressures for ensuring conservation with the need to create jobs and generate incomes. The literature offers rich accounts of this variety of strategies, and many studies explore the relationship between these approaches and the cosmologies of the participating communities, particularly in community-managed forests, which comprise 71% of the nation's forests (e.g. Bray, Merino-Pérez and Barry, 2005; Cronkleton, Bray and Medina, 2011; Barkin and Fuente, 2013; Stevens et al., 2014).[15]

The movement to reassert indigenous identities in Mexico was further strengthened in the aftermath of the 1994 uprising in Chiapas by the Zapatista Army of National Liberation (EZLN) (Muñoz, 2008).[16] Since then the activity and visibility of indigenous peoples throughout Mexico has increased, along with a gradual recognition of their importance in the population, because of, and in spite of, the growing intensity of repressive actions by the state and other actors, including private corporations given concessions in these territories, and organized groups in various parts of the society.[17] While a recounting of the initiatives being implemented in these communities would be too lengthy for inclusion here, suffice it to say that the discussion of many of them within the framework of the National Indigenous Congress, and the increased circulation of information and meetings among members are contributing to strengthen the resolve and ability of members to carry their projects forward.

In connection with their efforts to gain recognition and elaborate local management strategies, control of water resources has been particularly contentious as communities try to assert their rights to adequate supplies and protect their sources. We are accompanying a number of communities in their efforts to reinforce control in their territories by developing systems for managing water resources and organizing to impede encroachment by national and state-level authorities trying to limit their historical access. These movements are now inextricably combined with others in opposition to large-scale construction projects for dams designed to harness waters for electricity generation or for long-distance transfer between water basins to supply urban areas where ageing infrastructure and excessive growth in consumption are causing shortages due to a lack of administrative and technical capabilities of dominant bureaucracies. As a result, many communities that have historically been able to satisfy their own needs and even share surpluses with neighbouring communities are now finding themselves involved in coalitions with others defending their water sources, along with ecologists who are generally arguing that the engineering and public works approaches of the public sector are inappropriate and simply postponing the day of reckoning with regard to the need for a more ecologically informed approach to water management.

An interesting finding in our collaborations with communities involved with protecting water sources is the combination of traditional and leading-edge technologies applied to protect their natural sources – the streams and springs on which they depend. This combination of technologies with direct community involvement in water management contrasts sharply with the national water authorities' approach that eschews local diversity, preferring a homogenous administrative model conducive to centralized management and engineering solutions. In response to the great differences in local conditions, there are many examples of water-saving technologies being implemented by communities, such as installing composting toilets and separating grey from black water flows to allow for low-cost and passive biological processing conducive to restorative environmental practices. A particularly noteworthy project, Water Forever, transformed 1 million Ha of barren plateau and steep slopes using "appropriate" technologies to construct a large number of low-impact landscaping projects, including rock dams and ponds to channel surface flows and collect run-off, recreating underground aquifers and structures found in some of the oldest irrigation projects in the Western Hemisphere from the eleventh century. This project, which began in the 1980s, is noteworthy because it combines community-managed agroecological and agroindustrial activities and enterprises belonging to the participants, creating jobs and products that are proving attractive to consumers for their social, ecological and nutritional qualities (Hernández Garciadiego and Herrerías, 2008).[18] In Bolivia, the experience of the "Water War" of 2000 in Cochabamba is still vivid in people's memories as local water committees continue to organize actively while resisting the state's efforts to manage the commons (Fogelberg, 2013; Dwinell and Olivera, 2014).

These community-based management proposals embrace important parts of their members' collective existence but cannot provide for all of the needs of the community. Having adequate water supplies and sustainable models for forest management offer important points of departure for building stronger and more resilient communities. Unfortunately, recently the pressures on national governments to increase energy production from renewable sources are heightening the conflicts with indigenous communities threatened with being flooded out of their territories;[19] in Mexico, the refusal of the government to permit indigenous communities to undertake their own microhydroelectric power projects is clear evidence of the fear of the degree of independence that such activities would promote.

In spite of these obstacles and conflicts in the power and water sectors, numerous communities are undertaking productive activities to supply basic needs and create goods that can be traded for other products. Ongoing efforts are oriented towards identifying new activities that make use of available renewable resources to produce goods that might be advantageously exchanged with others to provide for these basic needs. The objective of this approach is to induce social dynamics that bring the producers together into stronger organizations that in turn become part of their communities.

As part of this effort, many groups are accompanying communities in introducing complementary activities and assisting them to modify technologies or introduce new ones that would strengthen their organizational capabilities to contribute to the collective wellbeing. The objective of these undertakings is to contribute to community efforts to strengthen their own capabilities to govern themselves. One of the most significant organizations engaged in accompanying people in strengthening their communities and enabling them to better meet the challenges of assuring a better style of life is Vía Campesina (VC). This group has a presence in 73 countries, representing more than 200 million members. Its purpose is to promote food production by using agroecological techniques to move groups of producers towards greater self-sufficiency. In 1996, VC expanded and redefined food sovereignty, associating it with the capacity to determine autonomously what to eat and how to produce it (Rosset, 2013).[20] Its achievements are best reflected in the somewhat controversial decision of the FAO to declare 2014 the International Year of Family Farming (CEPAL/FAO/IICA, 2014), where the organizations declare rather wistfully: "Countries look to family farming as the key to food security and rural well-being." VC also noted that this was the first time in its almost 60-year history that the organization made reference to the theme of agroecology, one of the principal strategies that can assure farmer control of agriculture and an appropriate response to the need for ensuring food security for societies.

Other social groups are actively engaged in activities that promote social, political and productive changes to contribute to improving their own lives as well as those of others while attempting to conserve and enhance environmental quality or sustainability. In Mexico, the local *Caracoles* in Chiapas are contributing to this objective, directly improving the lives of hundreds of thousands of its members while also portraying a model of social organization and change that continues to have a powerful effect on other communities as well as in other countries.[21] There is ample evidence that its activities are improving wellbeing, contributing to diversifying the economy, and increasing productivity in a region where perhaps as many as 500,000 people are participating; they have achieved a high level of self-sufficiency in food, health and education (Baronnet et al., 2011).

In South America, Andean communities are similarly involved in promoting collective strategies known as *Buen Vivir* (*Sumak Kawsay* is a Latinized version of an expression in Quechua).[22] Throughout the Americas, groups of communities are involved in mobilizations to defend their territories, cultures and societies from trespassing by people who lust after their resources or institutions that would erode the basis of their differences. There are groups such as Idle no More in Canada, the Haudenosaunee (Iroquois) Confederacy in eastern North America, the Landless Workers Movement (MST) in Brazil, the Mapuches in Chile, and numerous others throughout the region, as well as the National Indigenous Congress, the Network of Environmentally Affected Peoples and the Movement Against Mining in Mexico. Similarly, there is a coalition of indigenous peoples in the Americas and a series of international NGOs that are promoting strategies for better resource use, but most of the mobilizations are still defensive groupings helping to defend groups against others trying to take control of their resources, or organizing to forestall activities that might contaminate their lands or their waters (Vergara-Camus, 2014).

Accompanying these actions of resistance, many communities are involved in other constructive activities, promoting collaboration with university and civil society researchers who are helping to explain the value of the work, while contributing to diversifying economies and improving production in sustainable ways (Toledo, Garrido and Barrera Bassols, 2013; Toledo and Ortiz-Espejel, 2014). One application that has proved particularly illustrative involves the inclusion of unsalable avocados that were causing an environmental burden in diets to fatten hogs in backyard settings, resulting in metabolic changes to produce low-cholesterol meat, improving incomes as they are being marketed at a premium in local markets. In this case, as in others based on a similar paradigm, indigenous women were especially benefiting, as they implemented the projects and were soon recognized for their leadership capabilities (Barkin, 2012; Fuente and Ramos, 2013).

In a different approach, scholar-activists are working with producers in diverse regions to protect and enhance production of a traditional Mexican alcoholic drink, mezcal, modifying the traditional planting and harvesting techniques of agaves, taking care of the forest, and enriching community life by promoting cooperative production that is contributing to raising incomes and rehabilitating ecosystems (Delgado-Lemus et al., 2014). In Guerrero, this work is part of an ambitious programme of the Grupo de Estudios Ambientales (Illsley et al., 2007) for collaborative promotion of local forms of *Buen Vivir* and ecosystem restoration that was awarded the Equator Prize in 2012 by the United Nations Development Programme (UNDP). In another region of Oaxaca, four communities continue to care for their mulberry trees, raising silk worms to produce the traditional thread that they then weave into highly attractive and fairly priced garments, displayed and marketed locally and through a well-curated textile museum; elsewhere, others are experimenting with new plantings of perennial indigenous cotton varieties (that were cultivated before the Spanish Conquest) that are ideal for handicraft weaving as an alternative to genetically modified cotton that currently dominates the industry. In Peru and more recently Bolivia, a well-established technical promotion and development organization, Pratec, is deploying effective approaches to community-based learning, improving production in the multiple ecologies of the Andean world, focusing on potatoes but carefully balancing its work to support broad-based, diversified progress (Gonzales, 2014).[23] Ecotourism is another, more controversial, activity because it involves an explicit opening of the community to outsiders who are frequently unable to comprehend the magnitude of the cultural and economic chasm that separates them from their hosts (Barkin, 2002).

Elsewhere, indigenous peoples, peasants and industrial workers are all exploring new routes to reorganize their workplaces and contribute to improving living standards for themselves and their communities. New production systems are being invented as workers occupy closed factories, continuing operations by changing management and incentive systems (Ness and Azzellini, 2011). In many cases the initiatives have not only placed the direct producers in control of the enterprises but also often created possibilities that include the community in decisions and incorporate the impact on the environment into the new decision-making calculus.[24]

## The prospects for alternative strategies for environmental justice

While these initiatives are changing the map of the Americas (Navarro and Bessi, 2014), many other developments are threatening to erode the possibilities for improving peoples' lives and taking better care of the environment. Throughout the hemisphere, much environmental governance involves attempts to minimize the deleterious social and ecological impacts of the aggressive activities that are the foundation of national and international development. Industrial work is intensifying and ever more alienating, and labour has fewer protections; natural resource concessions are opening up vast new territories to exploration and production, with terrible environmental impacts. The privatization of public services and the deterioration in the quality of those remaining in the public sector are a palpable threat to peoples in every country.

Even as indigenous communities are asserting their new-found rights to proceed with forestry and water-management activities, governments are encouraging large-scale initiatives by transnational corporations that threaten to upset the delicate balance of productive activities on which the communities depend for their livelihoods and for ecosystem balance. These projects pose fundamental questions about the ability of the communities to defend their territories, including their substantial cultural, social and productive heritage that entrenches them in their ecosystems. The conflicts continue to this day, posing apparently irresolvable differences and often resulting in violent encounters, as mines, ecotourism and other projects (and with the recent reforms, fracking and other forms of resource extraction) threaten the very existence of the communities. The communities generally reject the assumption that the sacrifices that this destruction entails can be compensated by monetary offers that would only force them onto a path of institutionalized marginalization as isolated individuals, a life of limited opportunities without the social support systems and safety nets that their communities offer.

The ongoing initiatives to strengthen or generate "niches of sustainability" by peasant and indigenous communities throughout the Americas are heartening and important. While the momentum in the global marketplace is clearly threatening social groups and environments everywhere, the continuing successful efforts of peasants and indigenous peoples to implement their own strategies for social and productive change that deliberately incorporate the environment in the process offer a window on the possibilities for making environmental justice a reality for increasing segments of the population. This will not happen where the capitalist structure of production and control dominates. Thus the implementation of local solutions that create regions for autonomous action will become even more significant and effective as the spaces dominated by the global market continue to suffer from deteriorating environments and heightened conflict.

## Notes

1. Although Thompson was describing the notion of class consciousness in post-war England, it seems appropriate to apply his analysis to indigenous struggles in the Americas.
2. It is noteworthy that the attempt to integrate this rich heritage with the challenges of assuring an acceptable quality of life and the conservation

of the ecosystems appears to be a common trait among communities from different cultures and regions. The rich and abundant literature systematizing the experiences of indigenous peoples who are continuing to defend their own ways of life and prevent their territories from being despoiled or wrought from them clearly demonstrates the possibility of shaping alternative strategies to address the same challenges as those espoused in the dominant discourses of environmental governance that remain tied to the institutions of the market economy.

3. The specification of "regional sustainability" reflects the importance of defining ecosystems in terms of natural rather than administrative or political boundaries. The communities are acutely aware of the importance of respecting natural constructs, such as the river basin, that require cooperation and alliances among communities to implement sustainable management strategies.

4. The significance of these other epistemologies is explored in important contributions to our understanding by colleagues who are involved in exchanges with peoples whose organizations and productive systems are guided by other cosmologies. For an introduction to this other literature, see the contributions of Boaventura de Sousa Santos. His *Una Epistemología del Sur: La reinvención del Conocimiento y la Emancipación Social* (2009) offers a clear enunciation of this approach. The seminal work of Robert Vachon among the Iroquois in North America (1995) and the tradition of Ivan Illich (1977, 1982, 1992) have now abundant heirs.

5. An important effort to systematize our knowledge of these movements is reported in Chapter 2, as well as by the research programme Environmental Justice Organizations, Liabilities and Trade (http://www.ejolt.org), which maintains an ongoing inventory of resistance movements.

6. Both of these projects are very large-scale proposals for infrastructure investments to facilitate the penetration of large-scale capitalist organizations into the less exploited but important and well-endowed regions (cf. http://www.proyectomesoamerica.org/ and http://www.iirsa.org/).

7. The Indigenous and Tribal Peoples Convention 169 (http://www.ilo.org/indigenous/Conventions/no169) guarantees this right and, when ratified by a nation, has the standing of a constitutional mandate. It is noteworthy that of the 22 countries that ratified the convention, 17 are in Latin America.

8. See Benno Glauser's insightful presentation of this problem in his exchanges with leaders of the Ayoreo people in Paraguay (in Venkateswar and Hughes, 2011: Chapter 1). In its seven chapters, this book offers a variegated picture of indigenous activism in many parts of the world.

9. The working group charged with preparing the American Declaration on the Rights of Indigenous Peoples was formed following a resolution of the Organization of American States (OAS) General Assembly in 1989. As of 2014 the declaration had yet to be approved, reflecting the profound differences between the competing interests in the hemisphere.

10. Chapter VII, Article 231 of the 1988 constitution, as summarized in the OAS document mentioned in the previous footnote. Elsewhere in Latin America, these territorial rights are constitutionally protected (Argentina, Bolivia, Colombia, Ecuador, Mexico, Guatemala, Paraguay, Peru and

Venezuela). Moreover, the newest constitutions, like those of Ecuador (1998), included environmental and gender components.

11. At the end of his life, Hardin himself was forced to acknowledge that he only examined the "tragedy" of regimes of open access, as those dominant today, and not the commons (*The Ecologist*, 1993: 13).
12. Luis Villoro (2003) offered an insightful analysis of the differences in the meanings of social contracts in differing social contexts.
13. Mexican laws give the government the right to expropriate common land for public works or public interest. In 2013 the constitution was amended to permit this faculty to be applied for the benefit of private operators.
14. The 2007 United Nations Declaration of the Rights of Indigenous Peoples (http://undesadspd.org/indigenouspeoples/declarationontherightsofindigenouspeoples.aspx) should serve to reinforce the 1992 amendment to Article 4 of the Mexican Constitution asserting the country's "pluricultural character". Unfortunately the legislative changes were not accompanied by adjustments in the legal structure to define the judicial relationship between the state and the dozens of indigenous peoples. Serious conflicts continue to arise because recent legislation (2013–2014) reinforces the state's right to appropriate resources on lands in territories recognized as belonging to many of these peoples in spite of their declared opposition in the terms of the ILO Convention.
15. The efforts to assume collective control of the forests began in the 1970s (Simonian, 1995). Today, Mexico's community forest movement is recognized as one of the most effective and sustainable in the world, encompassing more than one-quarter of the nation's land area with differing management strategies that are cited as exemplary. The MOCAF (Mexican Campesino Forest Producers Network) and the Mexican Civil Society Organization for Sustainable Forestry (http://www.mocaf.org.mx and http://www.ccmss.org.mx) continue to play an important role in coordinating their activities and providing information about their history and achievements.
16. Cf. http://enlacezapatista.ezln.org.mx.
17. The very definition of "indigenous" in the Census was modified in 2010 as a result of the inadequacy of the previous categorization, based on fluency in a native language. While Bonfil Batalla mentioned there being about 8 million in his path-breaking book (1987), the Census reported only 6 million in 1990. Today, however, there are about 18 or 20 million people who consider themselves indigenous (Toledo, 2014). The Mexican indigenous population is the largest of any country in the hemisphere; Bolivia, Ecuador and Guatemala have larger proportions.
18. This project continues to mobilize the participation of more than 100,000 people in a region that has been in operation for more than a quarter of century. By focusing on a range of activities that create numerous opportunities, requiring an ever-increasing range of skills, the region is encouraging people to remain, strengthening communities and improving people's welfare.
19. The scope and intensity of conflicts originating from paradigmatic clashes with regard to the appropriate model for managing water and its use is such that a whole issue of the UNDP's *Human Development Report* (2006) was dedicated to the theme. Similarly, UNESCO's 2013 *World Social Science Report* (2013) addresses the need for a new kind of social science occasioned by the

scope of the social impacts of environmental changes resulting from conflicting models of environmental management and the legitimate rights of indigenous peoples.
20. Cf. http://viacampesina.org.
21. Five Caracoles or Good Government Councils were established in 2003 to implement a local governance structure in Zapatista territory.
22. There is ample literature describing and evaluating this approach, and similar proposals for alternative strategies to improve the quality of life in a "sustainable" manner that emerged from indigenous cosmologies (e.g. Bretón, 2005, 2013; Huanacuni, 2011; Acosta, 2013; Lang, 2013).
23. The breadth of this creativity can hardly be captured in this discussion. For more details about the projects mentioned in this paragraph, consult the following webpages: http://geaac.org, http://www.equatorinitiative.org/index.php?option=com_winners&view=winner_detail&id=67&Itemid=683&lang=es, http://www.museodetexitoaxaca.org and http://www.pratec.org. Among the groups participating in our project, peasant and indigenous communities are engaged in urban agriculture, waste separation for reutilization, and rainwater harvesting. Near the centre of Oaxaca's capital city, one of these initiatives received a national prize for Local Management and Governance in 2012 (http://oaxaca.me/recibe-san-bartolo-coyotepec-premio-nacional-por-el-cuidado-ecologico).
24. A review of many of these initiatives, involving different organizational models and cooperation among producers that encompasses not just the productive aspects but also the governance institutions that are now incorporating whole communities into the management process (e.g. Lavaca, 2003; Rebón, 2004; Giarraca and Teubal, 2005; Sitrin, 2006; Webber, 2011; Bollier and Helfrich, 2012; Burbach, Fox and Fuentes, 2013; Piñeiro, 2013).

## References

Acosta, A. (2013) *El Buen Vivir: Sumak Kawsay, una Oportunidad para imaginar Otro Mundo* (Barcelona: Icaria Antrazyt).
Apffel-Marglin, F., Kumar, S. and Misra, A. (2010) *Interrogating Development: Insights from the Margins* (New Delhi, India: Oxford University Press).
Apffel-Marglin, F. and Marglin, S. (1996) *Decolonizing Knowledge: From Development to Dialogue* (Oxford, UK: Clarendon Press).
Baran, P.A. (1957) *The Political Economy of Growth* (New York: Monthly Review).
Barkin, D. (2002) "Indigenous Ecotourism in Mexico: An Opportunity Under Constructio", in D. McLaren (ed.), *Rethinking Tourism and Ecotravel* (Westport, CT: Kumarian Press), 125–135.
Barkin, D. (2012) "Communities Constructing Their Own Alternatives in the Face of Crisis", *Mountain Research and Development* 32(S1): S12–S22.
Barkin, D. and Fuente, M. (2013) "Community Forest Management: Can the Green Economy contribute to Environmental Justice?", *Natural Resources Forum* 37(3): 200–210.
Barkin, D. and Lemus, B. (2014) "Rethinking the Social and Solidarity Society in Light of Community Practice", *Sustainability* 6(9): 6432–6445.

Baronnet, B., Bayo, M.M. and Stahler-Sholk, R. J. (eds) (2011) *Luchas 'muy otras': Zapatismo y Autonomía en las Comunidades Indígenas de Chiapas*. (Mexico: Universidad Autónoma Metropolitana Unidad Xochimilco).
Bessi, R. and Navarro, S.F. (2014) The Changing Map of Latin America, http://www.truth-out.org/news/item/25227-the-changing-map-of-latin-america, date accessed 1 August 2014.
Bollier, D. and Helfrich, S. (2012) *The Wealth of the Commons* (Amherst, MA: The Levellers Press).
Bonfil Batalla, G. (1987 [1996]) *Mexico Profundo: Reclaiming a Civilization* (Austin, TX: University of Texas Press).
Bray, D.B. and Merino-Pérez, L. (2004) *La Experiencia de las Comunidades Forestales en México. Veinticinco Años de Silvicultura y Construcción de Empresas Forestales Comunitarias* (México: SEMARNAT, Consejo Civil Mexicano para la Silvicultura Sostenible A.C., Fundación Ford).
Bray, D.B., Merino-Pérez, L. and Barry, D. (eds) (2005) *The Community Forests of Mexico: Managing for Sustainable Landscape* (Austin, TX: University of Texas Press).
Bretón, V. (2005) "Los Paradigmas de la 'Nueva' Ruralidad a Debate: El Poyecto de Desarrollo de los Pueblos Indígenas y Negros del Ecuador", *European Journal of Latin American and Caribbean Studies* 78: 7–30.
Bretón, V. (2013) "Etnicidad, Desarrollo y 'Buen Vivir': Reflexiones Críticas en Perspectiva Histórica", *European Journal of Latin American and Caribbean Studies* 95: 71–95.
Burbach, R., Fox, M. and Fuentes, F. (2013) *Latin America's Turbulent Transitions: The Future of Twenty-First Century Socialism* (London: Zed Books).
CEPAL, FAO, IICA (2014) *Perspectivas de la Agricultura y del Desarrollo Rural en las Américas: Una Mirada hacia América Latina y el Caribe* (San José, CR: IICA), http://www.fao.org/docrep/019/i3702s/i3702s.pdf and http://www.fao.org/family-farming-2014/en/, date accessed 1 August 2014.
Colectivo Situaciones (2002) *19 y 20: Apuntes para el Nuevo Protagonismo Social* (Buenos Aires: Colectivo Situaciones/Ediciones Mano a Mano).
Cronkleton, P., Bray, D.B. and Medina, G. (2011) "Community Forest Management and the Emergence of Multi-Scale Governance Institutions: Lessons for REDD+ Development from Mexico, Brazil and Bolivia", *Forests* 2(2): 451–473.
Delgado-Lemus, A., Casas, A. and Tellez, O. (2014) "Distribution, abundance and traditional management of *Agave potatorum* in the Tehuacán Valley, Mexico: Bases for sustainable use of non-timber forest products" *Journal of Ethnobiology and Ethnomedicine* 10: 63.
Dietrich, W. (ed.) (2011) *Peace Studies: A Cultural Perspective* (Houndmills, UK: Palgrave Macmillan).
Dwinell, A. and Olivera, M. (2014) "The Water Is Ours Damn It! Water Commoning in Bolivia", *Community Development Journal* 49(S1): i44–i52.
The Ecologist (1993) *Whose Common Future? Reclaiming the Commons* (London: Earthscan Publications).
Esteva, G. (2010) "From the Bottom-up: New Institutional Arrangements in Latin America", *Development* 53(1): 64–69.
Esteva, G. (2014) "Commoning in the New Society", *Community Development Journal* 49 (Suppl. 1): i144–i159.

Fogelberg, K. (2013) "From Adopt-a-Project to Permanent Services: The Evolution of Water for People's Approach to Rural Water Supply in Bolivia", *Water Alternatives* 6(2): Article 6.

Fuente, M. and Morales, F.R. (2013) "El Ecoturismo Comunitario en la Sierra Juárez-Oaxaca, México: Entre el Patrimonio y la Mercancía", *Otra Economía. Revista Latinoamericana de Economía Social y Solidaria* 7(12): 66–79.

Giarraca, N. and Teubal, M. (eds) (2005) *El Campo Argentino en la Encrucijada: Estrategias y Resistencias Sociales, Ecos en la Ciudad* (Buenos Aires: Alianza Editorial).

Giarraca, N. and Teubal, M. (eds) (2009) *La Tierra es Nuestra, Tuya y de Aquel... Las Disputas por el Territorio en América Latina* (Buenos Aires: GEMSAL).

Gonzales, T. (2014) "Kawsay (Buen Vivir) y Afirmación Cultural: Pratec-Naca, un Paradigma Alternativo en los Andes", in B. Marañon (ed.), *El Buen Vivir y Descolonialidad: Critica al Desarrollo y la Racionalidad Instrumentales* (México: UNAM-Instituto de Investigaciones Económicas).

Hardin, G. (1968) "The Tragedy of the Commons", *Science* 162: 1243–1248.

Hernández Garciadiego, R. and Herrerías Guerra, G. (2008) "El Programa Agua para Siempre: 25 Años de Experiencia en la Obtención de Agua mediante la Regeneración de Cuencas", in L. Paré, D. Robinson and M. González Ortiz (eds), *Gestión de Cuencas y Servicios Ambientales. Perspectivas Comunitarias y Ciudadanas* (México: Instituto Nacional de Ecología).

Huanacuni, F. (2011) *Buen Vivir/Vivir Bien. Filosofía, Políticas, Estrategias y Experiencias* (Lima: Coordinadora Andina de Organizaciones Indígenas – CAOI).

Illich, I. (1977) *Towards a History of Needs* (New York: Pantheon Books).

Illich, I. (1982) *Gender* (New York: Pantheon Books).

Illich, I. (1992) "A Plea for Research on Lay Literacy", *In the Mirror of the Past. Lectures and Addresses, 1978–1990* (London: Marion Boyars), 159–181.

Illsley Granich, C., González, J.A., Aguilar, J., Rivero, M.A., Frenk, G.A., León, M.D., Martínez, A.G., Marielle, C. and Ceniceros, A.J. (2007) "El Grupo de Estudios Ambientales, AC: Entre la Acción Social y la Consolidación Institucional", in A.J. Bebbington (ed.), *Investigación y Cambio Social. Desafíos para las ONG en Centroamérica y México* (Guatemala: School of Environment and Development et al.), 25, 175–208.

Lavaca Collective (2007) *Sin Patrón: Stories from Argentina's Worker-Run Factories* (Chicago: Haymarket Books).

Lang, M. (ed.) (2013) *Alternativas al Capitalismo/Colonialismo del Siglo XXI* (Quito: Fundación Rosa Luxemburg – Grupo Permanente de Trabajo sobre Alternativas al Desarrollo).

Linebaugh, P. (2008) *Magna Carta Manifesto: Liberties and Commons for All* (Berkeley: University of California Press).

Mariotti, D., Comelli, M., Petz, M.I., Wahren, J., Giarracca, N. and Teubal, M. (2007) *Tiempos de Rebelión: "Que se vayan Todos", Calles y Plazas en la Argentina: 2001–2002* (Buenos Aires: GEMSAL).

Martínez Luna, J. (2010) *Eso que llaman Comunalidad* (México: Conaculta).

McDermott, M. (ed.) (2014) "Introduction", *Common Development Journal*, Special Supplement: Commons Sense, New Thinking About an Old Idea 49 (Suppl. 1).

Muñoz Ramírez, G. (2008) *The Fire and the Word: A History of the Zapatista Movement* (San Francisco, CA: City Lights Press).

Navarro, S.F. and Bessi, R. (2014) "Across Latin America, a Struggle for Communal Land and Indigenous Autonomy", http://www.truth-out.org/news/item/24981-across-latin-america-a-struggle-for-communal-land-and-indigenous-autonomy, date accessed 1 August 2014.

Ness, I. and Azzellini, D. (2011) *Ours to Master and to Own: Workers' Control from the Commune to the Present* (Chicago: Haymarket Books).

Organization of American States, Committee on Juridical and Political Affairs (OAS-CJPA) (2003) *Working Group to Prepare the Draft American Declaration on the Rights of Indigenous Peoples* (OEA/Ser.K/XVI, GT/DADIN/doc.113/03 rev. 1, 20 February 2003), Report of the Rapporteur, "Traditional Forms of Ownership and Cultural Survival, Right to Land and Territories", http://www.oas.org/council/CAJP/Indigenous%20documents.asp#2003, date accessed 10 September 2014.

Ostrom, E. (1985) "Are Successful Efforts to Manage Common-Pool Problems a Challenge to the Theories of Garret Hardin and Mancur Olson?", Working Paper, *Workshop in Political Theory and Policy Analysis* (Indiana University).

Ostrom, E. (1986) "An Agenda for the Study of Institutions", *Public Choice* 48(1): 3–25.

Ostrom, E. (1990) *Governing the Commons: The Evolution of Institutions for Collective Action* (Cambridge, UK; Cambridge University Press).

Panikkar, R. (1979) *Myth, Faith and Hermeneutics* (Ramsey, NJ: Paulist Press).

Panikkar, R. (1995a) *Invisible Harmony: Essays on Contemplation and Responsibility* (Minneapolis, MN: Fortress Press).

Panikkar, R. (1995b) *Cultural Disarmament: The Way to Peace* (Louisville, KY: Westminster John Knox Press).

Piñeiro Harnecker, C. (2013) *Cooperatives and Socialism: A View from Cuba* (London: Palgrave Macmillan).

Rebón, J. (2004) *Desobedeciendo al Desempleo: La Experiencia de las Empresas Recuperadas* (Buenos Aires: Ediciones Picaso/La Rosa Blindada).

Robles Hernández, S. and Cardoso Jiménez, R. (2008) *Floriberto Díaz. Escrito, "Comunalidad, Energía Viva del Pensamiento Mixe"* (México: UNAM, Programa Universitario México Nación Multicultural) (La Pluralidad Cultural en México 14).

Rosset, P.M. (2013) "Re-thinking Agrarian Reform, Land and Territory in La Via Campesina", *Journal of Peasant Studies* 40(4): 721–775.

Santos, B.S. (2009) *Una Epistemología del Sur: La Reinvención del Conocimiento y la Emancipación Social* (México: Siglo XXI editores, CLACSO).

Simonian, L. (1995) *Defending the Land of the Jaguar: A History of Conservation in Mexico* (Austin, TX: University of Texas Press).

Sitrin, M. (2006) *Horizontalism: Voices of Popular Power in Argentina* (San Francisco: AK Press).

Smith, L.T. (2012) *Decolonizing Methodologies: Research and Indigenous Peoples*. 2nd Ed. (London: Zed Books).

Stephen, L. and Hale, C.R. (eds) (2013) *Otros Saberes: Collaborative Research on Indigenous and Afro-Descendant Cultural Politics* (Santa Fe, NM: SAR Press, LASA).

Stevens, C., Winterbottom, R., Reytar, K. and Springer, J. (2014) *Securing Rights, Combating Climate Change: How Strengthening Community Forest Rights Mitigates Climate Change* (Washington, DC: World Resources Institute).

Tenet, G. (2000) *Global Trends 2015: A Dialogue About the Future with Nongovernment Experts* (Washington, DC: National Intelligence Council), http://www.fas.org/irp/cia/product/globaltrends2015/index.html, date accessed 1 August 2014.

Thompson, E.P. (1959) "Commitment in Politics", in C. Winslow (ed.) (2014) *E.P. Thompson and the Making of the New Left* (London: Lawrence and Wishart).

Toledo, V.M. (2014) "México: La Batalla Final es Civilizatoria", Series of four articles, 22 July, 5 August, 2 September, 30 September, *La Jornada* (México).

Toledo, V.M., Garrido, D. and Barrera Bassols, N. (2013) "Conflictos Socio-Ambientales, Resistencias Ciudadanas y Violencia Neo-Liberal en México", *Ecología Política* 46: n115–n124.

Toledo, V.M. and Ortiz Espejel, B. (2014) *México, Regiones que Caminan Hacia la Sustentabilidad: Una Geopolítica de las Resistencias Bioculturales* (Puebla: Universidad Iberoamericana, Campus Puebla, Conacyt).

UNESCO (2013) *The World Social Science Report 2013* (Paris: International Social Science Council (ISSC), UNESCO and the Organisation for Economic Co-operation and Development (OECD)).

Vachon, R. (1995) "Guswenta or the Intercultural Imperative: Towards a Re-enacted Peace Accord Between the Mohawk Nation and the North American Nation-States (and Their Peoples)", *Interculture Journal* XXVIII (2, 3, 4), issues 127, 128, 129.

Venkateswar, S. and Hughes, E. (2011) *The Politics of Indigeneity, Dialogues and Reflections on Indigenous Activism* (London: Zed Books).

Villoro, L. (2003) *Creer, Saber, Conocer* (México: Siglo XXI).

Walljasper, J. (2010) *All That We Share: A Field Guide to the Commons* (New York: The New Press).

Webber, J. and Spronk, S. (2011) "The Bolivarian Process in Venezuela: A Left Forum", *Historical Materialism* 19(1): 229–266.

Zibechi, R. (2008) *Autonomías y Emancipaciones: América Latina en Movimiento* (México: Bajo Tierra Ediciones).

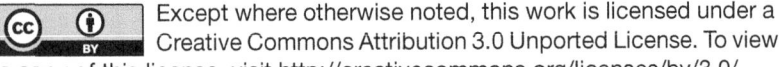

Except where otherwise noted, this work is licensed under a Creative Commons Attribution 3.0 Unported License. To view a copy of this license, visit http://creativecommons.org/licenses/by/3.0/

OPEN

# 11
# Community Consultations: Local Responses to Large-Scale Mining in Latin America

*Mariana Walter and Leire Urkidi*

## Introduction

This chapter studies the emergence and spread of community consultations in large-scale metal mining projects in Latin America. These consultations are different from the free, prior and informed consent (FPIC)-related consultations, or *consulta previa*, that are fostered by national governments. From Tambogrande (Peru) in June 2002 to Mataquescuintla (Guatemala) in November 2012, 68 consultations/referenda have been conducted in Peru, Argentina, Guatemala, Colombia and Peru. In all cases the result has been a large opposition to mining projects. This process is occurring in a context of growing pressures to extract mineral ores in Latin America and an increasing number of related socioenvironmental conflicts (see Chapter 2). The particularity of these consultations is that these are not commissioned by national governments as part of official procedures to consult communities but instead are promoted by environmental justice movements (EJMs), usually with the support of local governments.

The emergence and spread of consultations in Latin America remains poorly studied. Studies addressing mining consultations/referenda have focused on the first four cases: Tambogrande, Esquel, Sipakapa and Majaz/Río Blanco (Muradian, Martinez-Alier and Correa, 2003; Subies et al., 2005; Haarstad and Floysand, 2007; De Echave et al., 2009; McGee, 2009; Walter and Martinez-Alier, 2010; Fulmer, 2011; Urkidi, 2011; Bebbington, 2012a); along with the wave of consultations in Guatemala (Holden and Jacobson, 2008; Rasch, 2012; Trentavizi and Cahuec, 2012).

Nevertheless, the cases that followed, their connections and the institutional features of consultations have received poor scholarly attention. This research is born from the curiosity of understanding how and why these consultations have emerged and spread, and how community consultations are challenging the governance of mining activities.

Analysing the cases of community consultations conducted in Latin America from 2002 to 2012, we claim that these consultations (1) emerge in the context of environmental justice struggles and criminalization; (2) aim to reclaim the right of affected populations to participate, in empowering forms, in high-stakes decision-making that affect their lands and livelihoods; and (3) are a hybrid institution, the product of a dynamic multiscalar process where non-state and state actors, and formal and informal institutions, are mobilized to challenge the centralized governance of extractive activities.

## Struggles over the governance of mining activities in Latin America

As mentioned in Chapter 1, there is an ongoing shift in views that frame resource regulation from those that are led by state-based institutions of resource management (government) to a wider environmental governance perspective. The governance approach addresses the myriad of actors and institutions that guide the ways in which (global) environmental issues are addressed across different scales (Bulkeley, 2005).

State-centred frames are increasingly unsatisfactory and anachronistic to understanding different ways in which regulation is constructed and reconstructed. Recognizing the different spatial grammars at play becomes necessary in order to understand the emergence of hybrid forms of environmental governance and their implications (Bulkeley, 2005). Hybrid forms of governance challenge the conventionally recognized social roles of markets, states and, more recently, communities, as new dynamics and alliances are formed. Hybrid governance entails the formation of complex political spaces: networks of social, economic and cultural relations, actors connecting from distant locations, sharing networks with common social and political objectives.

In this chapter we refer to hybrid governance as a process of institutional bricolage where different (non-state and state) actors shape institutions that combine formal and informal components in a multiscalar dynamic. We conceive scale as an epistemological, not an ontological, entity. Leitner, Seppard and Sziarto (2008: 159) conceptualize scale "as a relational, power-laden and contested construction that actors strategically engage with, in order to legitimize or challenge existing power relations".

Hybrid institutions can be addressed from different theoretical perspectives. Instrumentalist approaches assume that actors are political and social entrepreneurs who actively use their social capital to build institutions that strive for optimal resource management. It is usually claimed that, to use social capital appropriately, institutions must be properly embedded in the cultural and social context from which the norms to support purposive decision-making are drawn (Ostrom, 1990). However, it has been claimed that concepts of embeddedness foster a functional and static conceptualization of culture and tradition that obscures the complex dynamics of institutional construction and evolution (Cleaver, 2001). Cleaver (2002: 17) claims that "the evolution of collective decision-making institutions may not be the process of conscious selection of mechanisms fit for the collective action task (as in Ostrom's model) but rather a messier process of piecing together shaped by individuals acting within the bounds of circumstantial constraint".

In her studies of institutions for common property resource management in Tanzania, Cleaver (2001, 2002, 2013) develops the concept of "institutional bricolage" as a process by which people consciously and unconsciously draw on existing social and cultural arrangements (rules, traditions, norms, roles and relationships) to shape institutions in patch-together institutions to change situations (Cleaver et al., 2013). In this dynamic, the resulting institution is a mix of modern and traditional, of formal and informal practices. Institutional bricolage offers a compelling approach to understanding the way in which hybrid institutions can be the result of a complex and dynamic assemblage process where contexts, conflicts, needs, scales, actors, and formal and informal institutions come into play to produce a particular hybrid institution.

## Environmental Justice Movements (EJMs)

Latin American anti-mining movements and organizations played a central role in the emergence and spread of consultations. In this section we outline some key features of this actor, its central demands and its scalar dynamics.

Latin American anti-mining movements have been framed as EJMs because they demand socioecological equity and fair decision-making processes in the governance of mining activities (Urkidi and Walter, 2011). Recently, questions of participation and voice have been at the forefront of environmental justice studies (Schlosberg, 2007). The concept of environmental justice was born in the 1980s in tandem with Afro-American social movements fighting environmental racism (Bullard, 1990). Since then, the concept has travelled among social movements and has been appropriated by other social groups and movements in the world. As a result, national and regional environmental justice networks have emerged in Latin America in recent decades (Carruthers, 2008). Mining concerns and anti-mining movements have a central place in these Latin American networks.

It has been pointed out that the concept of environmental justice entails a politics of scale because it refers to the spatial and social distribution of environmental impacts and economic benefits, and to the scales, institutions and agents that regulate environmental decisions (Kurtz, 2003). Some political geographers express criticism regarding EJMs' "militant particularism" (Harvey, 1996), according to which movements have to find a way to cross the problematic divide between actions that are profoundly embedded in place and local experience, on the one hand, and a wider movement and discourse on the other. According to this perspective, local loyalties and identity politics of resistance movements prevent engagement in wider and emancipating politics of scale. We claim, however, that EJMs tend to transcend place-based militant particularism (Kurtz, 2003). EJMs build strategies and discourses that transcend the particularities of local demands, acknowledging the structural roots of their struggles and establishing solidarity networks with other communities and groups (Urkidi and Walter, 2011). These networks have been key for anti-mining groups in Latin America, such as OCMAL and the No a la Mina platform in Argentina.

EJMs should not be seen as static but rather as learning and flexible movements that expand and contract in space as conflicts unfold and movements jump scales (Smith, 1996; Leitner, Seppard and Sziarto, 2008). There are different spatialities at play in contentious politics (e.g. scale, networks, place, mobility), and participants usually draw on several at once (Leitner, Seppard and Sziarto, 2008). The analysis of EJMs should also acknowledge this spatial complexity. We claim that these features of EJMs played a central role in the shaping of community consultations.

EJM concerns usually address three key dimensions of environmental justice: distribution, recognition and participation (Schlosberg, 2007). These can be seen as key lenses through which EJMs frame injustice. EJMs address not only inequity but also, and sometimes centrally, the political processes that construct environmental inequities. Anti-mining groups in Latin America frequently argue that the approval of mining projects involves the misrecognition of the material and cultural dependence on water and land of the affected populations and that it ignores the concerns expressed in local participatory stages, or that it lacks such spaces altogether (Muradian, Martinez-Alier and Correa, 2003; Haarstad and Floysand, 2007; Urkidi and Walter, 2011).

The main features of the procedures that govern mining activities are shared by most Latin American countries. Indeed, Latin American mining laws were developed under similar guidelines drafted by international financial institutions (e.g. the World Bank) (Chaparro, 2002; Bridge, 2004). The approval of mining projects is centralized in the national (or provincial, in the case of Argentina) government, and is based on the assessment of an environmental impact report. Participation arenas are set in relation to this technical document and are non-binding. Civil society actors can usually present allegations (e.g. online or on paper) and, sometimes, can express their views in front of a public audience where the technical document is presented. Usually, law requires that these concerns be addressed by the mining company when providing the final environmental impact assessment that has to be approved by the national government (usually by the mining or environmental departments). However, EJMs claim that participation in mining decisions is mainly "informative" and insufficient, when not secretive (Janhcke Benavente and Meza, 2010).

Projects affecting indigenous communities are under specific regulations. Most Latin American countries (all those studied in this chapter) have subscribed to the 169 ILO Convention, which requires the prior and informed consent of communities before decisions about activities that could affect them are made, a process that should follow customary procedures. This right is usually ignored or misapplied (Janhcke Benavente and Meza, 2010). However, even if put in practice, the way the 169 ILO Convention and other international documents (e.g. the UN Declaration on the Rights of Indigenous People) frame "consent" is ambiguous and does not necessarily imply a binding power to community views (McGee, 2009; Janhcke Benavente and Meza, 2010). As the cases presented in this chapter illustrate, and as pointed out by other studies (e.g. Janhcke Benavente and Meza, 2010), the way decisions regarding mining activities exclude or mistreat local actors, their values, concerns and institutions is fuelling unrest and frustration among the affected communities.

## The rise and spread of mining consultations in Latin America

In order to study the process of emergence and spread of Latin American mining consultations, we identified and analysed all cases of metal-mining consultations/referenda fostered by EJMs from 2002 (Tambogrande) to 2012 in Latin America. We considered those consultations/referenda that were not fostered by the central government or private companies as part of an official consultation process, and aimed to consult the local citizens at large whether or not a community/municipality/district was in favour of large-scale metal mining activities in their territory.

We reviewed and triangulated primary and secondary, and activist and academic, sources (e.g. newspapers, activist and government websites, reports, scientific papers). As the analysis unfolded, we identified the main commonalities and differences, and developed a series of hypotheses for the emergence and spread of consultations that made us revisit and expand our sources: an iterative process that led us to refine the findings outlined in this chapter.

We identified 68 metal-mining consultations in five Latin American countries: Peru (2002, 2007, 2008, 2009, 2012), Argentina (2003, 2012), Ecuador (2011), Colombia (2009) and Guatemala (57 municipal consultations from Sipakapa in 2005 to Mataquescuintla in 2012) (Tables 11.1 and 11.2). We grouped the cases into three main "travel paths" according to the connections and similarities of consultation cases, not their chronological order. In this vein we aim to identify how consultations have been transmitted from conflict to conflict as a useful participation institution. For each "travel path" we highlight the key elements of the leading case(s), identify how consultations emerged, their institutional features and the EJMs involved, and analyse the multiple spatialities at play in the transference of consultation experiences among EJMs.

The first travel path presents the main features of the first consultation case in Tambogrande (2002), the spread of the experience to other Peruvian communities and its arrival in Ecuador. The second travel path outlines the key features of the Argentinean process triggered by Esquel (2003). The third travel path addresses the Guatemalan wave of consultations born from Sipakapa (2005), and the arrival of this experience in Colombia. The case of Guatemala presents some particular features. While the first case of consultation (Sipakapa) occurred in the context of an active conflict, most of the following cases were part of a regional campaign to prevent the expansion of mining activities in the country.

Table 11.1 Mining consultations in the context of active mining conflicts, 2002–2012

|  | Consultation case | Conflict duration | Date of *consulta* | Mining project and mining company | Secret/ non-secret | Uses official voters list (Y/N) | Type of *consulta* (legal framework) municipal ordinance (MO) | Consulta | | | |
|---|---|---|---|---|---|---|---|---|---|---|---|
|  |  |  |  |  |  |  |  | Participation (% eligible voters) | % against mining | % in favour of mining | % whites/ null |
| Peru | **Tambogrande case** district of Tambogrande (Piura) | 1990–2003 | 01/06/02 | Tambogrande project, Manhattan Minerals (Canada, junior), gold and silver | S | Y | Consulta Vecinal (MO) | 27,015 (69%) | 93.85% | 1.98% | 4.17% |
|  | **Majaz/Rio Blanco case** Ayabaca and Pacaipampa district (Ayabaca municipality) and Carmen de la Frontera district (Huancabamba municipality), Piura. | 2002–today | 16/09/07 | Majaz project, Monterrico Metals (UK, junior), sold in 2007 to Zijin Mining (China), copper and molybdenum | S | Y | Pacaipampa (Consulta Vecinal, MO) Ayabaca (Consulta Vecinal, MO) Carmen de la frontera (Consulta Vecinal, MO) | 6,091 (71.47%) 8,873 (50.09%) 3,053 (59.26%) | 17,033 (94.54%) | 285 (1.58%) | 699 (3.88%) |

Table 11.1 (Continued)

| Consultation case | Conflict duration | Date of consulta | Mining project and mining company | Secret/ non-secret | Uses official voters list (Y/N) | Type of consulta (legal framework) municipal ordinance (MO) | Consulta Participation (% eligible voters) | % against mining | % in favour of mining | % whites/ null |
|---|---|---|---|---|---|---|---|---|---|---|
| **Candarave case** districts of Candarave, San Pedro, Cairani, Calacala; Talaca, Yucamani, Calientes and Pallata (Tacna, Atacama) | 1990s–today | 17/02/08 | Toquepala project, Southern Copper Corp. (US- Mexico), copper and molybdenum | S | Y | Consulta Vecinal (MO) | 3,478 (67%) | 3,215 (92%) | n.i. | n.i. |
| **Islay/Tía María case** districts of Cocachacra, Punta de Bombón, Dean Valdivia, Mejía, Islay-Matarani and Mollendo (Arequipa, Islay Province) | 2008–2011 | 27/09/09 | Tía María project, Southern Copper Corp., copper | S | Y | Cocachacra* (Consulta Vecinal, MO) | 3,131 (49%) | 2,916 (93%) | 139 (4.4%) | 76 (2.4%) |
| | | | | S | Y | Punta Bombón* (Consulta Vecinal, MO) | 2,004 (43%) | 1,883 (94%) | 71 (3.5%) | 50 (2.5%) |
| | | | | S | Y | Dean Valdivia (Consulta Vecinal, MO) | 2,304 (53%) | 2,211 (96%) | 52 (2.3%) | 41 (1.8%) |
| | | | | S | N | Mollendo (Consulta Popular) | 3,643 (n.i.) | 3,573 (98%) | 9 (0.3%) | 61 (1.7%) |
| | | | | S | ** | Mejía (Consulta Vecinal) | 272 (n.i.) | 245 (90%) | 26 (9.8%) | 1 (0.4%) |

295

| Country | Case | Period | Date | Project | | | Consulta type | Votes | | | |
|---|---|---|---|---|---|---|---|---|---|---|---|
| | | | | | S | N | Islay-Matarani (Consulta Popular) | 837 (n.i.) | 765 (91.4%) | 61 (7.3%) | 11 (1.3%) |
| | Kañaris case three districts (San Juan Bautista de Cañaris, Huacapampa, Congona), Lambayeque | 2004–today | 30/09/12 | Cañariaco project, Candente Copper (Canada, junior), copper, gold and silver | S | Y | Consulta Comunitaria (ILO 169) | 1,896 (47.4%) | 1,719 (95%) | 106 (6%) | 71 (4%) |
| Argentina | Esquel case (Chubut province) | 2001–today | 23/03/03 | Esquel project, Meridian Gold (US, junior), sold in 2007 to Yamana Gold (Canada, Junior), gold and silver | S | Y | Compulsory Consulta Popular (MO) | 13,845 (75%) | 11,046 (81%) | 2,561 (17%) | 277 (2%) |
| | Loncopue case (Neuquén province) | 2007–today | 02/06/12 | Lonco project, Corporación Minera de Neuquén (provincial Argentina) and Metallurgical Construction Corp (China), copper and molybdenum | S | Y | Compulsory and Binding Referendum (to approve MO) | 2,588 (72%) | 2,125 (82.08%) | 388 (15%) | 75 (2.9%) |
| Ecuador | Quimsacocha project (Vitoria del Portete and Tarqui), Azuay | 2004–today | 02/10/11 | Project Quimsacocha, Iam Gold (Canada), sold in 2012 to INV Metals (Canada, junior), gold, silver and copper | S | N | Consulta Comunitaria conducted by Juntas de Agua to its members | 1,037 (66.6%) | 958 (92.38%) | 47 (4.53%) | 18 (1.73%) |

Table 11.1 (Continued)

|  | Consultation case | Conflict duration | Date of consulta | Mining project and mining company | Secret/ non-secret | Uses official voters list (Y/N) | Type of consulta (legal framework) municipal ordinance (MO) | Consulta Participation (% eligible voters) | % against mining | % in favour of mining | % whites/ null |
|---|---|---|---|---|---|---|---|---|---|---|---|
| Colombia | Mandé Norte Project two municipalities (Carmen del Darién, Murindó), Chocó | 2007–today | 28/02/09 | Mandé Norte Project, Muriel Mining (US, junior), copper, gold and molybdenum | NS | N | Consulta Inter-etnica, ILO 169 | 799 (n.i.) | 100% | | |
| Guatemala | Sipakapa case (San Marcos) | 2003–today | 08/06/2005 | Martin project, Montana Exploradora (Goldcorp-Canada, senior), gold, silver and open pit | Both (depending on communities) | Y | Consulta Comunitaria (MO) | 2,564 (45%) | 95.50% | 1.40% | 1.60% |
|  | Minera San Rafael (Santa Rosa) and Mataquescuintla (Jalapa) | 2010–today | 29/05/11 11/06/11 10/07/11 11/11/12 | Escobal Project, Oasis (Minera San Rafael: Tahoe Resources Canada – 40% of Goldcorp Canada), silver, gold and others | S | Y | Mataquescuintla (Consulta Comunitaria MO) | 10,375 (53%) | 97% | 1.6% | 1.8% |

*Notes:* * In these consultations, two questions were asked. We only present in this table the answer regarding acceptance or rejection of mining activities.
** There is divergent information among sources regarding the role of the local government in this consultation.

Table 11.2 Guatemalan wave of preventative consultations against mining activities, 2005–2012

| Department/region | Projects/licences | No. of Consultas | Municipalities/dates | Participation | % saying no to mining | Types of consultas | Consequencies/results |
|---|---|---|---|---|---|---|---|
| San Marcos | Exploitation in San Miguel Ixtahuacan and Sipakapa (Goldcorp – Canada: gold, silver, others). Exploration licences in every municipality with consulta (Goldcorp's subsidiaries Canada: gold, silver, nickel, cobalt, polimetalics, rare earths, others) | 11 | Sipakapa 18/05/2005, Comitancillo 18/06/2005, Concepción Tutuapa 13/02/2007, Ixchiguan 13/06/2007, Sibinal 18/04/2008, Comitancillo 14/05/2008, Tacaná 16/05/2008, Tajumulco 13/06/2008, San José Ojetenam 11/07/2008, Tejutla 30/09/2008, San Cristobal Cucho 27/06/2009 | More than 60,000 people | 98 | Communitarian consultas. ILO 169 and Municipal Code. Non-secret vote. In Sipakapa just registered people (Tribunal Supremo Electoral), in the others all the community | In almost every consulta, support from communitarian and municipal governs. Non-binding consultations for national government. National government tried to regulate consultas in 2011, against consulted communities' wishes. Consultas meant the empowerment/of communities. Creation of networks against mining. Despite many exploration licences, only those very advanced projects have prospered after consultations (Sipakapa and San Rafael) |

Table 11.2 (Continued)

| Department/region | Projects/licences | No. of Consultas | Municipalities/dates | Participation | % saying no to mining | Types of consultas | Consequencies/results |
|---|---|---|---|---|---|---|---|
| Huehuetenango | Exploration licences in almost every municipality with consulta (Goldcorp's subsidiaries – Canada and Tenango S.A. Canada: gold, silver, others; Guatemala: copper; copper/cobalt; Minas de Guatemala: polimetalics; other companies). Few lead/zinc mines | 28 | Concepción Huista, Todos Santos Cuchumatán, San Juan Atitán, Colotenango, Santiago Chimaltenango 25–27/07/2006, Santa Eulalia 30/08/2006, San Pedro Necta 30/03/2007, San Antonio Huista 12/05/2007, Santa Cruz Barillas 23/06/2007, San Ildefonso Ixtahuacán 03/08/2007, Nentón 11/08/2007, San Sebastián Huehuet. 26/10/2007, San Miguel Acatán 01/12/2007, San Juan Ixcoy 13/05/2008, Tectitán 27/06/2008, Chiantla 13/07/2008, Jacaltenango 26/07/2008, Santa Ana Huista 06/08/2008, Aguacatán 03/10/2008, San Pedro Soloma 17/10/2008, Cuilco 25/10/2008, Santa Bárbara 28/11/2008, San Rafael Petzal 10/01/2009, San Rafael La Indep. 28/04/2009, San Mateo Ixtatán 21/05/2009, San Gaspar Ixchil 23/07/2009, San Sebastián Coatán 24/09/2009, Unión Cantinil 18/01/2010 | 377,615 people (without Tectitan: no data of participation) | 99 | Communitarian consultas. ILO 169 and Municipal Code. Non-secret vote. In six of them just registered people can participate, in the other 22 all the community can vote | |

| Department | Mining licences/companies | # | Location and date | People | % | Notes |
|---|---|---|---|---|---|---|
| Quiche | Some exploration licences (Nichromet Guatemala/Canada: nickel, cobalt, others) | 5 | Cunén 27/10/2009, Santa Cruz del Quiché 22/10/2010, Uspantán 29/10/2010, Sacapulas 20/05/2011, Chinique 14/03/2012 | 104,015 people | Almost 100 | Communitarian *consultas*. ILO 169 and Municipal Code. Non-secret vote. All the community |
| Quetzaltenango | Some exploration licences (Goldcorp's subsidiary – Canada: gold, silver, zinc; other companies) | 8 | Cajola 01/07/2011, San Miguel Siguilá 15/05/2011, Concepción Chiquirichapa 20/04/2011, San Martin Chile Verde 16/03/2011, Olintepeque 20/02/2011, San Juan Ostuncalco 18/02/2011, Huitan 22/11/2010 | 104,037 people (without Cajolá: no data) | 99 | Municipal *consultas*. ILO 169 and Municipal Code. Secret vote. Just registered people |
| Santa Rosa | Exploration licences (Tahoe Resources – Canada/USA: silver, gold, others –> Escobal project) | 3 | Nueva Santa Rosa 29/05/2011, Santa Rosa de Lima 10/07/2011, Casillas 10/07/2011 | 18,110 people | 98 | Municipal *consultas*. Secret vote (Tribunal Supremo Electoral). Under Municipal Code. Just registered people |
| Retalhuleu | Exploration licences (Mayan Iron Corp's subsidiary – Australia: iron sands, nickel, cobalt, rare earths, others), recognition licence (G4G Resources – Canada: iron sands) | 1 | Champerico 12/02/2012 | 16,699 people | 99 | Municipal *consulta*. Secret vote. Under Municipal Code |
| Jalapa | Exploration licences (Tahoe Resources Canada/USA: silver, gold, others: Escobal project; Goldcorp: gold, silver, others) | 1 | Mataquescuintla 11/11/2012 | 10,375 people | 97 | Municipal i. Secret vote. Under Municipal Code |

We explain the Sipakapa consultation in more detail and refer to the following cases as a regional process.

## Emergence and spread in Peru and Ecuador

### Tambogrande conflict (Piura)

Tambogrande is located in one of the poorest departments of Peru (Piura), with an arid climate that requires dams and irrigation canals (built with World Bank support) to sustain its agricultural export-oriented activities. The conflict was triggered by the Manhattan Minerals project, whose main deposit was located under the town of Tambogrande. Critical voices pointing to the environmental and social impacts of this activity, led by a local farmer and agrarian engineer who had emigrated from Lima, fostered the formation of the Frente de Defensa de Tambogrande y el Valle de San Lorenzo in 1999. This organization became the main local opposition to the project in collaboration with the local church and the National Coordinating Confederation of Communities Affected by Mining (CONACAMI) (Portugal Mendoza, 2005).

As the *Frente* was unable to engage in an exchange of views and concerns with the national government, local unrest rose (Portugal Mendoza, 2005). In March 2001, after a period of strikes, massive mobilizations and violent events in Tambogrande, the local leader Godofredo García Baca was shot dead by a hooded gunman (Muradian, Martinez-Alier and Correa, 2003). These events made the mining conflict nationally and internationally known (*The Economist*, 23 June 2001), thereby engaging new national and international support. Professionals from Piura and Lima constituted a working group to elaborate technical arguments and reports against the project, succeeding in involving transnational organizations and networks in the local struggle (Bebbington, Humphreys Bebbington and Bury, 2011).

Local tension was growing and social movements became concerned with a possible escalation of violence (Portugal Mendoza, 2005; Cabellos and Boyd, 2007; McGee, 2008). In this context, the *Frente*, its allies and Tambogrande's mayor – who was not clearly positioned before – agreed on the need to conduct a *consulta vecinal* (neighbours' consultation), a peaceful and democratic mechanism to channel local unrest and express local views (Portugal Mendoza, 2005; Subies et al., 2005; Bebbington, Humphreys Bebbington and Bury, 2011).

The municipality of Tambogrande issued the Municipal Ordinance No. 012-2001-MDT-C, which created the *consulta vecinal* as a mechanism for citizenship participation at the district level. The ordinance was based on international, national and municipal rights, and on laws regarding citizen participation (international treaties, national and municipal laws, constitutional articles and the Environment Code), setting the basic legal structure that would later be used in all following consultations in Peru. While ILO 169 was not referenced in the Tambogrande consultation ordinance (it was added in the following Majaz/Río Blanco municipal ordinances), it was used in activist discourses (Fulmer, 2011).

The National Office of Electoral Processes initially recognized the *consulta* and agreed to provide support. However, a formal complaint of unconstitutionality and illegality by the Ministry of Energy and Mines (MEM) reduced the final involvement of the office to advising and lending election materials (National Electoral Office, 2002). The technical advice of national and transnational groups and the financial collaboration of transnational organizations such as OXFAM were key to conducting the consultation (Portugal Mendoza, 2005; Bebbington, Humphreys Bebbington and Bury, 2011). Moreover, organizations such as the Mineral Policy Center, the Environmental Mining Council of British Columbia, OXFAM, and Friends of the Earth from Costa Rica and Ecuador contributed to building the legitimacy of the consultation by acting as observers, supporting and disseminating the experience (Muradian, Martinez-Alier and Correa, 2003).

On 2 June 2002, the *consulta* calling all district inhabitants was held and resulted in a massive rejection of the mining project (Portugal Mendoza, 2005). The participation mechanism followed the same procedures of a regular election (secret vote, registered voters, ballot boxes, etc.) (see Table 11.1). The consultation was not recognized either by the mining company or by the national government, which claimed that the EIA formal assessment was the legally binding decision-making process. The following month the *Frente* prevented three public audiences through organized protests. Finally, the public company revoked the Manhattan mining licence based on administrative grounds, thereby suspending the project. In November 2002 the president of the *Frente*, Francisco Ojeda, won the municipal elections (Portugal Mendoza, 2005).

*Majaz/Río Blanco conflict (Piura, Peru)*

As the Tambograde struggle was coming to an end, a new and relevant mining conflict was emerging nearby in the provinces of Ayabaca and Huancabamba (Piura Highlands) concerning the exploration of a copper-molybdenum mining deposit by a subsidiary of Monterrico Metals. The conflict of Tambogrande not only contributed to introducing mining scepticism in the region but was also a source of experience and support for local groups and authorities in this new struggle (Diez Hurtado, 2007; Bebbington, 2012a). For instance, the group of organizations and individuals supporting the *Frente* in Tambogrande – then formalized as Red Muqui – later in the conflict fostered the formation of the Majaz Support Group to create a bridge of experience, technical expertise and strategies among movements (Bebbington, 2012a).

The Majaz mining project was located in the peasant communities of Segunda y Cajas and Yanta (*comunidades campesinas*), lands that are administered under particular institutional arrangements legally recognized by the state (Bebbington, 2012a). The company did not comply with the required approval of the community assembly, triggering rejection and formal complaints (Bebbington et al., 2007).

In 2004, two "massive" mobilizations were conducted involving thousands of peasants concerned by the environmental (water), economic (agriculture, tourism) and social (land access) impacts of the mining project and its lack of recognition of local institutions. These protests resulted in police clashes, injuries and the death of two peasants, Remberto Herrero (April 2004) and Melanio García Gonzalez (July 2005) (Bebbington, 2012a). From 2004 to 2007, local activists denounced cases of activist kidnapping, tortures and persistent criminalization (discredit campaigns, unjustified imprisonment, legal prosecution) that even reached the UK justice courts (OXFAM, 2007, 2009; Cobain, 2009).

In 2005, mayors, local leaders and social organizations fostered the formation of the Frente por el Desarrollo Sostenible de la Frontera Norte del Perú (FDSFNP). The organization, critical of the mining project and the role of the national government, was composed of provincial and district government representatives, peasant communities, *rondas campesinas*, defence fronts from Huancabamba, Ayabaca, Tambogrande, and other anti-mining groups from the region.

Tension and distrust rose as negotiation attempts by the regional and national governments were failing and the government issued measures to limit public participation rights (Diez Hurtado, 2007; Red Muqui, 2009; Bebbington, 2012a). In this context, a consultation was promoted. As in Tambogrande, the consultation was seen as a peaceful channel of participation that would ease local tensions. The municipalities of Ayabaca and Huancabamba approved municipal ordinances, calling for a *consulta vecinal* (Bebbington, 2012a). The *consulta* resulted in a 94.5% rejection of mining activities in the district.

While in Tambogrande the national government minimized the weight of the consultation, in this instance it actively tried to prevent it. A vociferous campaign criminalized the consultation and its proponents, stating that the referendum was illegal, communist and politically manipulated by international NGOs that intended to delay the country's development (OXFAM, 2007; McGee, 2008). However, the Peruvian ombudsman and the human rights national council of the Justice Ministry declared that, even if this mechanism was non-binding, it was legal under constitutional law (OXFAM, 2009; Red Muqui, 2009; CISDE-ALAI, 2009). Moreover, the Majaz consultation led the national ombudsman of Peru to initiate a process of regulation of indigenous consultation rights. What is more, both in Majaz and Tambogrande (and in Esquel, Argentina), mining activities were halted and therefore became examples of successful cases.

*Toquepala expansion project (Candarave), Tía María project (Islay, Arequipa), Kañariaco project (Lambayeque) in Peru*

After these two consultations in Piura (North of Peru), there were three others on the south and central coast of Peru, where national organizations and networks played a key role in spreading the experience and providing support. The following consultation in Candarave (2008, Tacna region, Atacama Desert) is different from previous cases because it took place in an area with ongoing large-scale mining activities. The conflict that led to the consultation emerged when the mining company started negotiations to expand its water-use permits. Local and provincial governments, the irrigation users (*Junta de usuarios de riego*) and the local fronts of defence opposed new permits. They pointed to the need to decrease mining water use due to a regional water scarcity crisis that was affecting agricultural production and forcing peasant outmigration, and to the need to compensate for these impacts. In January 2008 the mayor of Candarave called for a *consulta vecinal* (Municipal Ordinance No. 001-2008-MPC/A) with the support of the provincial governor, local defence fronts and the Junta de Aguas. The *consulta* had observers from national and international NGOs who also provided technical support (Radio Uno, 2008). Consultation participants (67% of eligible voters) answered two questions: 92% rejected new mining activities, and 94% opposed the use of underground and superficial water for mining activities.

The fourth mining *consulta* in Peru occurred in 2009 in the province of Islay (Arequipa Department). Islay is a dry region inhabited by peasants and indigenous groups. The conflict emerged in 2008, with the

Southern Copper Peru Corporation Tía María large-scale copper mine project (Gutierrez Zeballos, 2011). Concerns regarding impacts on water availability and local livelihoods fostered the formation of the Frente Amplio de Defensa del Medio Ambiente y Recursos Naturales. This movement led to the organization of a regional front with the support of local groups, the mayor of Valdivia and national organizations such as the CONACAMI, Cooperacción, Red Muqui and the Coordinadora Andina de Organizaciones Indígenas (Gutiérrez Zeballos, 2011; Red Muqui, 2011).

On 27 September 2009, the six districts of Islay conducted a *consulta vecinal*. The provincial mayor refused to call for a provincial referendum. In some districts, consultations were called by local mayors who issued ordinances. In other districts, consultations were led by social movements, following the same procedures (CAOI, 2009; Gutiérrez Zeballos, 2011). The process was observed by a national congressman, members of the Flemish NGO Broederlijk Delen, and the Peruvian NGOs Transparencia Civil and CONACAMI (Márquez, 2009). The average turnout was 48.5% (considering the districts where voter lists were available), and 93–98% opposed the Tía María project.

The national government did not recognize the referendum and, some months later, called for a public audience to present the project's EIA. With the assistance of national and transnational organizations, around 3,000 technical comments on the EIA project were submitted. Moreover, a series of regional strikes were organized as dialogue spaces were perceived as sterile. These strikes were marked by hard police repression, activist criminalization, three deaths and more than 400 injuries (Gutiérrez Zeballos, 2011). In the midst of this violence, a report by the United Nations Office for Project Services, requested by the government and communities as an "independent" review, concluded that the EIA had serious deficiencies (UNOPS/PNUMA, 2011), forcing the MEM to suspend the project.

The fifth *consulta* of Perú took place in 2012 in the northern district of Kañaris (region of Lambayeque). The Kañariaco mining project was a large-scale copper mine, in exploration stages, owned by the junior Canadian company Candente Copper Peru SA. The project was located in a cloud forest area inhabited and cultivated by two Quechua-speaking communities (municipality of Kañaris, 2012). In an assembly in 2012, the community of San Juan de Kañaris decided to conduct a *consulta comunal* (community consultation) (Fedepaz, 2013). The mining company and the MEM claimed that a consultation had already been conducted following official procedures.

The community consultation followed the procedures of regular communal elections (secret, registered voters) without the support of local governments; the result was a 91% mining rejection (1,896 votes, 47.4% turnout). The process was supported and observed by CONCAMI, the Red Muqui and leaders of local organizations. The regional governor, the Ministry of Agriculture, and representatives of regional offices of development and production, and energy and mines, also participated as observers (Servindi, 2012).

When this consultation occurred, the national government was promoting a law to regulate indigenous consultation rights. The question of whether the Kañaris are peasant or indigenous, and hence entitled to FPIC according to ILO 169, triggered a wide debate (Greenspan, 2013). While the national ombudsman and transnational indigenous groups recognize the FPIC for Kañaris, the government denies this right and claims that the government consultation is the valid one. In 2013 the Candente mining company stopped mining exploration, pointing to low copper prices as the reason.

*Ecuador, Kimsakocha project (Azuay)*

In October 2011 the first mining community consultation of Ecuador took place. The conflict arose from an open-pit project owned by a junior Canadian company. Concerns rose regarding the impact on water resources among indigenous and peasant groups located downstream from the project area (Pérez Guartambel, 2012). The idea to conduct a consultation emerged in the context of growing pressure from the national government to promote mining activities in the country, in the midst of verbal and legal delegitimation and criminalization campaigns against Ecuadorian indigenous and anti-mining activists (interview with local activist, 2012). Moreover, local indigenous and peasant leaders were in contact with Latin American indigenous, anti-mining and human rights movements, in particular from Ecuador and Peru (interview with national anti-mining movement leader, 2012). In June 2011, local indigenous leaders led the organization of a continental peoples meeting with a strong emphasis on the impact of mining agendas on the environment and indigenous groups (Pérez Guartambel, 2012).

A community consultation was called by the Junta de Aguas, an indigenous and peasant organization that administers access to household water. The consultation was grounded in ILO 169, the UN Declaration on the Rights of Indigenous People and the Ecuadorian Constitution (Pérez Guartambel, 2012). The vote was carried out in the parishes of Victoria del Portete and Tarqui. The organization was led by local leaders of the Federation of Indigenous and Peasant Organizations of Azuay, with the support of national indigenous organizations (Ecuador Runakunapak Rikcharimuy/Movement of the Indigenous People of Ecuador (ECUARUNARI), La Confederación de Nacionalidades Indígenas del Ecuador/Confederation of Indigenous Nationalities of Ecuador (CONAIE)) and the mayor of Victoria del Portete. The consultation followed the Junta de Aguas election procedures: one vote per water right (a family can have more than one right). The vote was secret and for registered water right owners (head of family, not individuals). The consultation had national and international observers from organizations and the national ombudsman office. Days before the consultation, newspaper pages and leaflets calling people not to vote were distributed. There was a 67% turnout with a 92.3% opposition to mining activities. Provincial and national governments did not recognize the vote and led a strong, discrediting campaign.

### Argentina

*Esquel project (Chubut)*

The second consultation conducted in Latin America took place in Esquel in March 2003. The city of Esquel (28,089 inhabitants) is a main settlement of Argentinean Patagonia, an arid region also inhabited by Mapuche indigenous communities. In 2002 some 25% of the population were unemployed and 20% were under the poverty line. The arrival of Meridian Gold, a US junior mining company, with the intention to extract a gold and silver deposit located 6.5 km away from the city triggered the first mining conflict in the country.

The use of cyanide leaching techniques and the risks of water pollution in a water-scarce environment stirred initial concerns. The perception that the urgency to approve the project was undermining the quality of the technical assessment and was excluding local concerns led to the formation of a neighbours' assembly (Asamblea de Vecinos Autoconvocados (AVA)) opposed to the mine. The AVA brought together neighbours and organizations with different backgrounds, specialists in law, chemistry, medicine, geography, journalism and education, Mapuche groups and inhabitants of Esquel's poorer areas who became key information channels to marginal areas of the city. The movement deployed a range of strategies, from legal and administrative queries to mobilizations, technical arguments and advocacy networking. As the AVA jumped scales, contacting and obtaining the support of regional, national and international activists, organizations and networks, the

Esquel conflict started to be understood as part of an environmentally unjust process affecting many communities in Latin America (Urkidi and Walter, 2011).

Members of the AVA became acquainted with Tambogrande's consultation via the internet. The AVA also established contacts with the Mining Policy Center (now Earthworks), an NGO that supported the Tambogrande consultation and that would later finance (along with Greenpeace Argentina) the visit of an American hydrogeologist, who had also been in Tambogrande, to Esquel (Colao and Claps, 2005).

Two representatives of the local Deliberative Council, close to the AVA, presented a municipal ordinance proposal to call for a *consulta popular* (popular consultation/referendum) using a legal mechanism present in the provincial constitution. While the proposal was initially rejected, the mounting tension in Esquel fostered its approval by most political parties as a way to pacify local unrest.

A few days after the *consulta popular*, which resulted in an 81% rejection of the mining project (75% turnout), mining activities were halted and the Chubut legislature approved a provincial ban on open-pit mining. The Esquel case became a national referent (Svampa and Antonelli, 2009; Walter and Martinez-Alier, 2010). The AVA created an online platform (www.noalamina.org) that is still a key source of information for Argentinean and Latin American activists.

The Esquel case showed the strong political power that a non-binding consultation could have. In the years that followed, as mining investments were rising, more EJMs tried to foster similar consultations. In particular, the Government of the Province of Catamarca, the poorest province of Argentina where the oldest and largest mine operates (La Alumbrera), managed to stop at least three attempts of consultation in Tinogasta and Andalgalá in court.

*Lonco project (Neuquén)*

The second consultation in Argentina took place in the municipality of Loncopue. After a series of legal setbacks and different intimidation campaigns aimed at social movements and Mapuche indigenous communities, exploration activities were advancing without permits or consultation procedures. A local priest became involved and brought the matter to the town, connecting the urban movements with rural indigenous groups. A lawyer and anti-mining activist from Esquel, who was living in Loncopue, transferred his professional and activist experience to the emerging movement, advising and supporting the legal strategy (Yappert, 2009).

The call for a binding referendum to approve/reject a municipal law forbidding large-scale open-cast mining activities was fostered by Mapuche communities, neighbourhood assemblies, environmental groups and, as in Esquel, some politicians whose political parties were pro-mining at the provincial and national levels but who aligned themselves with anti-mining groups locally. With a 72% participation turnout, 82% voted in favour of a mining prohibition, but the provincial government presented a legal claim of unconstitutionality to disable the referendum (Yappert, 2009).

## Guatemala and Colombia

*Guatemala: Sipakapa, Escobal and the wave of consultas in West Guatemala*

The third Latin America bottom-up mining consultation after Tambogrande and Esquel (Argentina) occurred in Sipakapa (Guatemalan highlands) in June 2005. In 2003, Montana (now owned by the Canadian GoldCorp) obtained the exploitation permit for the Marlin gold mine in the municipalities of Sipakapa and San Miguel Ixtahuacan. These municipalities are inhabited by peasants who mostly identify themselves as indigenous. In Sipakapa, 87% live in relative poverty and 33% in absolute poverty (SEGEPLAN, 2002).

Research and interviews underline the fact that the first meetings held by the company with local groups and leaders were non-transparent, arbitrary and pro-mining (Van de Sandt, 2009; Urkidi, 2011). The opposition to mining in Sipakapa was born from the mistrust that arose among many community leaders in regard to information activities. Indigenous leaders met local priests and national groups (Movimiento de Trabajadores Campesinos, MadreSelva, Centro de Acción Legal Ambiental y Social de Guatemala (CALAS)) in order to get information about mining (Van de Sandt, 2009). These national organizations were already within Latin American networks (e.g. MadreSelva within OilWatch) and distributed information about the environmental impacts of mining activities. Local leaders from Sipakapa visited other gold-mining areas in Central America, such as Valle de Siria in Honduras, and got in touch with regional networks against mining (e.g. Central American Anti-Mining Network).

In December 2004 a community that blocked the passage of a truck heading to the mine in a neighbouring province was strongly repressed by police and military forces, resulting in the death of the peasant Raul Castro Bocel (Prensa Libre, 18 January 2005; Castagnino, 2006). The public resonance of these events forced the mayor of Sipakapa (in favour of mining) to arrange a public meeting to discuss the mining issue. This meeting led to a municipality agreement to conduct a consultation, based on the Municipal Code (2002) and ILO 169. The idea to conduct a consultation had been circulating since the beginning of 2004, born from an Italian priest who was acquainted with the Tambogrande experience (Van de Sandt, 2009).

The consultation was organized through the articulation of local, national and international organizations: the Municipal Development Council (Consejo Municipal de Desarrollo (COMUDE)), the parish and its catechists, the Linguistic Community of Sipakapa, the local justice of the peace, MadreSelva, the National Association of Maya Lawyers, the Catholic Church of San Marcos, and the Indigenous Advocacy of Human Rights, among others. National and international observers and human right activists were called in to verify the process. The Guatemalan Constitutional Court rejected an appeal of Montana to ban the consultation. On the same day of the consultation, flyers saying that the *consulta* was not going to occur were distributed in Sipakapa, presumably as a boycott by Montana.

However, 45% of the registered electorate took part in the consultation and 98% voted against mining. The voting was carried out in each community; some voted by a show of hands, others by secret ballot. In 2007 the Guatemalan Constitutional Court declared that the Sipakapa consultation was valid under ILO 169 and the Municipal Code, but that it was non-binding since such conventions and laws were imprecise and not coherent with the constitution, and also because mining activities were of national public interest. Hence the municipality of Sipakapa had no authority to decide on the matter (Xiloj and Porras, 2008).

The Marlin mine was in full operation in 2013, despite the consultation and different legal demands in relation to environmental impacts and the violation of human rights.[1] However, the process of Sipakapa was a milestone in the Guatemalan resistance against mining. The experience has been reproduced in 56 other consultations on metal mining in the country from 2005 to 2012 and more than 600.000 people have taken part in them, becoming one of the most relevant political processes of recent years in the country. A documentary on the Sipakapa *consulta* (Revenga, 2005) played a central role in spreading the experience throughout Guatemala and Latin America.

Some 52 of those 57 consultations occurred in western Guatemala and most of them in the highlands, as part of a regional campaign to reject mining activities. The Western People's Council (WPC), where the Huehuetenango Natural Resources Assembly had a central role, led the spread and organization of consultations. The WPC is a regional network organized in 2008 as a coalition of provincial organizations working in the defence of natural resources and local leaders of the municipalities that have held consultations. Its main objective is to develop a community-based strategy against mining. There are also national and international networks and NGOs[2] supporting the development of the consultations. However, one key characteristic of the Guatemalan process is the synergies between the anti-mining movement and the municipal governments in the organization of most consultations, and the active incorporation of local leaders in the regional network (Mérida and Krenmayr, 2010; Urkidi, 2011).

More recently, other cases of consultations that are not directly related to the WPC work are emerging in other areas of Guatemala. The consultation on the Escobal project in Santa Rosa is not part of the wave of consultations of western Guatemala, even if it has also been influenced by the Sipakapa experience. The context of Santa Rosa differs from the highlands, as most of its population are non-indigenous. There are, however, some *Xinca* communities. The conflict arose in 2010 when Tahoe Resources and Goldcorp were to start a metal mine in the area that might affect a nearby lake and its related water resources. A local committee was organized and, between 2011 and 2012, four consultations were developed in nearby towns with the support of the regional diocese, a national environmental organization (MadreSelva) and local governments. However, no consultation has been permitted in San Rafael Las Flores; the mine is in operation, the local population are highly divided, and violent events and criminalization processes have taken place over the last few years (OCMAL, 2013).

Apart from Sipakapa and Santa Rosa, the rest of the Guatemalan consultations are not associated with imminent mining projects but with exploration or research licences, so that they could be understood as preventative consultations. Indeed, no new exploration licences were granted in the country from 2008 to 2012. Table 11.2 presents more details about the cases of preventative consultations of Guatemala. The Guatemalan Government has not accepted community referendums and has proposed to regulate them with a specific law (Prensa Libre, 23/02/2011). The WPC defends that the current legal framework is sufficient to accept the consultations and their results, and that further regulations would just lead to more restrictive conditions for participation (Nisgua, 2011; Prensa Libre, 23/02/2011).

The Guatemalan anti-mining movement seeks to be inclusive in many senses, resulting in heterogeneous consulting processes. Mainly indigenous but also non-indigenous communities have been consulted (these last ones not appealing to ILO 169 but just to the Municipal Code (2002)), by secret ballot or by show of hands, in municipal or just communitarian *consultas*. In some cases, mainly in Huehuetenango, non-registered people have been able to take part in indigenous community meetings. This has led to greater participation of women than in other voting processes since women are proportionally less frequently registered than men in Guatemala (Mérida and Krenmayr, 2010). Such *consultas* have also spread to other extractive projects in Guatemala, such as hydroelectricity.

*Colombia, Mandé Norte project (Carmen de Darién, Chocó)*

Between 24 and 28 February 2009, the first community consultation on mining in Colombia took place. The conflict started with the arrival of Muriel Mining (Río Tinto and other companies), and the initial consultation activities led by the government and company to obtain the communities' approval to explore for copper, gold and molybdenum ores. Exploration sites were located in Afro-descendant and indigenous peoples' lands, including their homes and sacred areas, in the departments of Antioquia and Chocó. Indigenous and Afro-descendant communities started to search for information and contacted a national church organization working in the area. A support group was created, bringing information, documentaries (e.g. the Sipakapa case) and activists from other countries and communities to Carmen de Darién (Jahncke Benavente and Meza, 2010). Communities claimed that the official consultation process was not adequately conducted, excluding affected communities and endangering their livelihoods. As a reaction to local unrest, the national government militarized mining areas, intimidating and limiting community access (Jahncke Benavente and Meza, 2010; Movice, 2012).

Communities, inspired by the Sipakapa experience, promoted the organization as an interethnic consultation, following their own procedures (own language, registered, older than 14 years old). Human rights, indigenous, church and anti-mining organization representatives from Colombia, Paraguay, Honduras, Guatemala, Germany and Canada observed the process (CENSAT, 2009).

The consultation was grounded on international and national indigenous consultation rights, including the Colombian Constitution's special consideration for indigenous consultation rights. The legality and legitimacy of the process was confirmed by an important verdict (T-769, 2009) of the Colombian Constitutional Court, which led to the suspension of the project. Nevertheless, in the year that followed, campaigns to delegitimize local communities and further intimidation actions were conducted by the government in the area. In January 2010 the Colombian army conducted air bombings (Movice, 2012).

Consultation attempts have also been deployed by other non-indigenous communities in Colombia. During 2011, social movements in the department of Santander tried to conduct a popular consultation framed around the protection of water to stop gold-mining developments in upstream *Páramo* areas. This initiative was politically blocked (Comité por la defensa del agua y el páramo de Santurbán, 2012). Recently, in July 2013, the municipality of Las Piedras (Tolima region) conducted a popular consultation on mining activities, resulting in a 60% participation and 99% rejection of a large-scale mining project to be carried out by Anglo Gold Ashanti (EJOLT, 2013).

## Discussion

The cases of consultation analysed in this chapter represent an innovative governance experience that seeks to ensure inclusive participation in mining activities. Moreover, this governance perspective goes beyond local/global, formal/informal, state/non-state divides. These points lead to four aspects of consultations, which are elaborated in this discussion.

### Contexts: Conflicts, exclusion, criminalization and violence

The mining conflicts that led to consultations involved high-stake struggles. Mining disputes revolve around how the spatial and social distribution of uncertain benefits and impacts of mining activities are defined, and which are the legitimate scales of participation and decision-making to govern this activity. Consultations are neither the first nor the only action deployed by EJMs, but instead are promoted alongside a range of strategies (e.g. negotiations, mobilizations, legal and technical allegations, dissemination activities) aimed at influencing and challenging centralized mining governance institutions.

The discourses deployed by anti-mining movements in our cases reflect Schlosberg's (2007) key dimensions of environmental justice: recognition, distribution and participation. Anti-mining groups see the approval of mining projects as the misrecognition of their material and cultural dependence on land and water, and also as a disregard of their views and customary procedures (Muradian, Martinez-Alier and Correa, 2003; Haarstad and Floysand, 2007).

Social movements opposing mining activities claim that developing mining activities jeopardizes local (and supralocal) livelihoods. Communities in Peru, Guatemala, Colombia and Ecuador signal the risks to their livelihoods, which are dependent on agriculture, cattle and forests. Concerns about health also appear, with high relevance in Esquel (Argentina) regarding cyanide use. Worries about water quality, and availability for local economic activities and household use, are common to all studied cases.

While the affected communities signal such concerns as grounds to redraft or even stop a mining project and national mining plans, governments and companies claim that these decisions are not for local communities to make. Central governments argue that mining is an issue of national interest and experts within a national decision-making process should have the last word. Governments and mining companies frame local alarm as an exaggeration that undermines the positive impacts of mining. Moreover, critical communities' and EJM's views are being labelled by Latin American national governments as irrational, ignorant, anti-development, politically driven, promoted by foreigners' interests or by a radical, subversive environmentalism (Bebbington, 2012b), hand in hand with criminalization processes (OCMAL, 2011).

Official participation arenas become frustrating spaces given the partial information that is shared and the powerless participation modes they offer (Cole and Foster, 2001). As decision-making procedures are unable to address local communities' concerns, disputes form around these procedures and their decisions (Muradian, Martinez-Alier and Correa, 2003; Suryanata and Umemoto, 2005; Walter and Martinez-Alier, 2010; Urkidi and Walter, 2011). It is becoming increasingly common for EJMs to prevent or boycott public audiences, as these are seen as an empty requisite for project approval (Jahncke Benavente and Meza, 2010). There were cases of boycotts of public audiences in Tambogrande, Toquepala, Tía María, Esquel and Loncopue. Indigenous communities rejected and misrecognized the alleged consultation processes led by mining companies and governments in Peru, Colombia and Guatemala. In Ecuador and Argentina, indigenous communities claimed that formal consultation never occurred (Urkidi and Walter, 2011; Pérez Guartambel, 2012).

Furthermore, one of the findings of this research has been the role played by violence in the fostering of consultations. Human Rights claims have been identified as a particular root of Latin American EJMs (Carruthers, 2008). Mining referenda have emerged in contexts of repression and criminalization of activists, where concerns regarding the physical and psychological integrity of activists were rising. In this line, consultations can be seen as an innovative form of protest that aims to foster participation, promoting a democratic setting that protects its participants. These consultations have succeeded in pacifying local tensions, at least for a while.

While contexts of activist and protest criminalization and repression are not new in mining struggles, the particularity of these cases has been the ability of EJM to transform a risky protest environment into a democratic participation process. To do so, EJMs have constructed a hybrid participation institution.

### Community consultations: A hybrid institution

Latin American mining consultations/referenda are based on the claim that communities – whether indigenous or not – have the right to participate in high-stake decisions that affect their livelihoods, a right deemed legitimate by affected communities. This right is recognized in a variety of indigenous and non-indigenous, international, national and municipal norms and rights (Jahnchke Benavente and Meza, 2010; Fulmer, 2011). However, how participation is framed by regulations and actors varies widely, being mostly informing and non-binding. As analysed by Arnstein (1969) in his eight-rung participation ladder ((1) manipulation, (2) therapy, (3) informing, (4) consultation, (5) placation, (6) partnership, (7) delegated power and (8) citizen control), there are different levels of exclusion/involvement and empowerment. As pointed out by Arnstein, as we step down the ladder, frustration rises. Communities are struggling to climb this ladder.

Community consultations reclaim and rebuild the right of affected communities to participate, in meaningful and empowering ways, in decisions regarding high-impact activities that affect them. With this aim, in each context, communities strive for local participation rights appealing to, combining and reshuffling available regulations, rights and local traditions. This process of institutional bricolage draws on a particular mix of formal and informal, and modern and traditional, institutions according to the particular context.

For instance, communities are expanding and resignifying, in their discourse and practices, the way "consultation" is framed in ILO 169 – and the United Nations Declaration on the Rights of Indigenous People – forcing new debates about the convention's reach (McGee, 2008; Fulmer, 2011). ILO 169 asserts that consultations should be conducted by states. However, the studied consultations are not organized by the central government (Jahncke Benavente and Meza, 2010; Fulmer, 2011). Community consultations appeal to ILO 169 consultation rights, stretching the convention's reach according to what is considered just and legitimate by affected communities. In a similar vein, the way in which consultations appeal to national, municipal and international participation laws and rights in order to allow for a local referenda on mining challenges the national-government scale monopoly in mining decisions.

In each context, this hybrid institution is legitimized by reference to tradition and/or to the social perception of what are the acceptable ways of doing things (Cleaver et al., 2013). A relevant source of (internal and external) legitimacy of *consultas*/referenda is rooted in the procedures used to consult people that appeal to democratic values and to indigenous consultation rights. In most cases, communities put in place hybrid procedures that combine democratic participation institutions (e.g. official election procedures), indigenous customary rights, and experiences/lessons from previous consultations. In most consultations, including many indigenous communities in Guatemala, the consultation followed the same procedures as those of a regular election: formal call to vote, registered voters, the secret vote and the quality of the process as certified by external observers, as in Tambogrande. In Sipakapa, each of the 13 communities consulted chose its own procedure: some followed a traditional Western election format, while others voted by a show of hands or other formats. However, the consultation was called by the municipality and all members of the municipality could vote (even non-indigenous). In Sipakapa, indigenous customary votes were the most criticized by the government and by companies that claimed that their result could be manipulated (Fulmer, 2011). The consultation conducted by indigenous groups in Colombia followed the example of Sipakapa by merging procedures.

Some forms of (hybrid) governance that would include diverse social actors and visions a priori have been criticized because they continue to exclude disempowered groups (Ford, 2003; Cleaver et al., 2013). In contrast, consultations are organized by, and take into account, marginalized groups such as indigenous peoples, women and peasants. As a result, consultations usually stretch the reach of formal and informal institutions in order to foster local participation.

Consultations are more than the sum of existing regulations and rights but, while grounded on these, they reclaim their scope and meaning based on what is deemed legitimate and just by local communities. Moreover, the significance of community consultations is that communities are not only mobilizing and discursively struggling to contest the governance of mining activities but are also deploying innovative strategies to demand empowering and democratic participatory institutions. The community consultations studied here are a form of political mobilization, a form of protest grounded on democratic and indigenous, formal and informal institutions.

### The roles of movements, governments and state bodies

While EJMs have played a key role in the emergence and spread of consultations, a particular feature of community consultations has been the role played by local governments. Community consultations combine the formal and informal capabilities (i.e. rule-making, management, communication) and different forms of power (e.g. legitimacy, networks, resources, trust) of social movements and local governments.

Cases of consultations conducted without alliances with local governments are the exception. In some cases local governments rapidly align with social movements or even play a central role in the formation of movements critical of mining activities (e.g. Majaz, Toquepala, Guatemala's wave of consultations). In other cases, local governments change their position as conflicts unfold and finally allow or support consultations in order to preserve local governability or local power (e.g. Esquel, Sipacapa), sometimes adopting a position that differs from their national political parties.

However, the legitimacy of consultations is in dispute by different actors within states and governments. While national governments and mining departments reject, ignore or criminalize (define as illegal acts) these participatory events, some local and provincial governments – as well as national and regional departments, authorities and tribunals – recognize this participation institution (e.g. National Electoral Office, Constitutional Court, ombudsman, Human Rights National Councils, Ministry of the Environment).

The alliance with local governments was key to building the legitimacy of consultations (Red Muqui, 2009), framing them as a formal local (and democratic) participation institution, not a mere anti-mining social movement strategy (Muradian, Martinez-Alier and Correa, 2003). The fact that the first cases of consultations were conducted with the support of local ordinances contributed to building the grounds for legitimating the following wave of consultations, conducted with or without this formal support (e.g. some municipalities in the Tía María consultation in Peru and the Kimsakocha case in Ecuador). Moreover, the involvement of social movements reduced, in some places, the distrust that many rural communities have in relation to government bodies, including municipalities. In Guatemalan consultations, the fact that actors not directly related to the municipal government were also promoting the *consultas* was pointed out as a source of local trust and willingness to participate (interviews Guatemala 2009; Mérida and Krenmayr, 2010). We could also say that the legitimacy of consultations is, in part, both a cause and a consequence of the hybrid alliances formed between local governments and social movements.

The involvement of local governments and the diverse positions adopted within state and government bodies regarding community consultations reflect the heterogeneity of interests and values across these structures. This feature of consultations points to the need to further problematize the role of governments and the state in environmental governance frameworks. Hybrid institutions led by civil society, such as community consultations, do not necessarily aim to "bypass governments" (as pointed out by Delmas and Young, 2009) but, on the contrary, to anchor part of its legitimacy in some of its bodies (local governments).

Currently, the strength of the consultation's legitimacy grounded in its "legality" (i.e. formal institutional support) is becoming a weakness as the struggle is now revolving around the formalization of consultation rights (i.e. regulating consultation procedures) by central governments, with risks of co-optation, exclusion and denaturalization of the institution.

### A multiscalar institutional bricolage

Finally, we would like to point out that, while consultations could be framed as a hybrid institution that exemplifies a process of governance from "below" (Paterson, Humphreys and Pettiford, 2003), the strength and legitimacy of this institution is multiscalar. Analysing the spread of consultations in Latin America, we identify that this institution was fostered hand in hand with a diversity of spatial processes that have been key in its emergence, spread and legitimation in Latin America. Along these lines, consultations can be seen as the result of a dynamic multiscalar process of institutional bricolage.

Mining consultations are promoted by social movements composed of a myriad of groups, including indigenous and peasants' movements, farmers, (urban) professionals, local priests, teachers, community leaders and NGOs. As mining conflicts unfolded, these social movements engaged with networks and organizations (e.g. environmental, anti-mining, human rights, indigenous, Catholic) that move across multiple geographical scales. In the wave of consultations in Guatemala, national anti-mining networks fostered the participation of local actors and leaders. These networks circulate information, experiences and strategies, and promote the mobility of activists to learn and share experiences among communities, to Latin America and international forums, to foreign (e.g. UK courts in the Majaz case) and international tribunals (e.g. Sipakapa to the Inter-American Commission on Human Rights).

Additionally, among the EJMs and networks driving the spread of consultations, some were born from the first mining consultation experiences: Tambogrande, Esquel and Sipakapa. These first cases are relevant mining conflicts at national and transnational scales and have become milestones in the mining consultation processes in Latin America and in their own countries. Red Muqui, born from the Tambogrande conflict, was a key provider of information, experience and materials for the Majaz/Río Blanco case and following consultations. The "Noalamina" platform, coordinated by the Esquel anti-mining movement, is a key provider of information and resources for Latin American communities. In Guatemala, the great multiplication of mining consultations is partially grounded in the national and international repercussion of Sipakapa's experience. With the support of different national NGOs and associations, two regional networks were created around mining and hydropower conflicts (Huehuetenango Natural Resources Assembly and the Western Peoples Council). There has been an experience-sharing process, where new consultations have been organized by knowing and learning from previous ones, via these national and transnational organizations and networks (Red Muqui, 2009; Jahncke Benavente and Meza, 2010).

Organizations and networks have not only played a key role in spreading the experience of previous consultations but also provided logistical, technical and sometimes financial resources. A range of transnational actors have also supported consultations as observers, contributing to building the international legitimacy of these processes. OXFAM, Friends of the Earth, Greenpeace, the Mineral Policy Centre, Peace Brigades International, Nisgua, Catapa, Rigths Action in Sipakapa and Mining Watch are among the international observers that have been present in Latin American mining consultations.

Furthermore, as consultation experiences multiplied in Latin America, national and transnational networks have deployed efforts to systematize and strengthen the ongoing experience and its lessons, by organizing international events (e.g. Bi-national encounter Ecuador-Peru on Community Consultations, 28 February 2012) and elaborating reports (e.g. McGee, 2008; CISDE-ALAI, 2009; Jahncke Benavente and Meza, 2010; Mérida and Krenmayr, 2010; Duthie, 2012). National and transnational movements have also supported legal strategies – to defend the legality of consultations and condemn human rights abuses – at national and international tribunals (Constitutional Court case in Colombia, Interamerican Human Right Commission presentation of Sipakapa), thus systematizing and denouncing the growing number of criminalization cases (e.g. OCMAL, 2011).

When considering how consultations have travelled among Latin American communities, we point out that the internet and documentaries are powerful transporters of testimonies and experiences among distant people and places. While the role of the internet has been discussed in previous studies (Bickerstaff and Agyeman, 2009), we also found that documentaries are significantly contributing to social learning processes.

Sipakapa's documentary was a key source of inspiration in the organization of the Embera Katio indigenous consultation in Carmen de Darien (Colombia, 2009) (interview with Colombian activist, Jahncke Benavente and Meza, 2010). An indigenous leader that led the consultation of Ecuador also underscored the relevance of videos and documentaries to explain the implications of large-scale mining activities.[3] The documentaries on the Choropampa mercury spill in Cajamarca (Peru) and the cases of the Tambogrande and Sipakapa consultations have been widely distributed in the region (Choropampa: el precio del oro, 2002; Sipakapa no se vende, 2005; Tambogrande: mangos, muerte, minería, 2007). These and other documentaries have shown the impacts of large-scale mining activities and the strategies of anti-mining groups, contributing to a regional EJM learning process. In this regard we agree with Bickerstaff and Agyeman (2009) that there is a promising line of research to be explored in relation to the development of "assemblage" perspectives – coming from the actor-network theory (ANT) – when analysing how people, texts, machines, devices and discourses relate and collectively constitute environmental justice scales. How to conceptualize the role of these devices in processes of institutional bricolage could be explored in further detail.

Colombian activists highlight how Carmen de Darien's indigenous communities were moved to see – in the documentary on the Sipakapa consultation – other indigenous groups faced with similar struggles, telling similar histories and learning from their consultation experience (interview with Colombian activist). Documentaries played a central role in making affected communities acknowledge that their conflict was not local but simultaneously local, national, regional, global and structural. In this process, a common perspective is constructed and solidarity linkages are strengthened.

The construction, the spread and the sources of legitimacy of this hybrid institution (i.e. community consultations) are embedded in a complex and dynamic interplay of actors, discourses, networks and strategies that move among multiple scales. The political power of consultations is related to the ability of supralocal social movements to move and disseminate these events at multiple scales, creating new supports and reactions. Consultations, whether *vecinal, popular, comunitaria or inter-étnica*, are embedded in municipal, national and international norms and rights that are reclaimed by EJMs. In this regard, Latin American mining consultations are a multiscalar institution since they are constituted by (and constitutive of) actors, strategies, regulations and discourses rooted in different, multiple and changing scales.

## Conclusions

The process of meeting, consulting and voting is part of the functioning of many indigenous and peasant communities and organizations in Latin America. However, the mining consultations studied in this chapter, while nurtured and legitimated by these traditions, are something different. Mining consultations constitute a common institution in the current Latin American anti-mining protest cycle. Consultations reclaim and resignify the right of the local population and indigenous peoples to participate, in empowering ways, in high-stake decisions affecting their lands and livelihoods. Consultations are put forward not just as a form of protest but also as a decision-making event that challenges official decision-making institutions. Moreover, consultations show how we should move beyond analytical polarizations and try to understand the tensions and dynamics in the process of governance hybridization through cross-scale interactions, discourses and practices.

## Notes

1. In 2010 the Inter-American Commission on Human Rights ruled in favour of the precautionary closure of the project because of potentially harmful health and environmental impacts.
2. Mainly environmental and human rights associations and NGOs from Europe and Canada (CATAPA, Network in Solidarity with the People of Guatemala – NISGUA or Rights Action, among many others).
3. Interview conducted by Sara Latorre and Stalin Herrera with local leader, shared with us.

# References

Arnstein, S.R. (1969) "A Ladder of Citizen Participation", *JAIP* 35(4): 216–224.
Bebbington, A. (2012a) "Social Conflict and Emergent Institutions. Hypotheses from Piura, Peru", in A. Bebbington (ed.), *Extractive Industries, Social Conflict and Economic Development: Evidence from South America* (London: Routledge), 67–88.
Bebbington, A. (2012b) "Underground Political Ecologies: The Second Annual Lecture of the Cultural and Political Ecology Specialty Group of the Association of American Geographers", *Geoforum* 43: 1152–1162.
Bebbington, A., Connarty, M., Coxshall, W., O'Shaughnessy, H. and Williams, M. (2007) *Mining and Development in Peru. With Special Reference to the Rio Blanco Project* (Piura: Peru Support Group).
Bebbington, A., Humphreys Bebbington, D. and Bury, J. (2011) "Federating and Defending Water, Territory and Extraction in the Andes", in R. Boelens, D. Getches and A. Guevara Gil (eds), *Out of the Mainstream: The Politics of Water Rights and Identity in the Andes* (London: Earthscan), 307–327.
Bickerstaff, K. and Agyeman, J. (2009) "Assembling Justice Spaces: The Scalar Networking of Environmental Justice in North-East England", *Antipode* 41(4): 781–806.
Bridge, G. (2004) "Mapping the Bonanza: Geographies of Mining Investment in an Era of Neo-Liberal Reform", *Professional Geographer* 56(3): 406–421.
Bullard, R. (1990) *Dumping in Dixie: Race, Class, and Environmental Quality* (Boulder, CO: Westview Press).
Bulkeley, H. (2005) "Reconfiguring Environmental Governance: Towards a Politics of Scale and Networks", *Political Geography* 24: 875–902.
Cabellos, E. and Boyd, S. (2007) *Tambogrande: Mangos, Muerte, Minería*, Documentary, Guarango Association, Peru.
CAOI (Coordinadora Andina de Organizaciones Indígenas) (2009) Seis Distritos de la Provincia Arequipeña de Islay se pronunciaron a través del Voto, http://elecochasqui.wordpress.com/actualidad/setiembre-2009/caoi-seis-distritos-de-la-provincia-arequipena-de-islay-se-pronunciaron-a-traves-del-voto-rechazo-al-proyecto-minero-tia-maria-alcanza-el-97/, date accessed 10 January 2013.
Carruthers, D. (ed.) (2008) *Environmental Justice in Latin America: Problems, Promise, and Practice* (Cambridge, MA: MIT Press).
Castagnino, V. (2006) *Metal Mining and Human Rights in Guatemala: The Marlin Mine in San Marcos* (Guatemala: Peace Brigades International).
CENSAT (2009) Colombia: No a la Minería en Territorio Indígena, http://censat.org/component/content/article/314, date accessed 10 January 2013.
CISDE-ALAI (2009) *América Latina: Riqueza Privada, Pobreza Pública* (CISDE-ALAI).
Chaparro Avila, E. (2002) Actualización de la Compilación de Leyes Mineras de Catorce Países de América Latina y el Caribe, UUNN División de Recursos Naturales e Infraestructura ECLAC http://www.eclac.org/publicaciones/xml/6/10756/LCL1739-P-E.pdf, date accessed 10 January 2013.
Cleaver, F. (2001) "Institutional Bricolage, Conflict and Cooperation in Usangu, Tanzania", *IDS Bulletin* 32(4): 26–35.
Cleaver, F. (2002) "Reinventing Institutions: Bricolage and the Social Embeddedness of Natural Resource Management", *The European Journal of Development Research* 14(2): 11–30.

Cleaver, F. (2012) *Development Through Bricolage: Rethinking Institutions for Natural Resource Management* (London: Routledge).
Cleaver, F., Franks, T., Maganga, F. and Hall, K. (2013) *Beyond Negotiation?: Real Governance, Hybrid Institutions and Pastoralism in the Usangu Plains, Tanzania*, Working paper 61 (London: King's College).
Cobain, I. (2009) "British Mining Company Faces Damages Claims Alter Allegations of Torture in Peru", *The Guardian*, 18 October 2009.
Comité por la Defensa del Agua y el Páramo de Santurbán (2012) Queja Presentada ante la Oficina del Ombudsman y Asesor en Materia de Observancia, http://www.aida-americas.org/es/release/queja-interpuesta-contra-el-financiamiento-de-mina-de-oro-en-fragiles-humedales-colombianos, date accessed 10 August 2013.
Colao, D. and Claps, L.M. (2005) *Comunicación, Recursos Naturales y Comunidad en el Caso Esquel*, Graduate Thesis (Buenos Aires: Universidad de Buenos Aires).
Cole, L.W. and Foster, S.R. (2001) *From the Ground Up: Environmental Racism and the Rise of the Environmental Justice Movement* (New York: University Press London).
De Echave, J., Diez, A., Huber, L., Revesz, B., Lanata, X.R. and Tanaka, M. (2009) *Minería y Conflicto Social* (Lima: Instituto de Estudios Peruanos).
Delmas, M.A. and Young O.R. (eds) (2009) *Governance for the Environment* (New York: Cambridge University Press).
Diez Hurtado, A. (2007) "Ronderos y Alcaldes en el Conflicto Minero de Río Blanco en Piura, Perú", in J. Bengoa (ed.), *Territorios Rurales. Movimientos Sociales y Desarrollo Territorial Rural en América latina* (Santiago de Chile: RIMISP).
Duthie, K. (2012) *Local Votes and Mining in the Americas* (Mining Watch Canada).
Ejolt (2013) A Letter from Piedras, Tolima, Colombia: Local Referendum Against Anto Gold Ashanti, Environmental Justice Organizations, Liabilities and Trade Project, http://www.ejolt.org/2013/09/a-letter-from-colombia/, date accessed 10 October 2013.
Fedepaz (2013) Kañaris, Última Comunidad de Habla Quechua en la Costa Peruana, realiza Consulta Comunal, Fundación Ecuménica para el desarrollo y la Paz, http://www.fedepaz.org/index.php?option=com_content&task=view&id=194&Itemid=18, date accessed 20 August 2013.
Ford, L.H. (2003) "Challenging Global Environmental Governance: Social Movement Agency and Global Civil Society", *Global Environmental Politics* 3: 120–134.
Fulmer, A. (2011) "La Consulta a los Pueblos Indígenas y su Evolución como Herramienta de Negociación Política en América Latina. Los Casos de Perú y Guatemala", *Apuntes* 68: 37–62.
Greenspan, E. (2013) Peru backslides on Indigenous Rights, Oxfam America, http://politicsofpoverty.oxfamamerica.org/2013/05/08/peru-backslides-on-indigenous-rights/, date accessed 10 October 2013.
Gutiérrez Zeballos, P.J. (2011) "Las Razones de Lucha por el Valle del Tambo", in Cooperacción/Red Muqui/Frente Ambito de Defensa del Valle de Tambo/ Municipalidad Distrital de Dean Valdivia, *Valle de Tambo-Islay. Territorio, Agua y Derechos Locales en Riesgo con la Minería a Tajo Abierto*.
Haarstad, H. and Floysand, A. (2007) "Globalization and the Power of Rescaled Narratives: A Case of Opposition to Mining in Tambogrande, Peru", *Political Geography* 26(3): 289–308.

Harvey, D. (1996) *Justice, Nature and the Geography of Difference* (Cambridge, MA: Blackwell).
Holden, W. and Jacobson, D. (2008) "Civil Society Opposition to Nonferrous Metals Mining in Guatemala", *Voluntas* 19: 325–350.
Jahncke Benavente, J. and Meza, R. (2010) *Derecho a la Participación y a la Consulta Previa en Latinoamerica* (Lima: Fedepaz Muqui Miserer CIDSE).
Kurtz, H. (2003) "Scale Frames and Counter Scale Frames: Constructing the Social Grievance of Environmental Injustice", *Political Geography* 22: 887–916.
Leitner, H., Seppard, E. and Sziarto, K.M. (2008) "The Spatialities of Contentious Politics", *Transactions Institute of British Geographers* 33: 157–172.
Marquez, N. (2009) "Más del 90% de Pobladores de Cocachacra se pronunció en contra de la Instalación Extractiva y de la Utilización de Aguas Subterráneas", *La Jornada de Arequipa Journal,* http://www.jornaldearequipa.com/Tambo.htm, date accessed 10 August 2013.
McGee, B. (2008) An International Observer's Account of a Local Vote in Río Blanco Peru, Environmental Defender Law Center, http://www.edlc.org/resources/local-votes/observers-account/, date accessed 20 January 2013.
McGee, B. (2009) "The Community Referendum: Participatory Democracy and the Right to Free, Prior and Informed Consent to Development", *Berkeley Journal of International Law* 27(2): 570.
Mérida, A.C. and Krenmayr, W. (2010) *Sistematización de Experiencias 2008–2009: Tejiendo entre los Pueblos la Defensa del Territorio* (Guatemala: Asamblea Departamental por la Defensa de los Recursos Naturales Renovables y no Renovables de Huehuetenango).
Movice (2012) Pueblos Indígenas y la Protección del Territorio. Cartografía de Impulsos y Restricciones de Movimiento en el Territorio, http://www.conlospiesporlatierra.net/?p=1543, date accessed 10 August 2013.
Municipal Code (2002) "Decree 12–2002: Municipal Code Congress of the Republic of Guatemala".
Municipality of Kañaris (2012) Datos Económicos y Sociales, http://www.munikanaris.gob.pe/DatosDistrito.php, date accessed 20 August 2013.
Muradian, R., Martinez-Alier, J. and Correa, H. (2003) "International Capital Versus Local Population: The Environmental Conflict of the Tambogrande Mining Project, Peru", *Society and Natural Resources* 16: 775–792.
National Electoral Office (2002) Resolución RJ N°098–2002-J/ONPE, http://www.web.onpe.gob.pe/modResoluciones/descargas/RJN098–2002-J.pdf, date accessed 2 March 2002.
Nisgua (2011) Comunidades rechazan Iniciativa por Normar las Consultas Comunitarias, http://nisgua.blogspot.com.es/2011/03/comunidades-rechazan-iniciativa-por.html, date accessed 3 January 2014.
OCMAL (2011) *Cuando tiemblan los Derechos: Extractivismo y Criminalización en América Latina* (Quito: OCMAL-Acción Ecológica).
Ostrom, E. (1990) *Governing the Commons: The Evolution of Institutions for Collective Action* (New York: Cambridge University Press).
OXFAM (2007) Río Blanco: History of a Mismatch in Peru, Oxfam America, http://www.oxfamamerica.org/articles/rio-blanco-history-of-a-mismatch-in-peru/?searchterm=majaz, date accessed 10 January 2013.
OXFAM (2009) Oxfam calls for an Investigation of Alleged Torture of 28 in Peru, http://www.oxfamamerica.org/articles/oxfam-calls-for-an-investigation-of-alleged-torture-of-28-in-peru, date accessed 19 January 2013.

Paterson, M., Humphreys, D. and Pettiford, L. (2003) "Conceptualizing Global Environmental Governance: From Interstate Regimes to Counter-Hegemonic Struggles", *Global Environmental Politics* 3: 1–10.
Perez Guartambel, C. (2012) *Agua u Oro. Kimsakocha la Resistencia por el Agua* (Cuenca: Universidad Estatal de Cuenca).
Portugal Mendoza, C. (2005) "Gobernanza en el Acceso a la Actividad Minera a los Recursos Naturales Locales: El Caso Tambogrande", Grupo de Investigaciones Económicas.
Radio Uno (2008) Este Viernes remitirán Resultados de Consulta en Candarave al Ejecutivo y al Legislativo, http://www.radiouno.pe/noticias/3963/upt_banner.swf, date accessed 10 August 2013.
Rasch, E.D. (2012) "Transformations in Citizenship: Local Resistance against Mining Projects in Huehuetenango (Guatemala)", *Journal of Developing Societies* 28(2): 159–184.
Red Muqui (2009) "La Consulta Vecinal: Un Mecanismo de Democracia Directa para los Pueblos", in *America Latina: Riqueza Privada, Pobreza Pública* (Lima: CIDSE-ALAI).
Red Muqui (2011) "Valle de Tambo-Islay. Territorio, Agua y Derechos Locales en Riesgo con la Minería a Tajo Abierto", Red Cooperacción, Frente de Defensa del Valle de Tambo, Municipalidad Distrital de Dean Valdivia Muqui.
Revenga, A. (2005) *Sipakapa NO se vende. Sipakapa qal k'o pirk'ey xik*, Documentary. Caracol Producciones.
SEGEPLAN (2002) *Propuesta de Política para el Desarrollo Rural* (Guatemala: Secretaría General de Planificación y Programación de la Presidencia SEGEPLAN).
Servindi (2012) Perú: En Consulta Communal el 95% de Votantes de Kañaris rechazó Proyecto Minero Cañariaco, guamina.blogspot.com.es/2012/10/peru-en-consulta-comunal-el-95-de.html, date accessed 10 January 2013.
Schlosberg, D. (2007) *Defining Environmental Justice: Theories, Movements, and Nature* (New York: Oxford University Press).
Smith, N. (1996) "Spaces of Vulnerability: The Space of Flows and the Politics of Scale", *Critique of Anthropology* 16: 63–77.
Subies Grau, T., Beltrán, M.J., Mérida, J.M., Moreno, M., Salas, I., Sánchez Corominas, A., Soler, M. and Parea, M. (2005) "El Éxito de Tambogrande", *Ecología política* 30: 95–116.
Suryanata, K. and Umemoto, K. (2005) "Beyond Environmental Impact: Articulating the 'Intangibles' in a Resource Conflict", *Geoforum* 36: 750–760.
Svampa, M. and Antonelli, A. (eds) (2009) *Minería Transnacional, Narrativas del Desarrollo y Resistencias Sociales* (Buenos Aires: Biblos).
Trentavizi, B. and Cahuec, E. (2012) "Las Consultas Comunitarias de 'Buena Fe' y las Prácticas Ancestrales Comunitarias Indígenas en Guatemala", *Centro de Investigaciones Regionales de Mesoamérica – Oficina del Alto Comisionado para los Derechos Humanos de las Naciones Unidas*.
UNOPS/PNUMA (2011) Revisión "ad hoc" del Estudio de Impacto Ambiental, Proyecto Tía María, Informe de Observaciones y Requerimiento de Información Complementaria, http://xa.yimg.com/kq/groups/2122051/1296039927/name/Revisi%C3%B3n+del+Estudio+de+Impacto+Ambiental+del+Proyecto+T%C3%ADa+Mar%C3%ADa.pdf, date accessed 10 January 2013.
Urkidi, L. (2011) "The Defence of Community in the Anti-Mining Movement of Guatemala", *Journal of Agrarian Change* 11(4): 556–580.

Urkidi, L. and Walter, M. (2011) "Concepts of Environmental Justice in Anti-Gold Mining Movements in Latin-America", *Geoforum* 42: 683–695.
Van de Sandt, J. (2009) *Mining Conflicts and Indigenous Peoples in Guatemala* (The Hague: Cordaid – University of Amsterdam).
Walter, M. and Martinez-Alier, J. (2010) "How to Be Heard When Nobody Wants to Listen: The Esquel Mining Conflict", *Canadian Journal of Development Studies* 30(1–2): 281–303.
Xiloj, L. and Porras, G. (2008) *Diagnóstico sobre el Derecho a la Consulta que tienen los Pueblos Indígenas y sus Consecuencias Jurídicas y Políticas a Partir del Caso de la Explotación Minera en el Departamento de San Marcos* (Guatemala: Fundación Rigoberta Menchú).
Yappert, S. (2009) "Pedimos que pongan Freno a estos Proyectos de Muerte", *Rio Negro Newspaper* 5/5/2009.

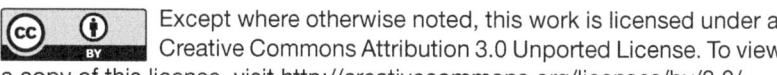
Except where otherwise noted, this work is licensed under a Creative Commons Attribution 3.0 Unported License. To view a copy of this license, visit http://creativecommons.org/licenses/by/3.0/

OPEN

# Afterword: From Sustainable Development to Environmental Governance

*Eduardo Silva*

This collection represents a milestone in the political ecology of environmental governance in Latin America. Given the enduring tension between the environment and development, its overarching purpose in my reading of the text is to elucidate the headway, obstacles and potential of our times to achieving the aspirations and goals of sustainable development, particularly in some of its early versions. Because this is a highly contested concept, and one that some of the contributors to this volume would most likely reject, I use it heuristically. It offers a platform from which to discuss the necessity of environmental governance that genuinely addresses environmental protection, social equity and broad-based political participation in the context of (largely) capitalist development; and this, in my view, is the central subject of the book.

Much has changed in the world since the Brundtland Commission first launched the idea of sustainable development in the mid-1980s: the end of the Soviet Bloc, free-market globalization, the rise of emerging market economies and profound transformations in manufacturing processes, just to name a few. Latin America transitioned to democratic governments, it benefited from sustained economic growth in the 2000s with some improvement in poverty alleviation, and it weathered a world recession. The region also witnessed an unexpected resurgence of the left. Left governments that would have been targets for decisive destabilization by the USA during the Cold War have successfully asserted their sovereignty, survived, and even thrived in some cases. This volume asks: What is the significance of these changes for advances in environmental governance that seriously addresses the problem of environmental sustainability beyond end-of pipe treatment, that supports and promotes sustainability of livelihoods by subordinate peoples, and that offers meaningful political participation in the context of the current stage of capitalist development?

The first significant contribution of this book is that it moves us substantially closer towards establishing an empirical baseline on where we stand with respect to these vital questions. In my reading, the overall tenor of the chapters is this. For all of the region's economic prosperity over the past decade and relative success in reducing poverty, pushing up the salience of the environmental issue area in the policy process with an emphasis on alternative technologies, livelihoods and meaningful participation from below has not gained much political traction. Indeed, in some cases we even see backsliding.

That said, the volume establishes that there have been advances in pushing up the salience of the environmental issue area in general. This is no small matter given the very low priority that environmental issues had for political leaders grappling with debt, economic restructuring, political instability and the imperatives of finding their way to a stable development model in the 1970s and 1980s. Today, no major stakeholders in government, business, civil society or academia take a cornucopian hands-off human ingenuity and the price system shall provide position (see Chapter 6). Although many believe in the imperative of development, they acknowledge that environmental concerns cannot be ignored.

The terrain then logically shifts to questions regarding the contested meanings of development, social equity, environmental sustainability and broad-based political participation; the competing conceptualizations of relations of domination and subordination in society embedded in those meanings; and their social, political, economic and cultural distributional consequences. In short, the book starts from the right place: the classic questions of what type of development, and development for whom.

The collection's political ecology approach points us in a fruitful direction to tackle these questions. The editors start from the analytic distinction that I, and others, made long ago between market-based and livelihood or bottom-up grassroots development models of sustainable development (Silva, 1994, 1997). To this they add a neodevelopmental model born of the left turn in Latin American governments, which features a return of the state and economic nationalism in development policy. The ascent of left governments initially fuelled expectations that the alternative approaches to the environment and development would, at last, find favourable political support after decades of market dominance. The book, in an unflinching yet realistic analysis, accounts for hopes dashed.

The volume's more significant contribution, however, is that it reveals a more complex reality. It shows how those left governments straddle various approaches, how they juggle or mix them. On the one hand, the neodevelopmental state plays a larger role in the economy and society than a neoliberal state, but it accepts market dictates such as GM organisms, carbon swaps, a focus on public and private nature conservation, and the expansion of megaextractive projects in mining, agribusiness and energy. On the other hand, some states under left government rule challenge the market-oriented approach that dominates the international climate-control regime. Moreover, although left governments fight anti-megaproject protests, they do not always ram them through no matter what the consequences. At times they cancel projects. More optimistically, the book also reminds us that states are not unitary rational actors. Agencies, departments and even a ministry or two may support projects and policies that advance alternative, livelihoods approaches to the environment and development. Since states operate in the context of democratic regimes, organized subaltern groups and their communities can ally with supportive agencies to survive and, in a few instances, grow. At minimum the struggle continues.

In addition to these advances in the characterization of contemporary environmental governance in Latin America, we also gain an elegant picture of the logic behind its current construction in chapters 2, 4 and 5. Latin American states, their governments, elites, class structures and, of course, economies are part of a world capitalist system, and generally in a subordinate position within it. Absent revolutionary breakdowns of existing polities, market-oriented and supportive development policies in general, and environmental policies specifically, are logically the norm and incremental reforms at the margin are about the most we can expect. The state is organized accordingly.

The collection makes it clear that it is within this context that we need to understand unquestionable advances in institutional capacity-building in the environmental issue area. The environment is firmly on the policy agenda. All Latin American states have built up ministries, agencies and departments for the environment. The technical knowledge and the national and international networks of their professional staff have expanded by leaps and bounds over the past 30 years. Their infrastructural reach has intensified. But they are at the service of states and dominant elites that must respond to the vicissitudes of world markets for their continued reproduction and development. Thus environmental protection regimes have improved by instituting environmental impact reporting, advances in end-of-pipe treatment, adoption of polluter pays principles, and nature conservation. However, we should not be surprised that they are consonant with market or neodevelopmental approaches to environmental sustainability that often disregard questions of social sustainability or that, at best, only seek to compensate – usually poorly and belatedly – for losses to adversely affected communities.

The book drives home a further telling point. How do at least nominally progressive governments juggle these contradictions? By using the commodity boom to substantially expand their welfare effort in social assistance, education, health and infrastructure development. Many of these welfare policies are now reaching not only the urban poor but, significantly, the rural poor. Some governments have also concentrated on substantially raising minimum wages. Thus previously marginalized peoples, at least statistically, have been lifted out of poverty and seen the life chances of their children improve. Judging by electoral returns, governments successfully argue that the aggressive expansion of resource extraction makes these compensatory policies possible.

In short, we gain a concise, well-defined specification of Latin American environmental governance regimes understood as the principles, norms, rules and procedures that infuse governing environmental institutions, their managers, their policies and their operations. This is an important step forward. It clarifies policy agendas and bounds policy prescriptions. It also sets up the relationship of official institutions to society: who's in, under what conditions, and who's out.

This significant contribution to Latin American political ecology is developed in the first half of the volume. The opening chapters, however, stress that Latin American environmentalism is also infused with a widespread, deep-seated concern for livelihoods, environmental justice and alternative production models that is unique to the region (see Chapter 1). These concerns cannot simply be ignored and swept under the carpet. Note to technocratic policy-makers: Resistance to policies that do so is to be expected. To believe otherwise is wishful thinking at best, or an act of willful, arrogant domination at worst.

The second half of the book explores the multifaceted reality of resistance to market and neodevelopmental policies and their lack of social and environmental justice and sustainability. But it does so within the context of a more holistic conceptualization of environmental governance. This permits us to begin to discern the larger significance of fragmented, local resistance for the construction of a more inclusive form of environmental governance in Latin America.

Herein lies the second major contribution of the volume to understanding Latin American environmental politics. Governance is generally understood as how the state organizes systems for rule, meaning that citizens obey its maxims, institutions and policies (Weber, 1978). It is also about the state effectively controlling citizen behaviour and instructing its citizens in how to behave properly (Foucault, 1997). In sum, the focus is on the state's capacity to foster and maintain order with a minimum of overt conflict and coercion. Effective governance systems channel societal tensions in ways that minimize open conflict. Ineffective ones fail to do so. Either way, resistance conflict and institutional change are external to the governance regime.

The editors understand this and, as I argued above, do us the service of specifying for the first time how Latin American states and their elites organize for rule in the environmental issue area. But they go further: they argue that resistance, conflict and coercion are not external to governance; they are an integral part of governance. Governance is a dialectic process that involves order and resistance. Understood this way, we can get at the potential sources of institutional change in Latin American environmental governance. This is the crucial issue for those interested in the question of how to open up policy space for concerns about livelihood, social and environmental justice, and alternative, small-scale, ecofriendly environmental policies. Even more importantly, this approach to environmental governance opens up a path to figuring out the broader, cumulative policy and political impacts of fragmented, localized, heterogeneous and territorially marginal protests. I will come back to this theme later.

The book, correctly, underscores two major paths to change. That they are both represented in one volume is rare and permits us to appreciate more the strengths and weaknesses of each. One argues that communities of marginal, subaltern social groups – of which there are many – should disengage from the policy process (see Chapter 10). They would be better off devoting their energies and scarce resources to creating their own alternative worlds by practising the principles of solidarity, social justice, freedom, autonomy, and alternative knowledge and production modes they espouse (see Chapter 3). Their multiplication coupled with networking among themselves will corrode the dominant society and force change, either because the dominant society collapses under its own weight or because it must adapt to avoid being overtaken. We see this position most clearly in Chapter 10. Questions remain about how those networks might expand, how their experiences can overcome resource deprivation in the midst of more or less functioning capitalist societies, and what the "tipping points" might be. Eventually, these experiences would also have to engage in politics and articulate with other actors, but which and how remains underexamined.

The other potential path for change involves engagement with established political systems at the international, national or subnational scale or combinations of them to effect institutional and policy reforms. For example, national left governments more connected to their social bases may seek to modify international regimes that favour market-oriented policies in ways that accommodate alternative approaches, as seen in Chapter 8 about REDD. By the same token, communities may attempt to use REDD as an opportunity to promote projects with a more alternative, livelihoods cast as described in Chapter 9 about community forestry in Mexico. Perhaps most promising of all was the experience depicted in Chapter 11 of communities making multiscalar alliances with local government, national government agencies and international organizations to organize consultations on unwanted megamining projects promoted by national governments, transnational corporations and multilateral lending institutions. Their application of national and international standards for conducting such extraofficial consultations was inspiring, as was the fact that such campaigns at times contributed to stopping the project.

The weaknesses of engagement with politics are well known. How can one avoid co-optation as leadership loses touch with its social bases? Will engagement merely result in cosmetic or symbolic changes or simply serve to disarticulate environmental justice movements (EJMs)? Will political allies betray such movements? What opportunities must EJMs seek to create in the face of unfavourable structural conditions? How can they scale up successful experiences? These are the perennial hard questions to which we have no good answers. This book, and similar research, suggests that networking analysis, which is gaining increasing attention, might be a fruitful way to go (von Bülow, 2010).

This brings me to a third major contribution of this volume to pushing the frontiers of Latin American political ecology. It offers a cogent and readily apprehensible approach to multiscalar analysis. All too often, analyses of problems that involve multiple scales simply devolve into descriptions of a bewildering multiplicity of actors with different interests and power resources. It becomes difficult to understand the relationship between them and how outcomes are affected. This book, because of the careful groundwork it has laid with respect to the specification of dominant governance regimes and its relationship to sources of resistance, overcomes those weaknesses. We understand the logic behind the multiscalar relationships that are forming, breaking apart and reconfiguring, thus we are able to easily follow them and apprehend their significance for the larger questions that stand at the centre of this collection. This is a manageable method that travels well in new and evolving situations.

This is a significant advance because a sign of our times is the increasing complexity of our multiscalar world, its fragmentation, heterogeneity, compartmentalization and segmentation. These characteristics permeate the environmental issue area, exacerbated by its intersection with economic, political and social structures. It is also one of the few arenas in our contemporary stage of capitalism where private property rights are hotly contested. This book reminds us that careful specification of ideal types can help us to make sense of the mixed types that exist in reality and the logical consequences for social action that follow from their characteristics. Hybrid types and bricolage cease to be ad hoc descriptive, and often confusing, assemblages, and they acquire real analytical bite.

In sum, this volume makes valuable empirical, conceptual, analytical and methodological contributions to the political ecology of environmental governance. It establishes an empirical benchmark from which to assess change or a lack thereof by carefully specifying actors, interests, power, and the structures and ideational frameworks in which they exist and operate. It provides a theoretical framework grounded in the everyday reality of the vast sea of marginalized peoples caught on the receiving end of harsh systems of domination, systems that are characterized with equal attention to reality. Consequently, the book constitutes an important advance from earlier efforts in Latin American political ecology (Painter and Durham, 1995).

Where do we go from here? This collection points us in fruitful directions. Its specification of the concept of environmental governance and its political ecology operationalization offers an innovative common framework for analysis. But what do we expect from improved environmental governance? Here is a question that merits further exploration. The book proposes that at minimum it should provide spaces for alternative, more ecofriendly approaches to inclusive development. That opens up another persistent problem: how to integrate subaltern social groups in a manner that does not compromise their autonomy, and with alternative forms of ecofriendly, low-impact, small-scale economic production that includes conservation measures.

Of course, examples of poster projects abound, but the issue is not so much technical as political. What are the conditions under which those isolated experiences might thrive and expand? As I mentioned previously, one path the book points to is the disengagement of marginalized peoples from the political process with a focus on building the alternative world they dream of. This suggests more work on the conditions that facilitate this process and its scaling-up through networking, as well as thinking about tipping points for the dominant society.

Engagement with politics by organized subaltern groups is the other path. I have already suggested the types of research needed to advance further on this front. The larger question, to expand on the point above, involves the conditions under which, in the context of contemporary capitalism, spaces for progressive environmental governance can be created, supported and scaled up to a meaningful proportion. Those who argue for disengagement clearly think it is an impossible proposition. The burden of proof is on those who focus on social movement resistance that engages authorities over time, seeking reforms to existing arrangements.

In an earlier comment I submitted that the book offers some starting points. I would like to expand on that by suggesting that a tighter focus on the outcomes of resistance might be useful. Given the decentred nature of much resistance, asking what the cumulative effects of the multitude of local actions are could be a fruitful research agenda. Understanding the causal mechanisms behind those results would also be advantageous. The classic approach has been to focus on proposed or enacted legislation. However, we must transcend that approach. One research strategy might be to analyse the effects of protest on the different stages of the policy process, such as agenda-setting, initiation, formulation, implementation and evaluation. Going beyond that relatively superficial level, one could also research deeper changes in the distribution of power. This includes significant changes in access to power by subordinate groups, often requiring institutional reforms; diffusion of new values; and improvements in movement resources to support the consolidation of new channels of access to power or new values.

By the same token, we need to improve our understanding of the causal mechanisms that influence those outcomes. Chapter 11 suggests that networking across multiple scales and the appropriation of existing principles, norms, rules and practices for other purposes might be an avenue. We can build on that and other works that focus on networks (von Bülow, 2010), but we must also expand our analytical tools to distinguish between direct effects and those mediated by third parties (Silva, 2013). That would advance analysis and contribute to the formulation of innovative strategies and policy proposals. Expanding these lines of research can help us with another vital issue. An oft-stated goal is to develop the participation of subaltern social groups that demand alternative forms of ecofriendly, small-scale economic production in the policy process. But what constitutes "real" participation and how do we get it? Yes, it must be deliberative and binding, and protest seems to be the only way to advance on this issue, but there is much more to it than that. A more social scientific formulation to guide us might be: How do we specify the dimensions of the concept and the consequences of different participatory mechanisms for containing political tensions, and in whose interest? What combinations of multiscalar action and ideational innovations are conducive to positive results? In this vein it might be useful to think about plausible alternatives or improvements to existing institutions, which might be informal rather than formal (Rodrik, 2007). The alternative consultations analysed in Chapter 11 are an example of this.

This book's approach to environmental governance unequivocally moves the discussion about environment, development and social justice past its current point of relative stagnation. To the extent that it highlights participation, conflict and resistance to environmentally and socially damaging development models it suggests that negotiation is another critical feature of environmental governance. Analysis frequently stresses maximum demands and outcomes benchmarked against them. But we also need to explore intermediate outcomes. Negotiation is about compromise, not capitulation. This raises additional questions worth further research. What are "exploited" actors willing to settle for? What is a second-best option? Where and what are those spaces for negotiation, formal or informal? What role does protest play in the negotiation process? Raising these questions in no way means losing sight of the potential for divisiveness or co-optation of movements when they begin to negotiate with governments or companies.

I conclude with the following brief reflection. Latin American political economy in general (and this collection in particular) focuses on natural resource extraction and the ever-present penetration of capitalism into rural or frontier regions where customary and collective forms of social organization are still relatively strong. Hence the environment is an issue area in which property rights and conceptions about economic, political and social organization are contested in places far from the urban centres. Moreover, with increasing frequency, indigenous cosmologies, presuming their affinity with nature, are invoked. What about environmental problems in urban areas? Do appeals to indigenous cosmology help or hinder? How does it travel to urban centres? These questions notwithstanding, is there a way to harness a dispossessed urban social subject around environmental issues in ways that facilitate alliances with rural actors? We know from history that rural-urban alliances of popular sectors tend to have greater power in pushing more radical reforms. Is it possible to construct such alliances for environmental governance along the lines this book advocates? It might be wishful thinking but the question is worth asking for the problem of environment and development is not only a rural one; it affects all. We know very little about the urban side of the equation. Perhaps we should re-examine our assumptions.

## References

Brundtland Commission (1987) *Our Common Future* (New York: Oxford University Press).
Painter, M. and Durham, W.H. (eds) (1995) *The Social Causes of Environmental Destruction in Latin America* (Ann Arbor: The University of Michigan Press).
Rodrik, D. (2007) *One Economy, Many Recipies: Globalization, Institutions, and Economic Growth* (Princeton: Princeton University Press).
Silva, E. (1994) "Thinking Politically About Sustainable Development in the Tropical Forests of Latin America", *Development and Change* 25 (4): 697–721.
Silva, E. (1997) "The Politics of Sustainable Development: Native Forest Policy in Chile, Venezuela, Costa Rica, and Mexico", *Journal of Latin American Studies* 29 (2): 457–493.
Silva, E. (2013) "Protest and Policy Change", paper presented at the American Political Science Convention, Chicago, Illinois, August 29–31.
Von Bülow, M. (2010) *Building Transnational Networks: Civil Society and the Politics of Trade in Latin America* (New York: Cambridge University Press).
Weber, M. (1978) *Economy and Society* (Berkeley: University of California Press).

Except where otherwise noted, this work is licensed under a Creative Commons Attribution 3.0 Unported License. To view a copy of this license, visit http://creativecommons.org/licenses/by/3.0/

# Index

agriculture, 29–30, 37, 68, 86–8, 90–103, 140, 152, 194, 242
alternative
 development, 177
 model, 11, 263, 278
Amazon, 9, 14, 31, 32, 35, 38, 40, 42, 49, 74–5, 80, 121, 153, 155, 205, 225, 228, 263
 Fund, 209–14
Andes, 36–7, 50, 77, 116
Argentina, 8, 13, 30–9, 44–7, 61–79, 140, 151–8, 165–72, 187, 193–7, 211, 237, 268, 287–95, 303–7
autonomy, 89, 91, 144–5, 222, 250, 262–8, 332

BINGOs, 215–28
biodiversity, 1, 14, 20, 34, 42, 47, 50, 51, 60, 65, 70, 74, 77, 80, 86–7, 94, 103, 126, 127, 179, 186, 195, 199, 206, 224, 227, 236, 242
biomass, *see* conflicts
Bolivia, 8, 11, 13, 34, 47–9, 51, 52, 74, 91, 113–32, 140, 145, 148, 156, 157, 209, 212, 218–22, 235, 251, 275, 278
Brazil, 13, 17, 30, 32, 34–6, 40, 42, 45–8, 51, 63, 64, 74–9, 91, 115, 141, 146, 153, 155, 209–2, 235, 237, 238, 251, 266, 268, 277
*Buen vivir*, 10, 11, 13, 38, 52, 124, 125, 262, 276, 277

carbon
 emissions, 1, 21, 205, 206, 211–13, 218, 222, 236
 markets, 115, 206, 208, 216–29
 sequestration, 205, 225
Chile, 30, 35, 39, 44, 45, 47, 48, 91, 115, 121, 157, 165, 166–8, 172–7, 212, 277
climate change, 35, 42, 46–50, 77, 91, 103, 127, 157, 165, 169, 170, 180, 186, 205, 209, 211, 213, 219, 220, 225, 236
Colombia, 30, 34, 41, 44, 47, 48, 61, 63–5, 74, 76, 78, 79, 137, 166, 167, 172, 212, 215–19, 224–7, 237, 287, 292, 296, 308, 311–19
commodities, 9, 60, 62, 64–6, 70, 74–8, 149
consensus, 49
commons, 18, 51, 96, 234–50, 268
community forestry, *see* forest
companies, 6, 9, 13, 49, 59, 69, 74, 91, 114, 117, 121, 123, 125, 132, 138, 141, 147–55, 164, 167–9, 174, 195, 217, 223, 251, 292, 313
conflicts
 biomass, 69
 distribution, 59, 60, 66, 78
 mining, 74, 293, 300, 301, 306, 312, 317–18
 (socio) environmental, 3, 7, 9, 17, 47, 51, 60, 68, 73, 77–8, 123, 132, 164–5, 287
consultation, 76, 225–6, 266, 270, 287–320
corporate social responsibility, 6
cosmologies, 9, 13, 258, 272–3, 335
Costa Rica, 211–12, 215, 217–19, 227, 301

decision making, 2–20, 74, 87, 125, 205, 269, 278, 289, 301, 313
deforestation, 4, 30, 34, 66–7, 96, 154, 206–29, 234–49
development
 eco, 40, 178
 institutional, 251
 resource-based, 115
 rural, 86–103
 sustainable, 38, 40, 51
discourse
 analysis, 168
 model, 170–81

distribution, *see* conflicts
diversity, 3, 7, 12, 32, 70, 74, 93, 96, 126, 131, 191, 196, 207, 240, 257, 274, 317

ecodevelopment, *see* development
ecological debt, 42, 46, 48, 52
Ecuador, 8, 30, 35, 38, 44, 48, 52, 61–4, 74–5, 80, 113–32, 140, 147–9, 165–78, 292, 305–6
environmental
　conflict, 3, 7, 9, 17, 47, 51, 60, 68, 73, 77–8, 123, 132, 164–5, 287
　crisis, 37, 41, 146, 153, 176, 188
　history, 29, 52
　impact, 47, 71, 73, 139, 159, 164, 173, 290–1, 309, 329
　justice, 43–51, 58, 60, 219, 257, 272–3, 287, 290, 312
　movements, 46, 140, 145, 157, 189, 190, 198
　policy, 87, 93–5, 127, 170
environmentalism, 29, 32–52, 86, 173, 313, 329
extractive industries, 79, 87, 113–18, 181
extractivism, 8, 41, 44, 52, 58, 64
　post, 43, 51

farming, 86, 87, 91–3, 97, 276
fisheries, 66, 69
forest
　community, 235, 331
　conservation, 154, 206, 228, 249–50
　governance, 141, 205–19, 234–5, 245–6
　policies, 237, 238, 240

Glocal, 16, 18
governance
　environmental, 2–21, 29, 40, 48, 58, 79, 89, 101, 113, 118, 126, 131–2, 137–59, 165–6, 190, 197, 205–7, 259, 278, 326–35
　hybrid environmental, 207
government, 2–20, 38–51, 113–32, 139–41, 145–9, 316–17
Gualeguaychu, 194

hybrid institutions, *see* institutions

indigenous
　knowledge, *see* knowledge
　people, 73–5, 91, 115, 123, 154, 164, 177, 207, 217–18, 225–7, 265–7, 277–9, 305
institutional development, *see* development
institutions
　environmental, 38, 329
　hybrid, 207, 289, 317
intercultural dialogue, 259

knowledge
　autochthonous, 87
　indigenous, 37, 86–103
　scientific, 88–103

large-scale mining, 287–320
Latin American
　conservationism, 35
　environmentalism, 29, 37, 42, 329
local
　communities, 4, 7, 9, 10, 15, 18, 69, 91, 126, 127, 150, 167, 219–20, 235–7, 246, 251, 312–13
　decision-making, 18

Matanza-Riachuelo, 196–8
Mesoamerica, 36, 50, 264
Mexico, 30, 35, 40–5, 86–103, 158, 212, 234–52, 264–79
mining conflicts, *see* conflicts
movements
　anti-mining, 289–90, 312
　environmental, 46, 140, 145, 157, 189, 190, 198

natural resources
　extraction, 334
nature-society relations, 1, 4, 8
networks, 7, 16, 45, 51, 97, 98, 150, 153, 155, 213, 222, 224, 288, 303, 306, 310–20
NGOs, 77, 207–29, 310

oil revenues, 120, 122, 123

338    *Index*

participation, 3, 14–15, 20, 94, 150, 179, 192, 245, 291, 311–17
peasants, 13, 20, 86–103, 278, 302
Peru, 29, 36, 47, 73, 211, 226, 278, 287, 292, 300–5
phased approach, 201–14, 207, 209
policy
  model, 4
  public, 35, 169, 181, 189, 229
political
  ecology, 45, 48, 60, 77, 90, 116, 142, 327, 332
  transformations, 7
politics, 48, 131, 139, 157, 179, 238, 260, 290, 330
popular environmentalism, 42–51
post-extractivism, 51
poverty, 40, 87, 132, 186–200
property rights, 86, 91, 234, 236, 332
public policies, 80, 169, 181, 229

quality of life, 119, 189, 191, 193, 282
  *see also Buen vivir*

raw materials, 44, 64, 70, 79, 80, 235
REDD, 141, 154, 205–29
referenda, 130, 287, 313–15
rentier states, 113–32
resisting strategy, 212, 218

resource-based development, *see* development
responsible mining, 170, 173
rural development, *see* development

scale, 1, 3, 6–18
scientific knowledge, *see* knowledge
self-sufficiency, 102, 262, 276
social
  change, 5
  exclusion, 3, 186, 198
  inclusion, 13, 20
  justice, 3, 7, 259, 261, 330
  metabolism, 33, 48, 58–80
  rights, 191
strategic actors, 164–81
strategy
  accommodating, 211, 215, 227
  assertive, 211, 212
surplus, 63, 270–4
sustainable consumption, 164–81
sustainable development, *see* development

transnational social movements, 7

water consumption, 171, 177
water-energy-mining complex, 165

 Except where otherwise noted, this work is licensed under a Creative Commons Attribution 3.0 Unported License. To view a copy of this license, visit http://creativecommons.org/licenses/by/3.0/